AF788201

Springer Proceedings in Mathematics & Statistics

Volume 70

For further volumes:
http://www.springer.com/series/10533

Springer Proceedings in Mathematics & Statistics

This book series features volumes composed of select contributions from workshops and conferences in all areas of current research in mathematics and statistics, including OR and optimization. In addition to an overall evaluation of the interest, scientific quality, and timeliness of each proposal at the hands of the publisher, individual contributions are all refereed to the high quality standards of leading journals in the field. Thus, this series provides the research community with well-edited, authoritative reports on developments in the most exciting areas of mathematical and statistical research today.

Wael Bahsoun • Christopher Bose • Gary Froyland
Editors

Ergodic Theory, Open Dynamics, and Coherent Structures

Springer

Editors
Wael Bahsoun
Department of Mathematical Sciences
Loughborough University
Leicestershire, UK

Christopher Bose
Department of Mathematics and Statistics
University of Victoria
Victoria, BC, Canada

Gary Froyland
School of Mathematics and Statistics
University of New South Wales
Sydney, Australia

ISSN 2194-1009 ISSN 2194-1017 (electronic)
ISBN 978-1-4939-0418-1 ISBN 978-1-4939-0419-8 (eBook)
DOI 10.1007/978-1-4939-0419-8
Springer New York Heidelberg Dordrecht London

Library of Congress Control Number: 2014935178

Mathematics Subject Classification (2010): 37-XX, 47A35, 37C30

© Springer Science+Business Media New York 2014
This work is subject to copyright. All rights are reserved by the Publisher, whether the whole or part of the material is concerned, specifically the rights of translation, reprinting, reuse of illustrations, recitation, broadcasting, reproduction on microfilms or in any other physical way, and transmission or information storage and retrieval, electronic adaptation, computer software, or by similar or dissimilar methodology now known or hereafter developed. Exempted from this legal reservation are brief excerpts in connection with reviews or scholarly analysis or material supplied specifically for the purpose of being entered and executed on a computer system, for exclusive use by the purchaser of the work. Duplication of this publication or parts thereof is permitted only under the provisions of the Copyright Law of the Publisher's location, in its current version, and permission for use must always be obtained from Springer. Permissions for use may be obtained through RightsLink at the Copyright Clearance Center. Violations are liable to prosecution under the respective Copyright Law.
The use of general descriptive names, registered names, trademarks, service marks, etc. in this publication does not imply, even in the absence of a specific statement, that such names are exempt from the relevant protective laws and regulations and therefore free for general use.
While the advice and information in this book are believed to be true and accurate at the date of publication, neither the authors nor the editors nor the publisher can accept any legal responsibility for any errors or omissions that may be made. The publisher makes no warranty, express or implied, with respect to the material contained herein.

Printed on acid-free paper

Springer is part of Springer Science+Business Media (www.springer.com)

Preface

Ergodic theory, as a mathematical discipline, refers to the analysis of asymptotic or long-range behaviour of a dynamical system, that is, a map or flow on a state space, using measure-theoretic or probabilistic methods. A close cousin to smooth dynamics (the study of differentiable actions on a smooth manifold) and to topological dynamics (comprising a continuous action on a topological space), there is a well-established and rich synergy between the three fields. Indeed, many important applications bring tools from all the three fields to bear in the study of a particular dynamical system.

As a quickly maturing mathematical field, both theoretical developments and applications in the physical sciences, engineering, and computer science are flourishing within the arena of modern research in ergodic theory. Driven by these new theoretical tools and a growing breadth of natural applications, computational aspects are now a central challenge to researchers in the field.

An open dynamical system is a natural extension of the traditional (closed) dynamical system. In an open system, the state space is no longer deemed to be invariant under the dynamical action, but some orbits are allowed to 'escape' depending on location and time. An everyday example is the dynamics of a ball on a billiard table; when the ball falls in a hole in the table, the orbit is terminated. As introduced by Yorke and Pianigiani in the late 1970s, the abstract concept of an open system leads immediately to the notion of a conditionally invariant measure and escape rate along with a host of detailed questions about how mass escapes or fails to escape from the system under time evolution.

Perhaps ironically, concepts from open systems have recently been used to analyse traditional, closed systems. For example, in many closed systems, relaxation to equilibrium is by no means uniform throughout the state space. There may be regions that remain 'almost invariant' for long periods of time, mixing with the rest of the space at quantifiably slower rates than the other parts of the system. These 'almost invariant sets' become key features determining the asymptotics of the system. One particularly fruitful idea is to study almost invariant sets as open subsystems of the larger closed dynamical system, wherein the escape rate determines the rate of mixing and relaxation to equilibrium.

In some realistic applications, time-varying parameters governing the flow or transformation on the state space necessitate modelling by a non-autonomous system. While the ergodic theory of non-autonomous systems parallels that of autonomous dynamics in many ways, there are important differences. Stable and unstable foliations, a foundation of geometric analysis for an autonomous map or a flow, become equivariant, time-dependent structures. Other dynamical objects such as Lyapunov exponents and Oseledets subspaces can be used in alternative ways to describe non-autonomous or random dynamics. Invariant or almost invariant objects arising in autonomous dynamics have non-autonomous analogues called coherent structures. These are features that move around in the state space under time evolution but that may still represent barriers to mixing and relaxation to 'equilibrium', a concept that also has to be reinterpreted compared to the autonomous setting.

From April 9 through April 15, 2012 a group of more than 40 researchers in ergodic theory gathered at the Banff International Research Station in Banff, Alberta, Canada to exchange cutting-edge developments in the field.[1] Thirty-five research talks were given during the course of the workshop, covering theoretical, applied, and computational aspects of both open and closed, autonomous and non-autonomous dynamics. After the workshop, a number of participants volunteered to expand on their presentations and contribute chapters to this volume. Each contribution was rigorously peer-reviewed before inclusion in this volume. We briefly outline the resulting contributions:

- Balasuriya considers time-dependent flows where the time dependence enters as a perturbation of an autonomous flow. He describes how to analytically estimate the perturbed stable and unstable manifolds, which may be regarded as Lagrangian coherent structures. He then uses Melnikov theory to quantify flux across these perturbed manifolds.
- Bandtlow and Jenkinson consider the spectrum of transfer operators of real-analytic expanding maps acting on holomorphic functions of the interval and other finite-dimensional spaces. They particularly consider the open setting where mass is leaving the phase space and prove bounds for each spectral point of the corresponding open operators.
- Bandtlow, Jenkinson, and Pollicott specialise the previous chapter to the setting of piecewise real-analytic expanding Markov maps of the interval, with escape through a Markov hole. They show that the leading spectral point of the transfer operator, which quantifies the escape rate, can be approximated using derivative information from all periodic points of increasing period. The invariant measure on the survivor set is also estimated.
- Basnayake and Bollt describe a method of extracting a flow field from a movie of observations. Under the assumption of smooth time dependence of the flow field,

[1]Materials from this workshop, including abstracts and some videos of the presentations, are available at the BIRS website, www.birs.math.ca; search on workshop code 12w5050.

they introduce a multi-step method to enforce smooth behaviour. As an example, they extract a flow field from a movie of sea surface temperature and calculate Lagrangian coherent structures in the form of finite-time Lyapunov exponent fields.

- Bose and Murray extend their earlier work on estimating absolutely continuous invariant measures (ACIMs) to the open dynamics setting. In general, an open system may support a continuum of escape rates and an infinity of absolutely continuous conditional invariant measures (ACCIMs). Their approach, based on maximum entropy and convex optimisation, allows one to prescribe the desired escape rate and find the corresponding ACCIM.
- Bruin studies a map on a Euclidean $(d-1)$-dimensional triangle that arises from a simple d-dimensional subtractive algorithm. For $d = 2$ the map becomes the well-known Farey map on $[0, 1]$; for $d = 3$, Bruin shows that the map is dissipative but at the same time ergodic with respect to two dimensional Lebesgue measure and even exact. This paper contributes to a long and historically important development of number theoretic applications of ergodic theory dating back to Renyi in the 1950s with his foundational work on the continued fraction expansion. At the same time, the tools used are up to date, bringing modern notions such as distortion, Schwarzian derivative, and random walks to bear on the problem.
- Bunimovich and Webb consider piecewise differentiable expanding Markov maps of the interval with a Markov hole. Their focus is on estimating the survival and escape probabilities after a finite number of iterations of the open system. They provide explicit upper and lower bounds for these probabilities in terms of eigenvalues of the transition matrices of induced Markov chains.
- Demers studies billiard dynamical systems with a variety of holes. Using transfer operator and Young tower techniques, he proves the existence of a natural escape rate and corresponding absolutely continuous conditional invariant measure (ACCIM). He then considers the question of stability as the size of the hole goes to zero and shows the limiting ACCIM is the SRB (or physical) measure for the closed system. Finally he shows that the escape rate also arises via a variational principle.
- Froyland and Padberg-Gehle give an overview of transfer operator methods for finite-time almost-invariant and coherent sets. Their chapter unifies the autonomous and time-dependent methodologies and then focuses on three aspects, namely the flow direction, the flow duration, and the level of diffusion present. They show that the coherent structures produced by the transport-based transfer operator approach are very natural from a geometric dynamical point of view.
- Haydn, Winterberg, and Zweimüller consider a general ergodic process and the return time and hitting time distributions corresponding to a sequence of sets of decreasing size. They first show that as the size of the sets approaches zero, limiting return and hitting time distributions exist. Further, they show that if one induces the original ergodic process via return times to a fixed set of positive

measure, the limiting distributions of the return and hitting times of the original and induced systems coincide.

The contributions to this book represent a broad cross section of the topics represented at the April 2012 workshop and, in turn, make a fine collection of sample papers for researchers who may be looking to broaden their outlook in modern aspects of the field. The editors wish to thank all the workshop participants for their contributions, but especially those participants who took the time to write up their work as a submission to this book and the referees who helped to hone the author's contributions into the high-quality research papers you will find in the following pages.

Finally, none of this would have been possible without the remarkable support offered by the Banff International Research Station and its staff. BIRS is indeed one of only a handful of first-class mathematical research venues in the world; if you have a chance to go there, do not hesitate!

Leicestershire, UK — Wael Bahsoun
Victoria, BC, Canada — Christopher Bose
Sydney, Australia — Gary Froyland
July 2013

Contents

List of Contributors

Sanjeeva Balasuriya School of Mathematical Sciences, University of Adelaide, Adelaide, SA, Australia

Oscar F. Bandtlow School of Mathematical Sciences, Queen Mary University of London, London, UK

Ranil Basnayake Department of Mathematics, Clarkson University, Potsdam, NY, USA

Erik M. Bollt Department of Mathematics, Clarkson University, Potsdam, NY, USA

Christopher Bose Department of Mathematics and Statistics, University of Victoria, Victoria, BC, Canada

Henk Bruin Faculty of Mathematics, University of Vienna, Vienna, Austria

Leonid A. Bunimovich ABC Math Program and School of Mathematics, Georgia Institute of Technology, Atlanta, GA, USA

Mark F. Demers Department of Mathematics and Computer Science, Fairfield University, Fairfield, CT, USA

Gary Froyland School of Mathematics and Statistics, University of New South Wales, Sydney, NSW, Australia

Nicolai T. A. Haydn Department of Mathematics, University of Southern California, Los Angeles, CA, USA

Oliver Jenkinson School of Mathematical Sciences, Queen Mary University of London, London, UK

Rua Murray Department of Mathematics and Statistics, University of Canterbury, Christchurch, New Zealand

Kathrin Padberg-Gehle Institute of Scientific Computing, Technische Universität Dresden, Dresden, Germany

Mark Pollicott Mathematics Institute, University of Warwick, Coventry, UK

Benjamin Z. Webb Laboratory of Statistical Physics, Rockefeller University, New York, NY, USA

Nicole Winterberg FB Mathematik, Universität Salzburg, Salzburg, Austria

Roland Zweimüller Fakultät für Mathematik, Universität Wien, Vienna, Austria

Chapter 1
Nonautonomous Flows as Open Dynamical Systems: Characterising Escape Rates and Time-Varying Boundaries

Sanjeeva Balasuriya

Abstract A Lagrangian coherent structure (LCS) in a nonautonomous flow can be viewed as an open dynamical system, from which there is time-dependent escape or entry. A difficulty with this viewpoint is formulating a definition for the time-dependent boundary of the LCS, since it does not correspond to an entity across which there is zero transport. Complementary to this is the question of how to determine the escape rate—the time-dependent fluid flux—across this purported boundary. These questions are addressed within the context of nonautonomously perturbed two-dimensional compressible flow. The LCS boundaries are thought of in terms of time-varying stable and unstable manifolds, whose primary locations are quantified. A definition for the time-varying flux across these is offered, and computationally tractable formulæ with a strong relation to Melnikov functions are provided. Simplifications of these formulæ for frequently considered situations (incompressibility, time-periodic perturbations) are demonstrated to be easily computable using Fourier transforms. Explicit connections to lobe areas and the average flux are also provided.

1.1 Introduction

A fairly recent inclusion into the theory of open dynamical systems is Lagrangian Coherent Structures (LCSs) in nonautonomous flows in fluids. LCSs are time-varying (unsteady) entities in a flow which remain 'almost' coherent as time evolves yet have some fluid exchange with their surroundings. They are prevalent in observational, experimental, and computational fluid mechanics and exist across

---

S. Balasuriya (✉)
Department of Mathematics, Connecticut College, New London, CT 06320, USA

School of Mathematical Sciences, University of Adelaide, Adelaide, SA 5005, Australia
e-mail: sanjeevabalasuriya@yahoo.com

W. Bahsoun et al. (eds.), *Ergodic Theory, Open Dynamics, and Coherent Structures*, Springer Proceedings in Mathematics & Statistics 70, DOI 10.1007/978-1-4939-0419-8_1,
© Springer Science+Business Media New York 2014

the gamut of spatial scales. Examples of LCSs include eddies in the ocean [2, 29, 31, 44] and the atmosphere [121, 123, 124], vortices at intermediate spatial scales [33, 56, 74–76], nonmixing patches of fluid in nanofluidic devices [119], and coherent shear jets [30, 34]. The term LCS was originally coined by Haller and collaborators [59, 66], to emphasise the *Lagrangian* nature of these entities, that is, the fact that these are associated with following fluid particles as opposed to the *Eulerian* approach of studying the time variation of instantaneous flow properties (vorticity field, rate of strain, pollutant concentration). If attempting to understand *transport* (of fluid, of pollutants such as oil or volcanic ash, of passively flowing organisms such as plankton, of heat energy which has strong implications on the weather and climate, etc), the Lagrangian viewpoint has clear importance [30, 34, 71, 79, 96, 101, 103, 112, 115, 123, 124]. Each LCS can be thought of as an open dynamical system, with most of the fluid particles within the LCS remaining within the LCS, yet with a smaller fraction exiting (or entering) the LCS as time evolves. Thus, the escape rate associated with the open dynamical system is the transport across the 'boundary' of the LCS. The obvious issues which immediately emerge are:

(i) How is the boundary of the LCS defined?
(ii) How can the time-dependent escape rate across the boundary be quantified?

These questions are intimately related and are an intensive area of current study [5,36,52,62, e.g.] due to their importance in transport across multiple spatial scales. In this chapter, I will answer these questions within a specific limited setting.

While the above questions could be posed in any dimension, two- and three-dimensional flows are the usual focus because of the fluid mechanical applications. Three dimensions would seem to be the most natural setting, but two-dimensional flows continue to attract vigorous research. An obvious reason is that these are more tractable than three-dimensional (3D) flows, but two-dimensional (2D) flows are also relevant in a variety of situations. In oceanic flows in particular, the dominant fluid motion when away from land masses and bottom topography tends to be on isopycnal (constant density) surfaces; most flow is on nearly horizontal 2D sheets [100]. This is sometimes called the *barotropic* assumption [100, 105], and thus studying flow in 2D is important in oceanography. Another situation of the 2D approximation being valid is when the spatial scales of two of the dimensions are hugely separated from the spatial scale of the third dimension, such as flow in long channels or in certain types of micro/nanofluidic devices. While many studies in the literature confine their attention to 2D incompressible (area-preserving) flow, it may be relevant to relax this *even if the fluid is incompressible*. As my first example of this claim, consider oceanic flows. The distance between the isopycnal surfaces need not remain constant; they may be close at some points in space, and far apart in other, and this relative positioning may also vary with time. Volume preservation in 3D will therefore not necessarily imply 2D area preservation within each isopycnal surface. As my second example of this claim, consider steady (autonomous) incompressible inviscid non-Beltrami flow in 3D: results due to Arnol'd [6–8] indicate that in this situation flow is confined to 2D surfaces which are topologically equivalent to cylinders or tori, except for surfaces which demarcate the transition from cylinders to tori.[1] The 2D surfaces flow is confined to are called *Lamb surfaces* [63, 120] and comprise level sets of the Bernoulli function. While in non-Beltrami flows with volume preservation and symmetry it is possible to reduce to a Hamiltonian structure on an abstract 2D manifold [63], area preservation is not necessary on the Lamb surfaces because of the possibility of compression/expansion in the normal direction of these surfaces. Thus, when relaxing steadiness or inviscidity (i.e. when permitting the flow to be nonautonomous or viscous), insisting on area preservation in 2D is not reasonable, even if the 3D flow was incompressible. Based on these arguments, there is cause for studying 2D flows within the *compressible* setting for incompressible 3D fluids, as it is of course also relevant when the (2D or 3D) fluid is compressible as well.

For 2D *autonomous* flows, compressible or not, it is easy to identify boundaries of LCSs and quantify transport across them. The reason is that the geometry here is confined to two dimensions only, and hence, in the long term, fluid particles approach either fixed points (stagnation points), periodic orbits, or *heteroclinic cycles* (a set of fixed points which are connected together by heteroclinic trajectories between the fixed points, forming a structure which is topologically equivalent to a circle) by the Poincaré-Bendixson Theorem [4, 9, 58]. This implies well-defined coherent motion associated with entities which remain spatially fixed, which moreover can be identified easily using either Lagrangian *or* Eulerian diagnostics. The paradigmatic example of such a coherent entity is a family of nested periodic trajectories which are eventually bounded by a heteroclinic cycle. These can be thought of as *vortices* (or *eddies*, in the geophysical literature) and are ubiquitous in applications. I show a variety of such coherent structures in Fig. 1.1, which illustrates the curves followed by particle trajectories. The Taylor-Green cellular flow [1, 3, 10, 11, 18, 39, 106, 116, 118] is one example, which is also the $\epsilon = 0$ flow (i.e. steady form) of the popular double-gyre model [35, 45, 49, 51, 81, 93, 114, e.g.]. This is displayed in Fig. 1.1a, and here each square cell is a coherent vortex structure, which is separated from adjacent square cells by the four heteroclinic trajectories along the sides of the square. Each heteroclinic trajectory is simultaneously a branch of the stable manifold of one fixed point and a branch of the unstable manifold of another. I shall often refer to this as a *heteroclinic manifold*, and thus the boundary of each square cell is a heteroclinic cycle comprising a collection of heteroclinic manifolds. The LCS boundary may even be a *homoclinic* manifold, that is, a connection from a fixed point to itself. For example, Fig. 1.1b shows an idealised model for a steady eddy [20, cf.]; the region of rotational motion is bounded by such a homoclinic manifold. A particular example of this occurs in the two-gyre Duffing model [5, 16, 17, 21, e.g.] as shown in Fig. 1.1c, in which case there are *two* such homoclinic manifolds. Another example, pictured in Fig. 1.1d, is the cats-eye structures in one-mode Bickley jet or two-mode Rossby wave models [24, 38, 53, 102, 122, 127] loosely meant to represent the eddies alongside the cores

[1]Indeed, these separating surfaces can be rationalised as being boundaries of LCSs in this highly idealised situation.

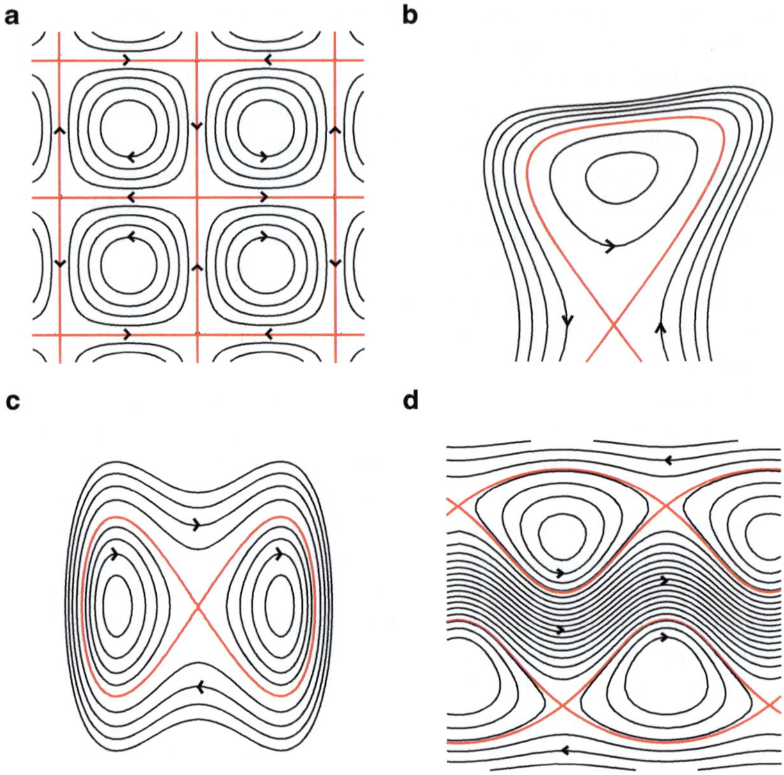

Fig. 1.1 Coherent structures and their boundaries (*heavy curves*) in autonomous flows: (**a**) Taylor-Green cellular flow, (**b**) an eddy, (**c**) two-gyre Duffing model, and (**d**) Bickley jet

of oceanic jets such as the Gulf Stream; in this case, each vortex is demarcated by two heteroclinic manifolds. Of note is that here the jet itself can be thought of as a coherent structure, and there is a heteroclinic manifold which demarcates the boundary between the jet and each of the cats-eye structures. This observation is generic: the boundaries of LCSs are heteroclinic cycles, and the escape rates across them is zero. In other words, 2D autonomous flows can be thought of as comprised of *closed* dynamical systems for each LCS, which is uninteresting from the perspective of this chapter.

Real flows are however *nonautonomous*; the associated velocity field is unsteady. This causes immediate difficulties, since heteroclinic manifolds are now moving with time and may also intersect each other in complicated ways[2] in each time slice. For any given stable and unstable manifolds, there may be infinitely many

[2]Transverse intersections are prohibited in autonomous flows, since the presence of such an intersection would violate uniqueness of trajectories—the trajectory passing through the intersection point will not be able to decide which manifold to follow.

intersections associated with transport which is potentially chaotic, a finite number of intersections or no intersections at all. This situation is compounded by the fact that all data from real flows is *finite time*, whereas infinite times are in reality needed to define stable and unstable manifolds, which are entities to which there is time-asymptotic exponential decay. How exactly could one *define* an LCS boundary in these situations? This question does not as yet have a completely agreed-on answer, but many *diagnostic* identification techniques have been developed, for which comprehensive reviews are available [32, 78, 99, 105]. The most popular technique is to use ridges of Finite-Time Lyapunov Exponent (FTLE) fields [33, 34, 37, 71–73, 103, 126], a method which continues to be developed [35, 65, 81, 113, 114] in order to achieve higher efficiency and because of the potential for false positives or negatives [46, 47, 60, 61]. A recently emerging class of diagnostics use Perron-Frobenius (transfer) and Koopman operators [43, 51, 53, 54, 80, 89, 91], while yet another is associated with averages along trajectories [70, 80, 85, 88, 92]. Several very recently developed methods [5, 36, 52, 62] illustrate an interesting diversity of approaches, indicating both the richness of this research area and the fact that a *full* understanding of the diagnostic methods used for LCSs is yet to be attained.

Despite this continuing development, many challenges remain. Firstly, what are the connections between each of these methods, and in what sense are they related to invariant manifolds? Secondly, are the methods *assured* of unambiguously identifying LCSs—that is, is there the potential for false positives or negatives? (Recently developed exact models in both two- and three-dimensional flows [17] may help in testing these two issues.) Thirdly, can the *infinite-time* necessity of idealised flows needed to define time-asymptotic decay towards/from invariant manifolds be reconciled with the *finite-time* nature of experimental, observational and numerical studies? Fourthly, if a particular diagnostic method can be used to identify LCSs, can the method be used to address the complementary question of quantification of transport from one LCS to another?

Methods for *quantifying* transport depend very much on the context in which the problem is posed. Two complementary aspects of transport are that of *advection* and *diffusion*, which are (somewhat confusingly) sometimes thought of as *transport* (or *stirring*) and *mixing*, respectively. The first of these relates to Lagrangian particle motion due to nonautonomous velocity fields and will be the main focus in this chapter. With regards specifically to transport from an LCS to another LCS, the results available tend to specialise towards specific scenarios. The idea of 'transitory' flows was used by Mosovsky and Meiss to quantify transport in incompressible flows from one LCS to another when the flow was nonautonomous only within a finite time interval [93, 94]. A highly popular idea due to Rom-Kedar and collaborators measures transport in the sense of crossing a 'pseudoseparatrix' for specifically incompressible time-periodic two-dimensional flows [109, 110, 128]. The transport measure in this case is an area of a lobe and builds on the theory of lobe areas in two-dimensional maps [40, 82, 83]. Under general time dependence, Haller and Poje [64] suggested measuring transport across a time-varying barrier consisting of a stable and an unstable manifold by measuring the flux of fluid which crosses a 'gate surface' connecting together these manifolds. A significant contribution here is to think of transport as a time-varying flux, but actually implementing this method is difficult in general due to the ambiguity of defining the gate surface.

The second aspect of transport, diffusion, incorporates the effects of small-scale randomness in many ways and is often associated with fluid mixing.[3] Quantifying diffusive effects tends to be associated with *global* mixing measures [55, 86, 95, 125, e.g.], and a standard numerical approach would be to quantify such a measure after using an advection-diffusion equation for passive scalars. What is particularly fascinating from the perspective of quantifying transport is the *interplay* between advection and diffusion, for which new insights are emerging [69, 108, 117, 122, e.g.]. From the perspective of this chapter, however, I will focus on advection, since it is instrumental in forming the gross LCS structures.

This chapter approaches the issues of quantifying transport between two-dimensional LCSs and demarcating the boundaries between them, under the constraint that the time variation of the flow occurs as a perturbation. Within this admittedly restrictive situation, a comprehensive theory on both identifying the boundaries and quantifying transport across them can be developed. Incompressibility is *not* a necessary ingredient in the theory. I begin in Sect. 1.2 by describing the setting in which the theory is to be developed. In Sect. 1.3, I adopt the attitude that the time-varying boundaries between the LCSs that we seek are stable and unstable manifolds and obtain leading-order expressions for their time-varying location. The development here is based on Melnikov theory which is well understood in certain situations [9,58,128, e.g.], but a couple of lesser known issues need to be addressed: compressibility and the tangential motion of invariant manifolds. I then in Sect. 1.4 develop a transport description for time-aperiodic perturbations in terms of an *instantaneous flux*, which is independent of the number of intersections between stable and unstable manifolds (which might be zero, a finite number or infinity), and is thus *not* dependent on lobe dynamics for which specific intersection patterns are usually required [50, 84, 110, 128]. I show how explicit expressions can be obtained for this instantaneous flux and in Sect. 1.5 outline the considerable simplifications which occur under additional conditions. I also describe how the instantaneous flux fits in with the well-known idea of lobe areas [109, 110, 128] which is relevant in time-harmonic incompressible flows. Finally, in Sect. 1.6, I discuss the implications and potential extensions of these results.

1.2 The Perturbative Setting

The focus shall be on systems in the form

$$\dot{\boldsymbol{x}} = \boldsymbol{f}(\boldsymbol{x}) + \varepsilon \boldsymbol{g}(\boldsymbol{x}, t) \tag{1.1}$$

[3] *Chaotic* mixing, on the other hand, can be thought of as a purely advective mechanism.

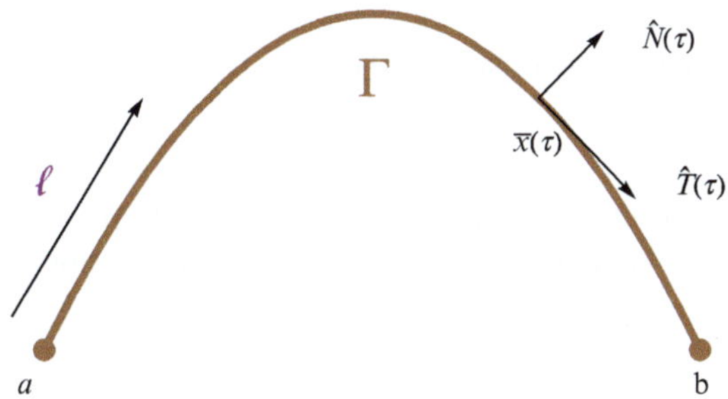

Fig. 1.2 The heteroclinic manifold Γ of (1.2), with a local coordinate structure at a general point $\bar{\boldsymbol{x}}(\tau)$

in which $\boldsymbol{x} \in \Omega$, a two-dimensional surface, and the parameter ε satisfies $0 \leq \varepsilon \ll 1$. The system (1.1) is a perturbation on the $\varepsilon = 0$ system

$$\dot{\boldsymbol{x}} = \boldsymbol{f}(\boldsymbol{x}) \tag{1.2}$$

for which certain hypotheses will be assumed.

Hypothesis 1. *The vector field $\boldsymbol{f} \in \mathrm{C}^2(\Omega)$.*

Hypothesis 2. *Equation (1.2) contains saddle-fixed points **a** and **b**, each possessing one-dimensional stable and unstable manifolds. A branch of the unstable manifold of **a** is assumed to coincide with a branch of the stable manifold of **b**, forming a heteroclinic manifold Γ.*

If $\boldsymbol{a}$ and $\boldsymbol{b}$ are the same point—which is permissible—Γ would be a homoclinic manifold. Figure 1.2 displays the required topological structure. In the unperturbed situation (1.2), Γ forms a barrier between the flows on the two sides and, depending on how the stable manifold of $\boldsymbol{a}$ and the unstable manifold of $\boldsymbol{b}$ behave (these are not pictured in Fig. 1.2), forms the boundary between two *closed* dynamical systems on the two sides of Γ. No transport occurs across Γ; this is zero escape rate from one dynamical system to another. Thus, Γ could be any of the heteroclinic manifolds illustrated in Fig. 1.1, which forms the boundary of an LCS. This behaviour changes when the perturbed flow (1.1) is considered, and characterising such transport between the *open* dynamical systems on the two sides of Γ, and defining the time-varying boundaries demarcating these systems, is the goal of the remainder of this chapter.

In order to do this, it helps to first introduce additional properties related to the unperturbed system (1.2), as indicated in Fig. 1.2. The interface Γ can be represented as a trajectory $\bar{\boldsymbol{x}}(t)$ of (1.2), such that

$$\lim_{t \to -\infty} \bar{\boldsymbol{x}}(t) = \boldsymbol{a} \quad \text{and} \quad \lim_{t \to \infty} \bar{\boldsymbol{x}}(t) = \boldsymbol{b}. \tag{1.3}$$

This representation is unique up to a shift (i.e. $\bar{\boldsymbol{x}}(t - \beta)$, for any real β, also has this property), since (1.2) posed with any initial condition on Γ has a solution $\bar{\boldsymbol{x}}(t)$ which satisfies these conditions. For a chosen initial condition $\bar{\boldsymbol{x}}(0)$, there is a solution $\bar{\boldsymbol{x}}(t)$ of (1.2) whose forward and backward trajectories span Γ. Let the parameter τ represent the position $\bar{\boldsymbol{x}}(\tau)$ on Γ. If ℓ is the arc length parametrisation along Γ such that $\ell = 0$ at $\boldsymbol{a}$ and $\ell = L$ (the length of Γ) at $\boldsymbol{b}$, then

$$\mathrm{d}\ell = |\boldsymbol{f}(\bar{\boldsymbol{x}}(\tau))| \, \mathrm{d}\tau, \qquad \text{or} \qquad \ell'(\tau) = |\boldsymbol{f}(\bar{\boldsymbol{x}}(\tau))|. \tag{1.4}$$

For each position $\bar{\boldsymbol{x}}(\tau)$ on Γ, I will set up a local orthogonal coordinate system by defining the *tangential* and *normal* unit vectors, respectively, by

$$\hat{\boldsymbol{T}}(\tau) = \frac{\boldsymbol{f}(\bar{\boldsymbol{x}}(\tau))}{|\boldsymbol{f}(\bar{\boldsymbol{x}}(\tau))|} \qquad \text{and} \qquad \hat{\boldsymbol{N}}(\tau) = \frac{J\,\boldsymbol{f}(\bar{\boldsymbol{x}}(\tau))}{|\boldsymbol{f}(\bar{\boldsymbol{x}}(\tau))|} \tag{1.5}$$

in which the 'rotation by $+\pi/2$' skew-symmetric matrix J is defined by

$$\mathrm{J} = \begin{pmatrix} 0 & -1 \\ 1 & 0 \end{pmatrix}.$$

These concepts are indicated in Fig. 1.2. The focus hereafter shall be on the perturbed situation (1.1). The perturbing vector field $\boldsymbol{g}$ will be assumed to satisfy:

Hypothesis 3. *The function* $\boldsymbol{g} \in \mathrm{C}^2(\Omega)$ *for any* $t \in \mathbb{R}$, $\boldsymbol{g} \in \mathrm{C}^2(\mathbb{R})$ *for any* $\boldsymbol{x} \in \Omega$, *and* $\boldsymbol{g}$ *and* $D\boldsymbol{g}$ *are both bounded in* $\Omega \times \mathbb{R}$.

The matrix $D\boldsymbol{g}$ mentioned above is the spatial derivative at fixed t. The formerly impermeable barrier Γ 'opens out' in the perturbed system (1.1), which needs characterisation. Defining the escape rates between the two open dynamical systems on the two sides of the perturbed version of Γ is a closely related question. The remainder of this chapter focusses on these issues.

Before getting to the results, I would like to point out several important subcases for which I will be able to express simpler results. These subcases are:

Subcase 1 (Time-periodic). *If there exists a constant* T *such that* $\boldsymbol{g}(\boldsymbol{x}, t+T) = \boldsymbol{g}(\boldsymbol{x}, t)$ *for all* $(x, t) \in \Omega \times \mathbb{R}$, *this situation shall be called the* time-periodic subcase.

Subcase 2 (Time-harmonic). *If the representation* $\boldsymbol{g}(\boldsymbol{x}, t) = \boldsymbol{h}(\boldsymbol{x}) \cos(\omega t + \phi)$ *for some smooth function* $\boldsymbol{h} : \Omega \to \mathbb{R}^2$ *and constants* $\omega \neq 0$ *and* ϕ *is possible, this shall be called the* time-harmonic subcase.

Subcase 3 (Incompressible). *The subcase in which* $\nabla \cdot \boldsymbol{f} = \mathrm{Tr}\, D\boldsymbol{f} = 0$ *shall be called the* incompressible (or area-preserving) subcase. *If so, there exists a Hamiltonian function (or stream function)* $H : \Omega \to \mathbb{R}$ *defined by*

$$\boldsymbol{f}(\boldsymbol{x}) = -J\, DH(\boldsymbol{x}) = -J\, \nabla H(\boldsymbol{x}, t), \tag{1.6}$$

such that H *is conserved along flow trajectories of (1.2),*[4] *and moreover* $\hat{\boldsymbol{N}}$ *in (1.5) is the unit vector in the direction of* ∇H.

[4] $\frac{dH(\boldsymbol{x}(t))}{dt} = DH(\boldsymbol{x}(t)) \cdot \dot{\boldsymbol{x}} = DH(\boldsymbol{x}(t)) \cdot [-JDH(\boldsymbol{x}(t))] = 0.$

Unless explicitly stated, the results that I will state will *not* assume a particular subcase. In Sect. 1.5, however, I shall describe simplifications which occur to my transport characterisations under various combinations of these subcases.

1.3 Boundaries Between Open Dynamical Systems: Invariant Manifolds

To take into account the explicit time dependence in (1.1), I shall represent it in the *augmented form*

$$\left.\begin{aligned} \dot{\boldsymbol{x}} &= \boldsymbol{f}\,(\boldsymbol{x}) + \varepsilon \boldsymbol{g}\,(\boldsymbol{x}, t) \\ \dot{t} &= 1 \end{aligned}\right\} , \qquad (1.7)$$

which now renders the system *autonomous* at the cost of expanding phase space to three dimensions. The $\varepsilon = 0$ version of (1.7) has trivial time dependence, whose important flow structures are shown in Fig. 1.3. The saddle point $\boldsymbol{a}$ of (1.2) becomes a trajectory $(\boldsymbol{a}, t)$ in the augmented system. This trajectory is *hyperbolic* and possesses two-dimensional stable and unstable manifolds, corresponding exactly to the one-dimensional stable and unstable manifolds of the saddle point $\boldsymbol{a}$ in the non-augmented system (1.2). For example, the branch of the one-dimensional unstable manifold of $\boldsymbol{a}$ which formed Γ becomes a two-dimensional surface in the augmented system. Trajectories on this two-dimensional surface are exponentially pushed away from $(\boldsymbol{a}, t)$ as time progresses. The hyperbolicity of $(\boldsymbol{a}, t)$ is associated exactly with such exponential rates of decay/growth and can be precisely characterised in terms of *exponential dichotomies* [27, 42, 97]. Similarly, the one-dimensional stable manifold of $\boldsymbol{b}$ of (1.2) corresponds to a two-dimensional stable manifold of the trajectory $(\boldsymbol{b}, t)$, which coincides with the two-dimensional unstable manifold of $(\boldsymbol{a}, t)$. Thus, the heteroclinic manifold in this situation can be thought of as $\Gamma \times \mathbb{R}$, and the intersection of this surface at *any* fixed time t gives exactly Fig. 1.2.

Now, when $0 < \varepsilon \ll 1$, the hyperbolic trajectories $(\boldsymbol{a}, t)$ and $(\boldsymbol{b}, t)$ of (1.7) perturb to nearby hyperbolic trajectories $\boldsymbol{A}_\varepsilon := (\boldsymbol{a}_\varepsilon(t), t)$ and $\boldsymbol{B}_\varepsilon := (\boldsymbol{b}_\varepsilon(t), t)$, respectively, and retain their stable and unstable manifolds. The proof of this fact is related to the persistence of exponential dichotomies under bounded perturbations

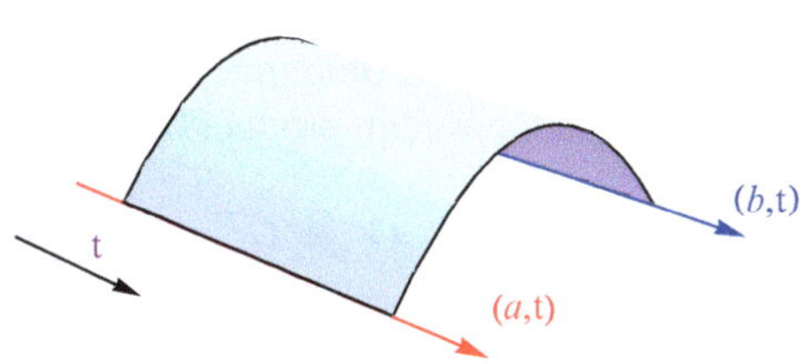

Fig. 1.3 Important structures associated with the augmented system (1.7) when $\varepsilon = 0$

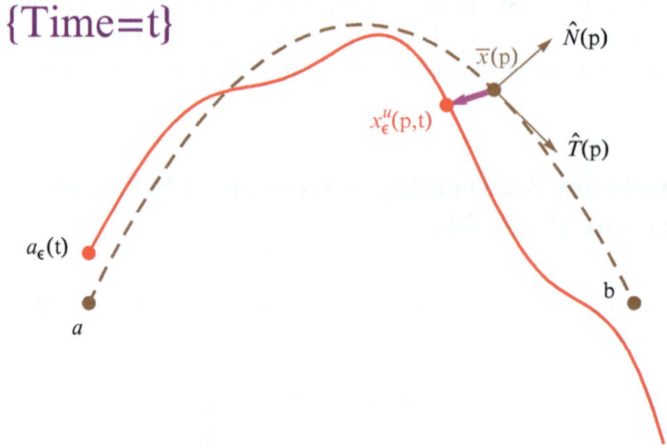

Fig. 1.4 Unstable manifold Γ_ε^u of the hyperbolic trajectory $(\mathbf{a}_\varepsilon(t), t)$ of (1.7) in a time slice t. The *dashed curve* shows the unperturbed unstable manifold Γ

[42, 129, 130]. (Related persistence results under slightly different assumptions are also available [48, 67, 97].) Now, the two-dimensional unstable manifold of $\boldsymbol{A}_\varepsilon$ and the two-dimensional stable manifold of $\boldsymbol{B}_\varepsilon$ are the entities of interest; when $\varepsilon = 0$ these coincided to form the impenetrable barrier $\Gamma \times \mathbb{R}$. The interesting issue is that these do *not* need to coincide when $\varepsilon \neq 0$, providing an avenue for exchange between regions above and below Γ in Fig. 1.2.

The perturbed versions of these manifolds will be characterised to leading order in ε based on recent results [16]. First, consider the unstable manifold of $\boldsymbol{A}_\varepsilon$, which shall be called Γ_ε^u. This must be attached to the hyperbolic trajectory $\boldsymbol{A}_\varepsilon$ and is a surface which is ε-close to a suitable restriction of $\Gamma \times \mathbb{R}$ in Fig. 1.3. The behaviour of Γ_ε^u as it gets close to the hyperbolic trajectory $\boldsymbol{B}_\varepsilon$ is not obvious. Since this second hyperbolic trajectory also possesses an unstable manifold, Γ_ε^u will get pulled away from $\boldsymbol{B}_\varepsilon$ as it gets closer to it and thus be subject to global flow properties. I will focus on the part of Γ_ε^u which remains $\mathcal{O}(\varepsilon)$-close and show how it can be characterised.

To do this, I fix a value of p and consider the point $\bar{\boldsymbol{x}}(p)$. I also fix a time value t—my intention is to represent the perturbed unstable manifold parametrically in terms of the *p*osition p and *t*ime-slice t. Now, the intersection of Γ_ε^u in the time slice t is pictured in Fig. 1.4. By smoothness in ε, there is a trajectory $\boldsymbol{x}_\varepsilon^u(p,t)$ of (1.1) which lies on Γ_ε^u in the time slice t, whose backward time trajectory remains $\mathcal{O}(\varepsilon)$-close to the backward time trajectory through $\bar{\boldsymbol{x}}(p)$ of the unperturbed system (1.2). Now I define the wedge product between vectors in $\mathbb{R}^2$ by

$$\boldsymbol{F} \wedge \boldsymbol{G} := \boldsymbol{G}^T J \boldsymbol{F} = \boldsymbol{G} \cdot (J \boldsymbol{F}) = F_1 G_2 - G_1 F_2 \,, \tag{1.8}$$

in which the final equality expresses the wedge product in component form. I also define the projected rate of strain:

$$R(\xi) := \hat{\boldsymbol{N}}^T(\xi)\left[(D\boldsymbol{f})^T(\bar{\boldsymbol{x}}(\xi)) + D\boldsymbol{f}(\bar{\boldsymbol{x}}(\xi))\right]\hat{\boldsymbol{T}}(\xi) \tag{1.9}$$

in which the superscript T represents the transpose, and the unit vectors $\hat{\boldsymbol{T}}$ and $\hat{\boldsymbol{N}}$ are defined in (1.5).

Theorem 1.1. *The part of the perturbed unstable manifold associated with the hyperbolic trajectory $\boldsymbol{A}_\varepsilon$ of (1.7) has a parametric representation with parameters $(p,t) \in (-\infty, P) \times (-\infty, T)$ for arbitrarily large P and T, given by*

$$\boldsymbol{x}_\varepsilon^u(p,t) = \bar{\boldsymbol{x}}(p) + \varepsilon\left[\frac{M^u(p,t)}{|\boldsymbol{f}(\bar{\boldsymbol{x}}(p))|}\hat{\boldsymbol{N}}(p) + \frac{B^u(p,t)}{|\boldsymbol{f}(\bar{\boldsymbol{x}}(p))|}\hat{\boldsymbol{T}}(p)\right] + \mathcal{O}(\varepsilon^2), \tag{1.10}$$

in which

$$M^u(p,t) = \int_{-\infty}^{p} \exp\left[\int_\tau^p \nabla\cdot\boldsymbol{f}(\bar{\boldsymbol{x}}(\xi))\,\mathrm{d}\xi\right]\boldsymbol{f}(\bar{\boldsymbol{x}}(\tau)) \wedge \boldsymbol{g}(\bar{\boldsymbol{x}}(\tau), \tau + t - p)\,\mathrm{d}\tau \tag{1.11}$$

and

$$B^u(p,t) = |\boldsymbol{f}(\bar{\boldsymbol{x}}(p))|^2 \int_0^p \frac{R(\tau)M^u(p,\tau+t-p) + \boldsymbol{f}(\bar{\boldsymbol{x}}(\tau))\cdot\boldsymbol{g}(\bar{\boldsymbol{x}}(\tau),\tau+t-p)}{|\boldsymbol{f}(\bar{\boldsymbol{x}}(\tau))|^2}\,\mathrm{d}\tau\,. \tag{1.12}$$

Proof. Details are available in [16]. □

Remark 1.1. There is freedom in determining $\boldsymbol{x}_\varepsilon^u(p,t)$ which is $\mathcal{O}(\varepsilon)$-close to the trajectory passing through $(\bar{\boldsymbol{x}}(p), t)$ of the unperturbed system, since any adjacent point on the unstable manifold also satisfies this requirement. This freedom is spurious if attempting to determine the unstable manifold in each time slice t, which is a one-dimensional curve. Hence, I have made a *specific* choice in (1.10), which is associated with choosing the specific lower limit 0 in the integral (1.12).

Remark 1.2. The limitation on P and T to be finite means that the *full* unstable manifold of $\boldsymbol{A}_\varepsilon$ is not given by (1.10); the expression loses control of the manifold as $p \to \infty$ or $t \to \infty$.

Remark 1.3. While the fact that the *normal* perturbation of the unstable manifold is related to the Melnikov-like function M^u has been implicitly understood in the time-periodic subcase, the tangential movement as characterised by $B^u(p,t)$ in (1.12) is a new finding [16]. The impact of this previously neglected quantity has been investigated in the situation of a time-aperiodically perturbed Duffing oscillator [16], and its effect on the location of the unstable manifold is nontrivial.

Remark 1.4. The improper integral over $(-\infty, p)$ in defining M^u in (1.11) converges; see Remark 1.15. Convergence issues of a boundary term are also important in deriving (1.11) [16].

An exactly analogous calculation can be performed to determine the location of the perturbed stable manifold of $\boldsymbol{B}_\varepsilon$ of (1.7):

Theorem 1.2. *The part of the perturbed stable manifold associated with the hyperbolic trajectory $\boldsymbol{B}_\varepsilon$ of (1.7) has a parametric representation with parameters $(p,t) \in (P,\infty) \times (T,\infty)$ for arbitrarily large $-P$ and $-T$, given by*

$$\boldsymbol{x}^s_\varepsilon(p,t) = \bar{\boldsymbol{x}}(p) + \varepsilon \left[\frac{M^s(p,t)}{|\boldsymbol{f}(\bar{\boldsymbol{x}}(p))|} \hat{\boldsymbol{N}}(p) + \frac{B^s(p,t)}{|\boldsymbol{f}(\bar{\boldsymbol{x}}(p))|} \hat{\boldsymbol{T}}(p) \right] + \mathcal{O}(\varepsilon^2), \quad (1.13)$$

in which

$$M^s(p,t) = -\int_p^\infty \exp\left[\int_\tau^p \nabla \cdot \boldsymbol{f}(\bar{\boldsymbol{x}}(\xi))\, \mathrm{d}\xi\right] \boldsymbol{f}(\bar{\boldsymbol{x}}(\tau)) \wedge \boldsymbol{g}(\bar{\boldsymbol{x}}(\tau), \tau + t - p)\, \mathrm{d}\tau \quad (1.14)$$

and

$$B^s(p,t) = |\boldsymbol{f}(\bar{\boldsymbol{x}}(p))|^2 \int_0^p \frac{R(\tau)M^s(p,\tau+t-p) + \boldsymbol{f}(\bar{\boldsymbol{x}}(\tau)) \cdot \boldsymbol{g}(\bar{\boldsymbol{x}}(\tau), \tau+t-p)}{|\boldsymbol{f}(\bar{\boldsymbol{x}}(\tau))|^2}\, \mathrm{d}\tau\,. \quad (1.15)$$

Proof. Details are available in [16]. □

Remark 1.5. The leading-order stable and unstable manifolds given in Theorems 1.1 and 1.2 are parametrised by p and t. The former parameter enables thinking of each manifold segment as the graph of a function from Γ. The inability to represent the invariant manifolds in regions beyond Γ (e.g. the parts of the unstable manifold pictured in Fig. 1.4 which have progressed further than $\boldsymbol{b}$) may be seen as a serious shortcoming when attempting to quantify LCS barriers. However, as shown in Sect. 1.4, knowledge of the parts of the manifolds as given in Theorems 1.1 and 1.2 is sufficient for flux characterisation.

Remark 1.6. It is natural to pose the inverse problem: for a specified behaviour of the nonautonomous stable and unstable manifolds, is it possible to determine the requirements on g? Ongoing work [22] has made promising strides in this *manifold control* problem.

Together, Theorems 1.1 and 1.2 characterise the *two* entities to which the unperturbed heteroclinic manifold Γ splits under the influence of the time-dependent perturbation in (1.1). These are associated with 'flow separators', that is, of time-varying boundaries between the 'top' and 'bottom' open dynamical systems which were separated by Γ in Fig. 1.2. However, exactly how these separate the two regions requires interpretation since many sorts of behaviours of these invariant manifolds are possible. For example, in each time slice t, these two manifolds could intersect infinitely many times (the classical paradigm under *time-harmonic* perturbations [109, 110, 128]), only finitely many times [13] or not at all [24, 111, e.g.]. This question is intimately linked to how one characterises the transport (flux) between the regions originally separated by Γ, that is, the escape of trajectories from one open dynamical system to another.

1.4 Flux Quantification

The results of Sect. 1.3 helps locate stable and unstable manifolds in nonautonomously perturbed systems in the form (1.1). The question to be addressed now is: knowing the time variation of these manifolds as described in Theorems 1.1 and 1.2, is it possible to *quantify* a transport across them?

Within the context of two-dimensional *maps*, transport is often expressed in terms of lobes resulting from the intersection of stable and unstable manifolds [40, 82, 83, 109]. A typical picture is shown in Fig. 1.5. By evaluating how each lobe evolves under the map P, a good qualitative understanding of flux can be obtained. Time-periodic flows can be viewed within this framework, by considering a Poincaré map which samples the flow at the period of the flow [110]. Lobes are therefore used extensively in time-periodic flows, or when the time aperiodicity lends itself to defining a sequence of maps [18, 28, 50, 84, 107, e.g.]. There are however several difficulties with this interpretation. First, if there are *no* intersections between the stable and unstable manifolds, then there are no lobes present.[5] How would flux be quantified in this case? Second, the lobes will typically have different areas, and so which lobe(s) are the ones whose areas are to be used? Would two small lobes be considered to represent 'equivalent' transport to one lobe whose area is the sum of the smaller ones? This problem is compounded by the fact that there are typically infinitely many lobes present. Third, when one thinks of flux as a transport across 'something', what exactly is the 'something' in a situation

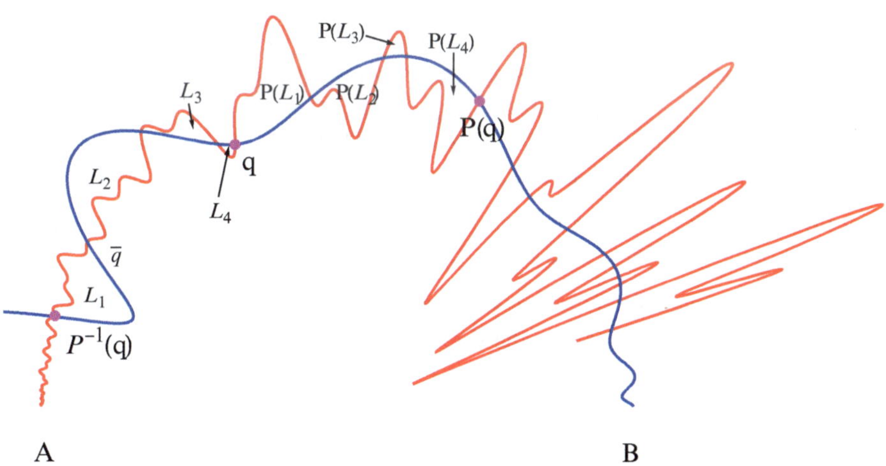

Fig. 1.5 A possible lobe structure resulting from the transverse intersection of a stable and unstable manifold for a map

[5] A concrete example with no lobes is presented in Remark 1.12. Viscosity-induced perturbations also result in no lobes [20, 24, 111].

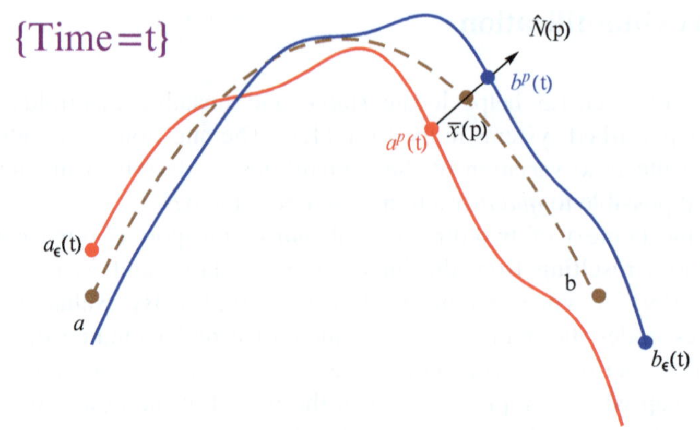

Fig. 1.6 A time slice t of the augmented system (1.7) for small $|\varepsilon|$. The pseudoseparatrix is indicated by the *curve* $\boldsymbol{a}_\varepsilon$–$\boldsymbol{a}^p$–$\boldsymbol{b}^p$–$\boldsymbol{b}_\varepsilon$, and the gate surface G is the *straight line* segment between $\boldsymbol{a}^p(t)$ and $\boldsymbol{b}^p(t)$, lying along the normal vector $\hat{\boldsymbol{N}}(p)$

like Fig. 1.5? In the instances in which the flow is incompressible and all lobes have equal areas, these questions can be resolved nicely by defining a pseudoseparatrix [109, 110, 128]. Indeed, there are pleasing formulæ using integrals of Melnikov functions to evaluate such lobe areas (originally developed by MacKay and Meiss [82]; see also [110, 128]). When the lobes are unequal or nonexistent, however, defining a flux becomes ambiguous.

To resolve this, I shall now define a flux which works not only in time-periodic situations such as those discussed above but also under general time-dependent perturbations. While the setting is assumed perturbative as in (1.1), some aspects of the following discussion apply more generally. For example, the gate surface in Definition 1.1 builds on an idea by Haller and Poje [64] to quantify transport associated with a stable and an unstable manifold under general time dependence.

Definition 1.1. Consider the flow of (1.7) intersected with a time slice t, and let $\bar{\boldsymbol{x}}(p)$ be a point lying on Γ (see Fig. 1.6):

(i) The *gate surface* G is the line segment between Γ_ε^u and Γ_ε^s, crossing through $\bar{\boldsymbol{x}}(p)$ and which lies along the direction $\hat{\boldsymbol{N}}(p)$. Let $\boldsymbol{a}^p(t)$ and $\boldsymbol{b}^p(t)$ be the points at which G intersects Γ_ε^u and Γ_ε^s, respectively.
(ii) The *pseudoseparatrix*, parametrised by (p, t), is the curve segment in the time slice t comprising the segments $\boldsymbol{a}_\varepsilon(t)$–$\boldsymbol{a}^p(t)$–$\boldsymbol{b}^p(t)$–$\boldsymbol{b}_\varepsilon(t)$, in which the first and the last segments are the curves lying along Γ_ε^u and Γ_ε^s, respectively, and the middle segment is the gate surface G.
(iii) The *instantaneous flux* at an instance in time t across the pseudoseparatrix is defined by

$$\Phi(p,t) := \operatorname{sgn}\left\{\left[\boldsymbol{a}^p(t) - \boldsymbol{b}^p(t)\right] \cdot \hat{\boldsymbol{N}}(p)\right\} \int_G \left[\boldsymbol{f}\left(\boldsymbol{x}(s)\right) + \varepsilon \boldsymbol{g}\left(\boldsymbol{x}(s), t\right)\right] \cdot \hat{\boldsymbol{T}}(p)\, \mathrm{d}s \tag{1.16}$$

in which $\boldsymbol{x}(s)$ is an arc length parametrisation of the gate surface G.

Remark 1.7. The pseudoseparatrix is a time-varying entity, and there is no transport across the segments $\boldsymbol{a}_\varepsilon(t)$–$\boldsymbol{a}^p(t)$ and $\boldsymbol{b}^p(t)$–$\boldsymbol{b}_\varepsilon(t)$ since these are segments of invariant manifolds. Thus, the time-varying transport across the pseudoseparatrix is limited to that across the gate surface. The integral in Definition 1.1 ensures that the instantaneous flux is measured as an *amount of fluid crossing the pseudoseparatrix per unit time*, hence justifying using the term 'flux'.

Remark 1.8. Since $\hat{\boldsymbol{T}}(p)$ is the normal vector to G at all points on G and is in the direction of the unperturbed flow, the integral in Definition 1.1 is positive to leading order in ε. The sign term outside ensures that the following convention is met: if the flux across the pseudoseparatrix crosses Γ in the direction of $\hat{\boldsymbol{N}}$, then the instantaneous flux is positive. If in the opposite direction, it is negative.

Let me elaborate further on the sign convention in Remark 1.8. At the instance of time pictured in Fig. 1.6, the transport occurs from left to right, or more properly, from the region above the pseudoseparatrix to below it. The positive direction across Γ is defined to be in the direction of $\hat{\boldsymbol{N}}(p)$. This remains unidirectional as p is varied; in Fig. 1.6, for example, it is always from below the pseudoseparatrix to above it. On the other hand, for the situation pictured in Fig. 1.6, the instantaneous transport across the gate surface leads to fluid from above the pseudoseparatrix travelling to below it. The sign term in (1.16) takes care of this; since $\hat{\boldsymbol{N}}$ is in the opposite direction to $\boldsymbol{a}^p(t) - \boldsymbol{b}^p(t)$, the flux at the instance pictured in Fig. 1.6 is negative. If the vectors in the sign term of (1.16) are in opposite directions (as will eventually occur when the manifold intersection point visible in Fig. 1.6 moves through the gate surface G), the transport direction is reversed, and $\Phi(p,t)$ will be positive.

In quantifying the instantaneous flux, the length of the vector from $\boldsymbol{b}^p(t)$ to $\boldsymbol{a}^p(t)$ in Fig. 1.6, that is, the length of the gate surface G, is seemingly key. This can be obtained in terms of a Melnikov function. Melnikov's original development [87] and standard expositions [9, 58, 128] are mostly geared towards proving that a perturbation leads to chaotic motion [68, 128, e.g.]. Melnikov methods have recently had a surprising application in determining wave speeds of travelling waves in reaction-diffusion equations in combustion and ecology [14,19,23,26,98]. While geometric methods are customarily used to derive the Melnikov function [9, 13, 16, 57, 58, 68, 128], alternative functional analytic methods have also been developed [27, 41, 97]. The most frequently used versions of Melnikov theory [9, 58, 128] specialise to both the time-harmonic and incompressible subcases. However, more general versions of Melnikov functions are available [11, 13, 16, 68, e.g.], and in particular:

Theorem 1.3. *The width of the gate surface at a location and time parametrised by (p,t) is given by*

$$\left[\boldsymbol{a}^p(t)-\boldsymbol{b}^p(t)\right]\cdot\hat{\boldsymbol{N}}(p)=\varepsilon\frac{M(p,t)}{\left|\boldsymbol{f}\left(\bar{\boldsymbol{x}}(p)\right)\right|}+\mathscr{O}(\varepsilon^2) \tag{1.17}$$

in which the Melnikov function *is*

$$M(p,t)=\int_{-\infty}^{\infty}\exp\left[\int_{\tau}^{p}\nabla\cdot\boldsymbol{f}\left(\bar{\boldsymbol{x}}(\xi)\right)\,\mathrm{d}\xi\right]\boldsymbol{f}\left(\bar{\boldsymbol{x}}(\tau)\right)\wedge\boldsymbol{g}\left(\bar{\boldsymbol{x}}(\tau),\tau+t-p\right)\,\mathrm{d}\tau\,, \tag{1.18}$$

or alternatively, in terms of the arc length parametrisation (1.4) on Γ,

$$M(p,t)=\int_{0}^{L}\exp\left[\int_{\tau(\ell)}^{p}\nabla\cdot\boldsymbol{f}\left(\bar{\boldsymbol{x}}(\xi)\right)\,\mathrm{d}\xi\right]g^{\perp}\left(\bar{\boldsymbol{x}}\left(\tau(\ell)\right),\tau(\ell)+t-p\right)\,\mathrm{d}\ell\,, \tag{1.19}$$

in which $\tau(\ell)$ is the relationship between a general position $\bar{\boldsymbol{x}}(\tau)$ and arc length ℓ as indicated by (1.4), and

$$g^{\perp}\left(\bar{\boldsymbol{x}}(\tau),\cdot\right):=\boldsymbol{g}\left(\bar{\boldsymbol{x}}(\tau),\cdot\right)\cdot\hat{\boldsymbol{N}}(\tau) \tag{1.20}$$

is the projection of $\boldsymbol{g}$ in the normal direction at each general point on Γ.

Proof. Subtracting (1.13) from (1.10) and considering only the normal term leads to (1.18). The expression (1.19) arises from the observation that $\boldsymbol{f}\wedge\boldsymbol{g}\mathrm{d}\tau=\boldsymbol{g}\cdot(J\boldsymbol{f})\,\mathrm{d}\tau=\boldsymbol{g}\cdot\hat{\boldsymbol{N}}\,|J\boldsymbol{f}|\,\mathrm{d}\tau=|\boldsymbol{f}|\,g^{\perp}\mathrm{d}\tau=g^{\perp}\mathrm{d}\ell$, in accordance with the connection between the arc length parametrisation and location given in (1.4). At a general position, this implies that $\ell'(\tau)=\left|\boldsymbol{f}\left(\bar{\boldsymbol{x}}(\tau)\right)\right|$, and $\tau(\ell)$ is the inverse relationship between τ and ℓ. □

Remark 1.9. While (1.18) was derived by Holmes [68] 30 years ago, establishment of the convergence of (1.11) and (1.14) when $\nabla\cdot\boldsymbol{f}\neq 0$, and the legitimacy of discarding boundary terms at $\pm\infty$—required for the subtraction in the proof—is more recent [16]. The basic reason for this convergence is explained in Remark 1.15.

Remark 1.10. In view of the expression (1.17), the existence of a transverse intersection of the stable and unstable manifolds in a fixed time slice t can be imputed using a straightforward implicit function theorem argument if $M(p,t)$ has a simple zero with respect to p. Alternatively, a transverse intersection near a fixed position $\bar{\boldsymbol{x}}(p)$ in some time slice can be imputed from a simple zero of $M(p,t)$ with respect to t. This duality occurs because each trajectory which is heteroclinic persists in every time slice.

Remark 1.11. The Melnikov function, being the $\mathscr{O}(\varepsilon)$-term of the distance function, is only able to quantify *primary* intersection points, as understood as follows from Fig. 1.6. In reality, in each time slice t, the unstable manifold emanating from $\boldsymbol{a}_\varepsilon(t)$ gets pushed out near $\boldsymbol{b}_\varepsilon(t)$ since there is an unstable manifold (not pictured in Fig. 1.6) emanating from $\boldsymbol{b}_\varepsilon(t)$ as well. Consequently, the $\mathcal{O}(\varepsilon)$ distance representation becomes meaningless in these parts of the manifold. In a global sense, the unstable manifold emanating from $\boldsymbol{a}_\varepsilon(t)$ may get pushed outside the $\mathcal{O}(\varepsilon)$-region but then return subsequently to intersect the stable manifold emanating from $\boldsymbol{b}_\varepsilon(t)$ elsewhere. The Melnikov function is not able to analyse such a possibility.

Remark 1.12. If $M(p,t)$ were strictly bounded away from zero for all (p,t), there would be no primary intersections of the manifolds. For example, suppose

$$\boldsymbol{g}(\boldsymbol{x},t) = J\boldsymbol{f}(\boldsymbol{x})\,[3+\cos\omega t]\;, \tag{1.21}$$

for which the flow (1.1) is time periodic. Then

$$\begin{aligned} M(p,t) &= \int_{-\infty}^{\infty} \exp\left[\int_\tau^p \nabla\cdot\boldsymbol{f}\,(\bar{\boldsymbol{x}}(\xi))\,\mathrm{d}\xi\right] |\boldsymbol{f}\,(\bar{\boldsymbol{x}}(\tau))|^2\,[3+\cos\omega\,(\tau+t-p)]\,\mathrm{d}\tau \\ &\geq \int_{-\infty}^{\infty} 2\exp\left[\int_\tau^p \nabla\cdot\boldsymbol{f}\,(\bar{\boldsymbol{x}}(\xi))\,\mathrm{d}\xi\right] |\boldsymbol{f}\,(\bar{\boldsymbol{x}}(\tau))|^2\,\mathrm{d}\tau > 0 \end{aligned} \tag{1.22}$$

where the inequalities are possible since the integrand is strictly positive. Thus, *even in the time-periodic case,* it is possible that there are no intersections.

Corollary 1.1. *Upon defining the scaled Melnikov function*

$$\tilde{M}(p,t) := \int_{-\infty}^{\infty} \exp\left[\int_\tau^0 \nabla\cdot\boldsymbol{f}\,(\bar{\boldsymbol{x}}(\xi))\,\mathrm{d}\xi\right] \boldsymbol{f}\,(\bar{\boldsymbol{x}}(\tau))\wedge\boldsymbol{g}\,(\bar{\boldsymbol{x}}(\tau),\tau+t-p)\,\mathrm{d}\tau\,, \tag{1.23}$$

the Melnikov function can be written as

$$M(p,t) = \tilde{M}(p,t)\exp\left[\int_0^p \nabla\cdot\boldsymbol{f}\,(\bar{\boldsymbol{x}}(\xi))\,\mathrm{d}\xi\right], \tag{1.24}$$

whose simple zeroes are exactly associated with simple zeroes of $\tilde{M}(p,t)$.

Proof. I split the integral inside the exponential in (1.18) into two integrals: one going from τ to 0 and the other going from 0 to p. The exponential of the second integral can then be pulled out of the outer integral, resulting in (1.24). □

Remark 1.13. The distance function (1.17) can be written in terms of the scaled Melnikov function as

$$[\boldsymbol{a}^p(t)-\boldsymbol{b}^p(t)]\cdot\hat{\boldsymbol{N}}(p) = \varepsilon\frac{\tilde{M}(p,t)\exp\left[\int_0^p \nabla\cdot\boldsymbol{f}\,(\bar{\boldsymbol{x}}(\xi))\,\mathrm{d}\xi\right]}{|\boldsymbol{f}\,(\bar{\boldsymbol{x}}(p))|} + \mathcal{O}(\varepsilon^2)\,. \tag{1.25}$$

In view of the fact that the other factors in the leading-order term are non-zero, simple zeroes of the function $\tilde{M}(p,t)$ relate to intersections of the stable and unstable manifolds. Thus, either the scaled or unscaled Melnikov function could be used without prejudice when seeking intersections of manifolds for small $|\varepsilon|$.

Remark 1.14. The advantage of working with $\tilde{M}(p,t)$ rather than $M(p,t)$ in the analysis of intersections is that $\tilde{M}$, unlike M, can be expressed in terms of just *one* variable $(t - p)$. However, I will retain both quantities since in Theorem 1.4 I show that M is critical to flux characterisation.

Remark 1.15. I now establish the convergence of the improper integral in $\tilde{M}$; this argument also works for convergence of the expressions (1.11) and (1.14) used for determining the unstable and stable manifolds. Upon defining

$$\mu(\tau) := \exp\left[\int_\tau^0 \nabla \cdot \boldsymbol{f}\,(\bar{\boldsymbol{x}}(\xi))\,\mathrm{d}\xi\right] \boldsymbol{f}\,(\bar{\boldsymbol{x}}(\tau)) \tag{1.26}$$

it is clear that the scaled Melnikov function can be written as

$$\tilde{M}(p,t) = \int_{-\infty}^{\infty} \mu(\tau) \wedge \boldsymbol{g}\,(\bar{\boldsymbol{x}}(\tau), \tau + t - p)\,\mathrm{d}\tau\,. \tag{1.27}$$

Now, $\boldsymbol{g}$ is bounded by hypothesis, and I now establish that $\mu(\tau)$ decays exponentially as $\tau \to \pm\infty$. Since $\boldsymbol{a}$ is a hyperbolic saddle fixed point of (1.2), I suppose the eigenvalues of $D\boldsymbol{f}\,(\boldsymbol{a})$ are α and β where $\alpha < 0 < \beta$, then as $\tau \to -\infty$, $\boldsymbol{f}\,(\bar{\boldsymbol{x}}(\tau)) \sim e^{\beta\tau}$, but $\exp\left[\int_\tau^0 \nabla \cdot \boldsymbol{f}\,(\bar{\boldsymbol{x}}(\xi))\,\mathrm{d}\xi\right] \sim \exp\left[\int_\tau^0 (\alpha + \beta)\,\mathrm{d}\xi\right] = e^{-(\alpha+\beta)\tau}$. Thus, $\mu(\tau) \sim e^{\beta t} e^{-(\alpha+\beta)t} = e^{-\alpha t}$ as $t \to -\infty$. A similar argument works as $\tau \to \infty$. This exponential decay in μ ensures that $\tilde{M}$ is well defined.

Using the distance function under both incompressibilty *and* time periodicity, and assuming that the Melnikov function has zeroes, a pleasing connection between the Melnikov function and lobe areas can be obtained [82, 109, 110, 128]. By (1.17), an intersection of manifolds occurs near zeroes of the Melnikov function. In a fixed time slice t, two adjacent zeroes in p (say p_1 and p_2) define the endpoints of a lobe. These correspond to points $\bar{\boldsymbol{x}}(p_1)$ and $\bar{\boldsymbol{x}}(p_2)$, which in turn are associated with arc length parameter values ℓ_1 and ℓ_2 using (1.4). Thus, using (1.17) [109, 110, 128],

$$\begin{aligned} \text{Lobe Area} &= \int_{\ell_1}^{\ell_2} \left| [\boldsymbol{a}^p(t) - \boldsymbol{b}^p(t)] \cdot \hat{\boldsymbol{N}} \right| \mathrm{d}\ell + \mathcal{O}(\varepsilon^2) \\ &= \varepsilon \int_{p_1}^{p_2} |M(p,t)|\,\mathrm{d}p + \mathcal{O}(\varepsilon^2)\,, \end{aligned} \tag{1.28}$$

which has also been obtained via an action formalism [77, 82]. This powerful expression has been frequently used in the incompressible and time-periodic subcases [28, 107, e.g.]. However, the stronger time-*harmonicity* condition is in actuality needed to show that the areas of *all* lobes are equal to leading order,[6] enabling (1.28) to be used as an unambiguous measure of flux.

[6]See Corollary 1.5 in Sect. 1.5.

Let me now return to the most general setting in which neither time periodicity nor incompressibility is assumed. While (1.28) remains a valid definition for the size of a *particular* lobe, establishing a connection to flux is difficult. I therefore revert to the *instantaneous flux* as defined in (1.16), which is valid even if there were no lobes. What is particularly pleasing is that for flows of the form (1.1), the instantaneous flux has a much stronger connection to the Melnikov function than, for example, in (1.28), where an *integral* of the Melnikov function is connected to a flux measure. In the general time-dependent setting, the Melnikov function *is* the leading-order flux:

Theorem 1.4. *The instantaneous flux $\Phi(p,t)$ associated with a time slice t and a gate surface drawn at $\bar{\boldsymbol{x}}(p)$ as shown in Fig. 1.6 is given by*

$$\Phi(p,t) = \varepsilon M(p,t) + \mathcal{O}(\varepsilon^2)\,, \tag{1.29}$$

in which the Melnikov function is defined in (1.18).

Proof. This argument was provided in [13], but the basic idea is that the $\varepsilon \boldsymbol{g}$ term appearing in (1.16) can be ignored when seeking leading-order information. Thus,

$$\begin{aligned}
\Phi(p,t) &= \operatorname{sgn}\left\{[\boldsymbol{a}^p(t) - \boldsymbol{b}^p(t)]\cdot\hat{\boldsymbol{N}}(p)\right\}\int_G [\boldsymbol{f}(\boldsymbol{x}(s)) + \varepsilon\boldsymbol{g}(\boldsymbol{x}(s),t)]\cdot\hat{\boldsymbol{T}}(p)\,\mathrm{d}s \\
&= \operatorname{sgn}\left\{[\boldsymbol{a}^p(t) - \boldsymbol{b}^p(t)]\cdot\hat{\boldsymbol{N}}(p)\right\}|\boldsymbol{f}(\bar{\boldsymbol{x}}(p))|\left|[\boldsymbol{a}^p(t) - \boldsymbol{b}^p(t)]\cdot\hat{\boldsymbol{N}}(p)\right| + \mathcal{O}(\varepsilon^2) \\
&= \operatorname{sgn}\{M(p,t)\}\,\varepsilon\,|M(p,t)| + \mathcal{O}(\varepsilon^2)
\end{aligned}$$

where the expression (1.17) has been used. □

Remark 1.16. Let $\rho_0(\boldsymbol{x})$ be an equilibrium scalar density advected by the steady flow (1.2). If $\rho(\boldsymbol{x},t) = \rho_0(\boldsymbol{x}) + \mathcal{O}(\varepsilon)$ is a scalar density advected by the perturbed flow (1.1), then by extending the above idea [13], the flux of ρ across the gate surface is

$$\text{Instantaneous Flux of Scalar} = \varepsilon\rho_0(\bar{\boldsymbol{x}}(p))\,M(p,t) + \mathcal{O}(\varepsilon^2)\,. \tag{1.30}$$

Remark 1.17. In view of (1.19), the instantaneous flux can be written as

$$\Phi(p,t) = \varepsilon\int_0^L \exp\left[\int_p^{\tau(\ell)} \nabla\cdot\boldsymbol{f}(\bar{\boldsymbol{x}}(\xi))\,\mathrm{d}\xi\right] g^{\perp}(\bar{\boldsymbol{x}}(\tau(\ell)), \tau(\ell)+t-p)\,\mathrm{d}\ell + \mathcal{O}(\varepsilon^2)\,, \tag{1.31}$$

In determining the instantaneous flux, note that it is not the normal component of $\boldsymbol{g}$ at that instance in time t that needs to be used, but this component at a time value which is shifted by $p - \tau(\ell)$. This reflects how the *Lagrangian* motion influences the flux, which is further evidenced by the time-evolving compressibility term of the interior integral in (1.31). If an Eulerian instantaneous flux definition were used, the temporal argument of $g^{\perp}$ would have been simply t, and the additional interior integral would be absent.

Theorem 1.4 provides an important method for *quantifying* the flux, as a time-dependent entity, for perturbed nonautonomous flows in the form (1.1). The leading-order term of this flux is *exactly* the Melnikov function, and hence, it is possible to think of the Melnikov function as the *flux function*. This flux function can of course be computed from (1.18). In some frequently considered situations (as in the subcases described in Sect. 1.2), the flux function can be simplified considerably, as shown in the next section.

1.5 Simplifications of Flux Formulæ in the Subcases

I present some simplifications of the flux formulæ in relevant subcases. Since the specialisation to these formulæ are often simple manipulations, many proofs in this section will be omitted.

Corollary 1.2. *Under the incompressible subcase (Subcase 3), the leading-order flux function can be written in either of the forms*

$$M(p,t) = \int_{-\infty}^{\infty} \nabla H\left(\bar{\boldsymbol{x}}(\tau)\right) \cdot \boldsymbol{g}\left(\bar{\boldsymbol{x}}(\tau), \tau + t - p\right) \mathrm{d}\tau \tag{1.32}$$

$$= \int_{0}^{L} g^{\perp}\left(\bar{\boldsymbol{x}}\left(\tau(\ell)\right), \tau(\ell) + t - p\right) \mathrm{d}\ell\,, \tag{1.33}$$

in which the (p,t) *dependence appears only in the combination* $(t-p)$.

Now I revert to compressible flow, but which is time periodic (Subcase 1) and so there exists T such that $\boldsymbol{g}\left(\boldsymbol{x}, t+T\right) = \boldsymbol{g}\left(\boldsymbol{x}, t\right)$ for all $(\boldsymbol{x}, t) \in \Omega \times \mathbb{R}$. The development I present here is a generalisation of the incompressible and time-periodic results in [11], for which I have also adopted a more natural parametrisation. I define $\omega = 2\pi/T$ (the natural frequency associated with the period T) and also the Fourier coefficients $\boldsymbol{g}_n(\boldsymbol{x})$ for $n \in \mathbb{Z}$ by

$$\boldsymbol{g}_n(\boldsymbol{x}) = \frac{\omega}{2\pi} \int_0^{2\pi/\omega} \boldsymbol{g}\left(\boldsymbol{x}, t\right) e^{-in\omega t}\, \mathrm{d}t\,. \tag{1.34}$$

Since smoothness in t is assumed, $\boldsymbol{g}$ has a complex Fourier series representation

$$\boldsymbol{g}\left(\boldsymbol{x}, t\right) = \sum_{n=-\infty}^{\infty} \boldsymbol{g}_n\left(\boldsymbol{x}\right) e^{in\omega t}\,. \tag{1.35}$$

I now define the functions

$$\lambda_n(\tau) := \exp\left[\int_\tau^0 \nabla \cdot \boldsymbol{f}\,(\bar{\boldsymbol{x}}(\xi))\,\mathrm{d}\xi\right] \boldsymbol{f}\,(\bar{\boldsymbol{x}}(\tau)) \wedge \boldsymbol{g}_n\,(\bar{\boldsymbol{x}}(\tau)) \tag{1.36}$$

for $n \in \mathbb{Z}$. By virtue of the argument in Remark 1.15, each λ_n has exponential decay as $\tau \to \pm\infty$. Thus, I can define the associated Fourier transforms:

$$\Lambda_n(\omega) := \mathscr{F}\{\lambda_n(\tau)\}(\omega) := \int_{-\infty}^{\infty} \lambda_n(\tau) e^{-i\omega\tau}\,\mathrm{d}\tau\,, \tag{1.37}$$

where in reality ω is a *constant.*

Theorem 1.5. *Under time periodicity (Subcase 1), the scaled Melnikov function (1.23) is T-periodic in both p and t and can be represented as a complex Fourier series in the variable $(t - p)$ by*

$$\tilde{M}(p,t) = \sum_{n=-\infty}^{\infty} \tilde{M}_n e^{in\omega(t-p)} \quad , \quad \tilde{M}_n = \Lambda_n(-n\omega)\,. \tag{1.38}$$

Proof. Using the definition of the scaled Melnikov function (1.23), I write

$$\begin{aligned}\tilde{M}(p,t) &= \int_{-\infty}^{\infty} \exp\left[\int_\tau^0 \nabla \cdot \boldsymbol{f}\,(\bar{\boldsymbol{x}}(\xi))\,\mathrm{d}\xi\right] \boldsymbol{f}\,(\bar{\boldsymbol{x}}(\tau)) \wedge \sum_{n=-\infty}^{\infty} \boldsymbol{g}_n\,(\bar{\boldsymbol{x}}(\tau))\, e^{in\omega(\tau+t-p)}\,\mathrm{d}\tau \\ &= \sum_{n=-\infty}^{\infty} e^{in\omega(t-p)} \int_{-\infty}^{\infty} \lambda_n(\tau) e^{in\omega\tau}\,\mathrm{d}\tau = \sum_{n=-\infty}^{\infty} \Lambda_n(-n\omega) e^{in\omega(t-p)}\,.\end{aligned}$$

□

Remark 1.18. The fact that $\tilde{M}(p,t)$ is T-periodic in both p and t under the time-periodic subcase is obvious from its definition (1.23) and is also reflected in (1.38). The unscaled Melnikov function $M(p,t)$ is however only guaranteed to be T-periodic in t, and not in p. On the other hand, since the zeroes of M and $\tilde{M}$ coincide (see Remark 1.13), intersection points of stable and unstable manifolds respect periodicity in both the position p and time slice t.

Remark 1.19. Since (1.38) converges very rapidly, efficient computation is possible using fast Fourier transform software. This rapid convergence follows from the fact that $\boldsymbol{g}$ and $\boldsymbol{f}$ are C^2 in the spatial argument, and thus, $\lambda_n \in C^2(\mathbb{R})$. Thus, $\Lambda_n(s)$ decays to zero at least as fast as $|s|^{-2}$ as $s \to \pm\infty$, and therefore, $\Lambda_n(-n\omega)$ decays at least as $|n|^{-2}$ as $n \to \pm\infty$. Sample computations demonstrating this rapid convergence are available [11].

Corollary 1.3. *In the setting of Theorem 1.5, suppose that incompressibility (Subcase 3) is also assumed. Then (1.36) becomes $\lambda_n(\tau) = \nabla H\,(\bar{\boldsymbol{x}}(\tau)) \cdot \boldsymbol{g}_n\,(\bar{\boldsymbol{x}}(\tau))$, and the Melnikov function acquires the form*

$$M(p,t) = \sum_{n=-\infty}^{\infty} \Lambda_n(-n\omega)\, e^{in\omega(t-p)} . \tag{1.39}$$

I now specialise Theorem 1.5 to the time-harmonic situation in which $\boldsymbol{g}(\boldsymbol{x},t) = \boldsymbol{h}(\boldsymbol{x})\cos(\omega t + \phi)$, without assuming incompressibility. I first define

$$\lambda(\tau) := \exp\left[\int_{\tau}^{0} \nabla \cdot \boldsymbol{f}\,(\bar{\boldsymbol{x}}(\xi))\, \mathrm{d}\xi\right] \boldsymbol{f}\,(\bar{\boldsymbol{x}}(\tau)) \wedge \boldsymbol{h}\,(\bar{\boldsymbol{x}}(\tau)) , \tag{1.40}$$

along with its Fourier transform $\Lambda(\omega) := \mathscr{F}\{\lambda(\tau)\}$.

Corollary 1.4. *Under time harmonicity (Subcase 2) the scaled Melnikov function has the representation*

$$\tilde{M}(p,t) = |\Lambda(\omega)| \cos[\omega(t-p) + \phi - \mathrm{Arg}[\Lambda(\omega)]] , \tag{1.41}$$

and therefore, the Melnikov function can be represented by

$$M(p,t) = \exp\left[\int_{0}^{p} \nabla \cdot \boldsymbol{f}\,(\bar{\boldsymbol{x}}(\xi))\, \mathrm{d}\xi\right] |\Lambda(\omega)| \cos[\omega(t-p) + \phi - \mathrm{Arg}[\Lambda(\omega)]] . \tag{1.42}$$

Proof. While Corollary 1.4 can be obtained as an easy corollary of Theorem 1.5, the following straightforward trigonometric calculation is worth stating. From (1.23),

$$\begin{aligned}
\tilde{M}(p,t) &= \int_{-\infty}^{\infty} \lambda(\tau) \cos[\omega(\tau + t - p) + \phi]\, \mathrm{d}\tau \\
&= \cos[\omega(t-p)+\phi] \int_{-\infty}^{\infty} \lambda(\tau) \cos\omega\tau\, \mathrm{d}\tau - \sin[\omega(t-p)+\phi] \int_{-\infty}^{\infty} \lambda(\tau) \sin\omega\tau\, \mathrm{d}\tau \\
&= \cos[\omega(t-p)+\phi]\, \mathrm{Re}[\Lambda] + \sin[\omega(t-p)+\phi]\, \mathrm{Im}[\Lambda] \\
&= |\Lambda| \left\{ \cos[\omega(t-p)+\phi] \frac{\mathrm{Re}[\Lambda]}{|\Lambda|} + \sin[\omega(t-p)+\phi] \frac{\mathrm{Im}[\Lambda]}{|\Lambda|} \right\} \\
&= |\Lambda| \{\cos[\omega(t-p)+\phi] \cos(\mathrm{Arg}[\Lambda]) + \sin[\omega(t-p)+\phi] \sin(\mathrm{Arg}[\Lambda])\} \\
&= |\Lambda| \cos[\omega(t-p) + \phi - \mathrm{Arg}[\Lambda]] ,
\end{aligned}$$

which is (1.41). This strategy was initially suggested in [11, 12] for dealing with the incompressible situation. Obtaining M in (1.42) is a simple application of (1.24).

□

Remark 1.20. Thus, $\Lambda(\omega) \neq 0$ is a sufficient condition for the presence of a heteroclinic tangle even when incompressibility is *not* assumed. One quick Fourier transform replaces the classical method of having to perform contour integration in the complex plane to prove the existence of simple zeroes of M [9, 58, e.g.].

Theorem 1.6. *Under incompressible (Subcase 3) and time-harmonic (Subcase 2) conditions, the leading-order flux function is itself harmonic in* $(t - p)$ *and is given by*

$$M(p,t) = |\Lambda(\omega)| \cos\left[\omega(t-p) + \phi - \text{Arg}[\Lambda(\omega)]\right] , \tag{1.43}$$

where

$$\Lambda(\omega) = \mathscr{F}\left\{\nabla H\left(\bar{\boldsymbol{x}}(\tau)\right) \cdot h\left(\bar{\boldsymbol{x}}(\tau)\right)\right\} . \tag{1.44}$$

Remark 1.21. Since the (t, p)-dependence in (1.43) appears directly as a harmonic function of $(t - p)$, the instantaneous flux is harmonic to leading order. This causes fluid to slosh back and forth from above to below Γ; in each period $T = 2\pi/\omega$ of the harmonic perturbation, the flux undergoes exactly one harmonic cycle as well, and so fluid sloshes back and then forth exactly once in each cycle. This result shows moreover that within a fixed time slice t, there are exactly two lobes between an intersection point $\boldsymbol{q}$ and its image under the T-periodic Poincaré map. The topological structure of the intersections between a point and its image must match the topological structure of how Melnikov function (1.43) intersects the p-axis in one period.

Corollary 1.5. *Under incompressible (Subcase 3) and time-harmonic (Subcase 2) conditions, the lobes caused through the intersection of stable and unstable manifolds in any time slice t have identical areas to leading order in* ε*, given by*

$$\text{Lobe Area} = \varepsilon \frac{2\,|\Lambda(\omega)|}{\omega} + \mathcal{O}(\varepsilon^2) . \tag{1.45}$$

Proof. I take a fixed time slice t. Suppose p_1 and p_2 are two adjacent values at which $M(p,t)$ has zeroes; given the form (1.43), $p_2 = p_1 + \pi/\omega$. By (1.17) this means that the perturbed stable and unstable manifolds intersect near $\bar{\boldsymbol{x}}(p_1)$ and $\bar{\boldsymbol{x}}(p_2)$ in this time-slide and thereby form a lobe. Using (1.43) on the lobe formula (1.28), I get [11, 12, 15, 18]

$$\begin{aligned}
\text{Lobe Area} &= \varepsilon \int_{p_1}^{p_2} |M(p,t)| \,\mathrm{d}p + \mathcal{O}(\varepsilon^2) \\
&= \varepsilon\,|\Lambda(\omega)| \int_{p_1}^{p_2} \left|\cos\left[\omega\,(t-p) + \phi\right]\right| \,\mathrm{d}p + \mathcal{O}(\varepsilon^2) \\
&= \varepsilon\,|\Lambda(\omega)| \int_{-\pi/(2\omega)}^{\pi/(2\omega)} \cos\left(-\omega p\right) \,\mathrm{d}p + \mathcal{O}(\varepsilon^2) ,
\end{aligned}$$

in which I have used the fact that harmonic functions have the same absolute area between any two adjacent zeroes and have thus moved the integral to convenient limit. Computing this integral leads to (1.45). Since independent of the two adjacent values p_1 and p_2, the answer is the same to leading order for all lobes. □

Remark 1.22. If, following [108, 110], I define the *average flux* as being one of these lobe areas divided by the time of the Poincaré map (which is $T = 2\pi/\omega$), then

$$\text{Average Flux} = \frac{1}{2\pi/\omega}\,\text{Lobe Area} = \varepsilon\frac{|\Lambda(\omega)|}{\pi} + \mathcal{O}(\varepsilon^2)\,. \tag{1.46}$$

Remark 1.23. If the flow is incompressible (Subcase 3) and time harmonic (Subcase 2), then one might consider three potential flux measures: the amplitude of the instantaneous flux (1.43), a lobe area (1.45) or the average flux (1.46). The leading-order term of *all* three measures are governed by the quantity $|\Lambda(\omega)|$.

1.6 Concluding Remarks

Extensions of the theory presented in this chapter to genuinely nonautonomous (not merely in a perturbative sense) flows will clearly be useful. Thinking of time-varying stable and unstable manifolds as flow barriers still remains attractive, but locating them in general is not easy from a theoretical perspective. While many diagnostic techniques have been proposed for locating boundaries between LCSs *numerically*, questions still remain as to whether the diagnostics necessarily identify stable and unstable manifolds. If the attitude that these stable and unstable manifolds *are* the important entities to identify, and even if assuming that such an identification can be done, the next issue is deciding how to quantify the transport 'across' them. Since the pseudoseparatrix defined in this chapter is dependent on the perturbative nature of the flow, their systematic extension remains problematic. The propensity for the manifolds to wander around space in complicated ways precludes putting in a *fixed* gate surface. A cautionary note is also provided by Remark 1.17, which shows that in quantifying *Lagrangian* transport, determining the flux across a curve at a fixed instance in time generally provides an invalid assessment of the particle transport. The history of the fluid crossing such a curve is necessary. Transport needs to be quantified in some form of Lagrangian frame (i.e. a frame which follows particle trajectories in some appropriate way), but deciding how to set up this frame under general time dependence is not obvious. After all, in a frame local to *each* fluid particle, that particle undergoes no net transport.

The other obviously valuable extension would be to three dimensions. There is certainly some insight in the literature into how time-varying two-dimensional invariant manifolds might result in different transport mechanisms [25, 36, 94, 104], but a comprehensive extension of the results of this chapter are still lacking. The concepts analogous to area preservation and Hamiltonian flows might be volume preservation and Liouville methods [63, 90, 94], but a theory which does not insist on incompressibility (such as presented here) would have applicability in the atmosphere. A significant difficulty in three dimensions is in deciding on a good interpretation of flux. For example, the corresponding gate surface (between a two-dimensional stable manifold and a two-dimensional unstable manifold, both of which are varying with time) should presumably be a two-dimensional entity, but how exactly should one define its boundary? (This was not a problem in the two-dimensional case; the one-dimensional gate surface ended when it met each of the manifolds—whether the manifolds intersected or not. If in a three-dimensional situation in which the two two-dimensional manifolds do *not* intersect, e.g. any presumptive two-dimensional gate surface connecting these two manifolds will have a direction in which it cannot be bounded.) On a similar note, how can intersections between two one-dimensional manifolds, or between a one-dimensional and a two-dimensional manifold, be used to characterise flux? Such difficulties make the study of three-dimensional nonautonomous flows intriguing, challenging, and (of course) highly relevant.

Acknowledgements Funding from Connecticut College's Research Matters fund for travel to the Banff International Research Station (BIRS) for the workshop associated with this volume and funding from BIRS while at Banff are gratefully acknowledged. I also express thanks to the Simons Foundation for a Collaboration Grant for Mathematicians. I also express thanks to the Simons Foundation for a Collaboration grant in Mathematics, and the Australian Research Council for grant FT130100484.

References

1. Abdullaev, S.: Structure of motion near saddle points and chaotic transport in Hamiltonian systems. Phys. Rev. E **62**, 3508–3528 (2000)
2. Aharon, R., Rom-Kedar, V., Gildor, H.: When complexity leads to simplicity: ocean surface mixing simplified by vertical convection. Phys. Fluids **24**, 056,603 (2012)
3. Ahn, T., Kim, S.: Separatrix map analysis of chaotic transport in planar periodic vortical flows. Phys. Rev. E **49**, 2900–2912 (1994)
4. Alligood, K., Sauer, T., Yorke, J.: Chaos: An Introduction to Dynamical Systems. Springer, New York (1996)
5. Allshouse, M., Thiffeault, J.L.: Detecting coherent structures using braids. Phys. D **241**, 95–105 (2012)
6. Arnold, V.I.: Sur la topologie des écoulements stationnaires des fluides parfaits. C. R. Acad. Sci. Paris **261**, 17–20 (1965)
7. Arnold, V.I.: Sur la géométrie differentielle des groupes de lie de dimension infinie et ses applications á l'hydrodynamique des fluides parfaits. Ann. Inst Fourier **16**, 316–361 (1966)
8. Arnold, V.I.: Mathematical Methods of Classical Mechanics. Springer, New York (1978)
9. Arrowsmith, D., Place, C.: An Introduction to Dynamical Systems. Cambridge University Press, Cambridge (1990)
10. Balasuriya, S.: Approach for maximizing chaotic mixing in microfluidic devices. Phys. Fluids **17**, 118103 (2005)
11. Balasuriya, S.: Direct chaotic flux quantification in perturbed planar flows: general time-periodicity. SIAM J. Appl. Dyn. Syst. **4**, 282–311 (2005)

12. Balasuriya, S.: Optimal perturbation for enhanced chaotic transport. Phys. D **202**, 155–176 (2005)
13. Balasuriya, S.: Cross-separatrix flux in time-aperiodic and time-impulsive flows. Nonlinearity **19**, 282–311 (2006)
14. Balasuriya, S.: Invasions with density-dependent ecological parameters. J. Theory Biol. **266**, 657–666 (2010)
15. Balasuriya, S.: Optimal frequency for microfluidic mixing across a fluid interface. Phys. Rev. Lett. **105**, 064501 (2010)
16. Balasuriya, S.: A tangential displacement theory for locating perturbed saddles and their manifolds. SIAM J. Appl. Dyn. Syst. **10**, 1100–1126 (2011)
17. Balasuriya, S.: Explicit invariant manifolds and specialised trajectories in a class of unsteady flows. Phys. Fluids **24**, 127101 (2012)
18. Balasuriya, S., Finn, M.: Energy constrained transport maximization across a fluid interface. Phys. Rev. Lett. **108**, 244503 (2012)
19. Balasuriya, S., Gottwald, G.: Wavespeed in reaction-diffusion systems, with applications to chemotaxis and population pressure. J. Math. Biol. **61**, 377–399 (2010)
20. Balasuriya, S., Jones, C.K.R.T.: Diffusive draining and growth of eddies. Nonlin. Proc. Geophys. **8**, 241–251 (2001)
21. Balasuriya, S., Padberg-Gehle, K.: Controlling the unsteady analogue of saddle stagnation points. SIAM J. Appl. Math. **73**, 1038–1057 (2013)
22. Balasuriya, S., Padberg-Gehle, K.: Nonautonomous control of stable and unstable manifolds in two-dimensional flows (2014, submitted)
23. Balasuriya, S., Volpert, V.: Wavespeed analysis: approximating Arrhenius kinetics with step-function kinetics. Combust. Theor. Model. **12**, 643–670 (2008)
24. Balasuriya, S., Jones, C.K.R.T., Sandstede, B.: Viscous perturbations of vorticity-conserving flows and separatrix splitting. Nonlinearity **11**(1), 47–77 (1998)
25. Balasuriya, S., Mezić, I., Jones, C.K.R.T.: Weak finite-time Melnikov theory and 3D viscous perturbations of Euler flows. Phys. D **176**(1–2), 82–106 (2003)
26. Balasuriya, S., Gottwald, G., Hornibrook, J., Lafortune, S.: High Lewis number combustion wavefronts: a perturbative Melnikov analysis. SIAM J. Appl. Math. **67**, 464–486 (2007)
27. Battelli, F., Lazzari, C.: Exponential dichotomies, heteroclinic orbits, and Melnikov functions. J. Differ. Equ. **86**, 342–366 (1990)
28. Beigie, D., Leonard, A., Wiggins, S.: Invariant manifold templates for chaotic advection. Chaos Solitons Fractals **4**, 749–868 (1994)
29. Beron-Vera, F., Olascoaga, M., Brown, M., Kocak, H., Rypina, I.: Invariant-tori-like Lagrangian coherent structures in geophysical flows. Chaos **20**, 017,514 (2010)
30. Beron-Vera, F., Olascoaga, M., Brown, M., Kocak, H.: Zonal jets as meridional transport barriers in the subtropical and polar lower stratosphere. J. Atmos. Sci **69**, 753–767 (2012)
31. Bettencourt, J., López, C., Hernández-García, E.: Oceanic three-dimensional Lagrangian coherent structures: a study of a mesoscale eddy in the Benguela upwelling region. Ocean Model. **51**, 73–83 (2012)
32. Boffetta, G., Lacorata, G., Radaelli, G., Vulpiani, A.: Detecting barriers to transport: a review of different techniques. Phys. D **159**, 58–70 (2001)
33. Borgogno, D., Grasso, D., Pegoraro, F., Schep, T.: Barriers in the transition to global chaos in collisionless magnetic reconnection. I. Ridges of the finite time Lyapunov exponent field. Phys. Plasmas **18**, 102307 (2011)
34. Branicki, M., Mancho, A., Wiggins, S.: A Lagrangian description of transport associated with front-eddy interaction: application to data from the North-Western mediterranean sea. Phys. D **240**, 282–304 (2011)
35. Brunton, S., Rowley, C.: Fast computation of finite-time Lyapunov exponent fields for unsteady flows. Chaos **20**, 017,503 (2010)
36. Budišić, M., Mezić, I.: Geometry of ergodic quotient reveals coherent structures in flows. Phys. D **241**, 1255–1269 (2012)

37. de la Camara, A., Mechoso, C., Ide, K., Walterscheid, R., Schubert, G.: Polar night vortex breakdown and large-scale stirring in the southern stratosphere. Climate Dyn. **35**, 965–975 (2010)
38. del Castillo Negrete, D., Morrison, P.: Chaotic transport by Rossby waves in shear flow. Phys. Fluids A **5**, 948–965 (1993)
39. Chandrasesekhar, S.: Hydrodynamics and Hydrodynamic Stability. Dover, New York (1961)
40. Channon, S., Lebowitz, J.: Numerical experiments in stochasticity and homoclinic oscillations. Ann. New York Acad. Sci. **357**, 108–118 (1980)
41. Chow, S.N., Hale, J.K., Mallet-Paret, J.: An example of bifurcation to homoclinic orbits. J. Differ. Equ. **37**, 351–373 (1980)
42. Coppel, W.A.: Dichotomies in Stability Theory. Lecture Notes in Mathematics, vol. 629. Springer, Berlin (1978)
43. Dellnitz, M., Junge, O.: Almost invariant sets in Chua's circuit. Int. J. Bif. Chaos **7**, 2475–2485 (1997)
44. d'Ovidio, F., Isern, J., López, C., Hernández-García, C., García-Ladona, E.: Comparison between Eulerian diagnostics and finite-size Lyapunov exponents computed from the altimetry in the Algerian basin. Deep-Sea Res. **56**, 15–31 (2009)
45. Duc, L., Siegmund, S.: Hyperbolicity and invariant manifolds for planar nonautonomous systems on finite time intervals. Int. J. Bifurcation Chaos **18**, 641–674 (2008)
46. Farazmand, M., Haller, G.: Computing Lagrangian coherent structures from variational LCS theory. Chaos **22**, 013,128 (2012)
47. Farazmand, M., Haller, G.: Erratum and addendum to "A variational theory for Lagrangian Coherent Structures". Phys. D **241**, 439–441 (2012)
48. Fenichel, N.: Persistence and smoothness of invariant manifolds for flows. Indiana Univ. Math. J. **21**, 193–226 (1971/1972)
49. Forgoston, E., Billings, L., Yecko, P., Schwartz, I.: Set-based corral control in stochastic dynamical systems: making almost invariant sets more invariant. Chaos **21**, 013,116 (2011)
50. Franco, E., Pekerek, D., Peng, J., Dabiri, J.: Geometry of unsteady fluid transport during fluid-structure interactions. J. Fluid Mech. **589**, 125–145 (2007)
51. Froyland, G., Padberg, K.: Almost-invariant sets and invariant manifolds: connecting probabilistic and geometric descriptions of coherent structures in flows. Phys. D **238**, 1507–1523 (2009)
52. Froyland, G., Padberg, K.: Finite-time entropy: a probabilistic method for measuring nonlinear stretching. Physica D **241**(19), 1612–1628 (2012)
53. Froyland, G., Lloyd, S., Santitissadeekorn, N.: Coherent sets for nonautonomous dynamical systems. Phys. D **239**, 1527–1541 (2010)
54. Froyland, G., Santitissadeekorn, N., Monahan, A.: Transport in time-dependent dynamical systems: finite-time coherent sets. Chaos **20**, 043116 (2010)
55. Gouillart, E., Dauchot, O., Thiffeault, J.L.: Measures of mixing quality in open flows with chaotic advection. Phys. Fluids **23**, 013604 (2011)
56. Green, M., Rowley, C., Smits, A.: The unsteady three-dimensional wake produced by a trapezoidal pitching panel. J. Fluid Mech. **685**, 117–145 (2011)
57. Gruendler, J.: The existence of homoclinic orbits and the method of Melnikov for systems in $\mathbb{R}^n$. SIAM J. Math. Anal. **16**, 907–931 (1985)
58. Guckenheimer, J., Holmes, P.: Nonlinear Oscillations, Dynamical Systems and Bifurcations of Vector Fields. Springer, New York (1983)
59. Haller, G.: Distinguished material surfaces and coherent structures in three-dimensional fluid flows. Phys. D **149**, 248–277 (2001)
60. Haller, G.: Lagrangian coherent structures from approximate velocity data. Phys. Fluids A **14**, 1851–1861 (2002)
61. Haller, G.: A variational theory for Lagrangian coherent structures. Phys. D **240**, 574–598 (2011)
62. Haller, G., Beron-Vera, F.: Geodesic theory of transport barriers in two-dimensional flows. Phys. D **241**, 1680–1702 (2012)

63. Haller, G., Mezić, I.: Reduction of three-dimensional, volume-preserving flows by symmetry. Nonlinearity **11**, 319–339 (1998)
64. Haller, G., Poje, A.: Finite time transport in aperiodic flows. Phys. D **119**, 352–380 (1998)
65. Haller, G., Sapsis, T.: Lagrangian coherent structures and the smallest finite-time Lyapunov exponent. Chaos **21**, 023,115 (2011)
66. Haller, G., Yuan, G.C.: Lagrangian coherent structures and mixing in two-dimensional turbulence. Phys. D **147**, 352–370 (2000)
67. Hirsch, M., Pugh, C., Shub, M.: Invariant manifolds. Bull. Am. Math. Soc. **76**, 1015–1019 (1970)
68. Holmes, P.: Averaging and chaotic motions in forced oscillations. SIAM J. Appl. Math. **38**, 65–80 (1980)
69. Hughes, C., Thompson, A., Wilson, C.: Identification of jets and mixing barriers from sea level and vorticity measurements using simple statistics. Ocean Model. **32**, 44–57 (2010)
70. Jimenez-Madrid, J., Mancho, A.: Distinguished trajectories in time dependent vector fields. Chaos **19**, 013,111 (2009)
71. Joseph, B., Legras, B.: Relation between kinematic boundaries, stirring and barriers for the Antarctic polar vortex. J. Atmos. Sci. **59**, 1198–1212 (2002)
72. Kasten, J., Petz, C., Hotz, I., Hege, H.C., Noack, B., Tadmor, G.: Lagrangian feature extraction of the cylinder wake. Phys. Fluids **22**, 091,108 (2010)
73. Katija, K., Beaulieu, W., Regula, C., Colin, S., Costello, J., Dabiri, J.: Quantification of flows generated by the hydromedusa Aequorea victoria: a Lagrangian coherent structure analysis. Mar. Ecol. Prog. Ser. **435**, 111–123 (2011)
74. Kelley, D., Ouellette, N.: Separating stretching from folding in fluid mixing. Nat. Phys. **7**, 477–480 (2011)
75. Kelley, D., Ouellette, N.: Spatiotemporal persistence of spectral fluxes in two-dimensional weak turbulence. Phys. Fluids **23**, 115,101 (2011)
76. Kourentis, L., Konstantinidis, E.: Uncovering large-scale coherent structures in natural and forced turbulent wakes by combining PIV, POD, and FTLE. Exp. Fluids **52**, 749–763 (2012)
77. Kovacic, G.: Lobe area via an action formalism in a class of Hamiltonian systems. Phys. D **51**, 226–233 (1991)
78. Krauskopf, B., Osinga, H., Doedel, E., Henderson, M., Guckenheimer, J., Vladimirsky, A., Dellnitz, M., Junge, O.: A survey of method's for computing (un)stable manifold of vector fields. Int. J. Bif. Chaos **15**, 763–791 (2005)
79. Lekien, F., Coulliette, C., Mariano, A., Ryan, E., Shay, L., Haller, G., Marsden, J.: Pollution release tied to invariant manifolds: a case study for the coast of Florida. Phys. D **210**, 1–20 (2005)
80. Levnajić, Z., Mezić, I.: Ergodic theory and visualization. I. Mesochronic plots for visualization of ergodic partition and invariant sets. Chaos **20**, 033,114 (2010)
81. Lipinski, D., Mohseni, K.: A ridge tracking algorithm and error estimate for efficient computation of Lagrangian coherent structures. Chaos **20**, 017,504 (2010)
82. MacKay, R., Meiss, J.: Relation between quantum and classical thresholds for multiphoton ionization of excited atoms. Phys. Rev. A **37**, 4702–4706 (1988)
83. MacKay, R., Meiss, J., Percival, I.: Transport in Hamiltonian systems. Phys. D **13**, 55–81 (1984)
84. Malhotra, N., Wiggins, S.: Geometric structures, lobe dynamics, and Lagrangian transport in flows with aperiodic time dependence, with applications to Rossby wave flow. J. Nonlin. Sci. **8**, 401–456 (1998)
85. Mancho, A., Hernández-García, E., Small, D., Wiggins, S., Fernández, V.: Lagrangian transport through an ocean front in the northwestern Mediterranean sea. J. Phys. Oceanography **38**, 1222–1237 (2008)
86. Mathew, G., Mezić, I., Petzold, L.: A multiscale measure for mixing. Physica D **211**, 23–46 (2005)
87. Melnikov, V.K.: On the stability of the centre for time-periodic perturbations. Trans. Moscow Math. Soc. **12**, 1–56 (1963)

88. Mendoza, C., Mancho, A.: Hidden geometry of ocean flows. Phys. Rev. Lett. **105**, 038,501 (2010)
89. Mezić, I., Banaszuk, A.: Comparison of systems with complex behavior. Phys. D **197**, 101–133 (2004)
90. Mezić, I., Wiggins, S.: On the integrability and perturbation of three-dimensional flows with symmetry. J. Nonlin. Sci. **4**, 157–194 (1994)
91. Mezić, I., Wiggins, S.: A method for visualization of invariant sets of dynamical systems based on the ergodic partition. Chaos **9**, 213–218 (1999)
92. Mezić, I., Loire, S., Fonoberov, V., Hogan, P.: A new mixing diagnostic and Gulf oil spill movement. Science **330**, 486–489 (2010)
93. Mosovky, B., Meiss, J.: Transport in transitory dynamical systems. SIAM J. Appl. Dyn. Syst. **10**, 35–65 (2011)
94. Mosovky, B., Meiss, J.: Transport in transitory three-dimensional Liouville flows. SIAM J. Appl. Dyn. Syst. **11**, 1785–1816 (2012)
95. Moura, A., Feudel, U., Gouillart, E.: Mixing and chaos in open flows. Adv. Appl. Mech. **45**, 1–50 (2012)
96. Olascoaga, M., Haller, G.: Forecasting sudden changes in environmental pollution patterns. Proc. Nat. Acad. Sci. **109**, 4738–4743 (2012)
97. Palmer, K.: Exponential dichotomies and transversal homoclinic points. J. Diff. Equ. **55**, 225–256 (1984)
98. Parks, H., Ermentrout, B., Rubin, J.: The dynamics of forced coupled network of active elements. Phys. D **240**, 554–567 (2011)
99. Peacock, T., Dabiri, J.: Introduction to focus issue: Lagrangian coherent structures. Chaos **20**, 017501 (2010)
100. Pedlosky, J.: Geophysical Fluids Dynamics. Springer, New York (1979)
101. Peng, J., Peterson, R.: Attracting structures in volcanic ash transport. Atmos. Environ. **48**, 230–239 (2012)
102. Pierrehumbert, R.: Chaotic mixing of tracer and vorticity by modulated travelling Rossby waves. Geophys. Astrophys. Fluid Dyn. **58**, 285–319 (1991)
103. Pierrehumbert, R., Yang, H.: Global chaotic mixing in isentropic surfaces. J. Atmos. Sci. **50**, 2462–2480 (1993)
104. Pouransari, Z., Speetjens, M., Clercx, H.: Formation of coherent structures by fluid inertia in three-dimensional laminar flows. J. Fluid Mech. **654**, 5–34 (2010)
105. Provenzale, A.: Transport by coherent barotropic vortices. Ann. Rev. Fluid Mech. **31**, 55–93 (1999)
106. Radko, T.: On the generation of large-scale structures in a homogeneous eddy field. J. Fluid Mech. **668**, 76–99 (2011)
107. Rodrigue, S., Eschenazi, E.: Lobe transport analysis of the Kelvin-Stuart cat's eyes driven flow. Chaos **20**, 013,101 (2010)
108. Rom-Kedar, V., Poje, A.: Universal properties of chaotic transport in the presence of diffusion. Phys. Fluids **11**, 2044–2057 (1999)
109. Rom-Kedar, V., Wiggins, S.: Transport in two-dimensional maps. Arch. Ration. Mech. Appl. **109**, 239–298 (1990)
110. Rom-Kedar, V., Leonard, A., Wiggins, S.: An analytical study of transport, mixing and chaos in an unsteady vortical flow. J. Fluid Mech. **214**, 347–394 (1990)
111. Sandstede, B., Balasuriya, S., Jones, C.K.R.T., Miller, P.D.: Melnikov theory for finite-time vector fields. Nonlinearity **13**(4), 1357–1377 (2000)
112. Sapsis, T., Peng, G., Haller, G.: Instabilities on prey dynamics in jellyfish feeding. Bull. Math. Biol. **73**, 1841–1856 (2011)
113. Senatore, C., Ross, S.: Detection and characterization of transport barriers in complex flows via ridge extraction of the finite time Lyapunov exponent field. Int. J. Num. Meth. Engin. **86**, 1163–1174 (2011)
114. Shadden, S., Lekien, F., Marsden, J.: Definition and properties of Lagrangian coherent structures from finite-time Lyapunov exponents in two-dimensional aperiodic flows. Phys. D **212**, 271–304 (2005)

115. Shadden, S., Lekien, F., Paduan, J., Chavez, F., Marsden, J.: The correlation between surface drifters and coherent structures based on high-frequency radar data in Monterey bay. Deep Sea Res. II Topic. Stud. Oceanography **56**, 161–172 (2009)
116. Shraiman, B.: Diffusive transport in a Rayleigh-Bénard convection cell. Phys. Rev. A **36**, 261–267 (1987)
117. Shuckburgh, E., Hayes, P.: Diagnosing transport and mixing using a tracer-based coordinate system. Phys. Fluids **15**, 3342–3357 (2003)
118. Solomon, T., Gollub, J.: Chaotic particle transport in time-dependent Rayleigh-Bénard convection. Phys. Rev. A **38**, 6280–6286 (1988)
119. Speetjens, M., de Wispelaere, H., van Steenhoven, A.: Multi-functional Lagrangian flow structures in three-dimensional ac electro-osmotic micro-flows. Fluid Dyn. Res. **43**, 035503 (2011)
120. Sposito, G.: On steady flows with Lamb surfaces. Int. J. Eng. Sci. **35**, 197–209 (1997)
121. Tallapragada, P., Ross, S., Schmale, D.: Lagrangian coherent structures are associated with fluctuations in airborne microbial populations. Chaos **21**, 033,122 (2011)
122. Tang, W., Walker, P.: Finite-time statistics of scalar diffusion in Lagrangian coherent structures. Phys. Rev. E **86**, 045201(R) (2012)
123. Tang, W., Chan, P., Haller, G.: Lagrangian coherent structure analysis of terminal winds detected by Lidar. part I: Turbulence structures. J. Appl. Meteor. Climat. **50**, 325–338 (2011)
124. Tang, W., Chan, P., Haller, G.: Lagrangian coherent structure analysis of terminal winds detected by Lidar. Part II: Structure evolution and comparison with flight data. J. Appl. Meteor. Climat. **50**, 2167–2183 (2011)
125. Thiffeault, J.L.: Using multiscale norms to quantify mixing and transport. Nonlinearity **25**, R1–R44 (2012)
126. Titaud, O., Brankart, J., Verron, J.: On the use of finite-time Lyapunov exponents and vectors for direct assimilation of tracer images into ocean models. Tellus Ser. A Dyn. Meteorol. Oceanography **63**, 1038–1051 (2011)
127. Weiss, J., Knobloch, E.: Mass transport by modulated traveling waves. Phys. Rev. A **40**, 2579–2589 (1989)
128. Wiggins, S.: Chaotic Transport in Dynamical Systems. Springer, New York (1992)
129. Yagasaki, K.: Invariant manifolds and control of hyperbolic trajectories on infinite- or finite-time intervals. Dyn. Syst. **23**, 309–331 (2008)
130. Yi, Y.: A generalized integral manifold theorem. J. Differ. Equ. **102**, 153–187 (1993)

Chapter 2
Eigenvalues of Transfer Operators for Dynamical Systems with Holes

Oscar F. Bandtlow and Oliver Jenkinson

Abstract For real-analytic expanding open dynamical systems in arbitrary finite dimension, we establish rigorous explicit bounds on the eigenvalues of the corresponding transfer operators acting on spaces of holomorphic functions. In dimension 1 the eigenvalue decay rate is exponentially fast, while in dimension d it is $O(\theta^{n^{1/d}})$ as $n \to \infty$ for some $0 < \theta < 1$.

2.1 Introduction

For an expanding map $T : X \to X$, the Perron-Frobenius operator $\mathscr{P}$ defined by

$$\mathscr{P} f(x) = \sum_{Ty=x} \frac{f(y)}{|T'(y)|}$$

and more general transfer operators $\mathscr{L}$ defined by

$$\mathscr{L} f(x) = \sum_{Ty=x} e^{\varphi(y)} f(y)$$

with potential function $\varphi : X \to \mathbb{R}$ are important objects in the thermodynamic formalism approach to ergodic theory.

Given a subset $H \subset X$, which we regard as a *hole* in X, it is natural to consider modified operators $\mathscr{P}_H$ and $\mathscr{L}_H$, defined by setting $\mathscr{P}_H f = \mathscr{P}(f\chi_{X\setminus H})$ and $\mathscr{L}_H f = \mathscr{L}(f\chi_{X\setminus H})$, in view of their connections with escape rate (see, e.g.

O.F. Bandtlow (✉) • O. Jenkinson
School of Mathematical Sciences, Queen Mary University of London, Mile End Road, London, E1 4NS, UK
e-mail: ob@maths.qmul.ac.uk; omj@maths.qmul.ac.uk

W. Bahsoun et al. (eds.), *Ergodic Theory, Open Dynamics, and Coherent Structures*, Springer Proceedings in Mathematics & Statistics 70, DOI 10.1007/978-1-4939-0419-8_2,

© Springer Science+Business Media New York 2014

[6, 9]) and various equilibrium measures supported by the survivor set $X_\infty = \cap_{n=0}^{\infty} T^{-n}(X \setminus H)$.

The purpose of this note is to describe, in the case where T is piecewise analytic and H is a suitable hole, explicit estimates on the spectral asymptotics of $\mathscr{P}_H$ and $\mathscr{L}_H$ when acting on various Banach spaces of holomorphic functions.[1]

Specifically, we take $X \subset \mathbb{R}^d$ to be compact and connected and $\mathscr{X} = \{X_i\}_{i\in\mathscr{I}}$ a finite partition (consisting of non-empty pairwise disjoint subsets of X, each one open in $\mathbb{R}^d$, whose union is dense in X). The map $T : X \to X$ is assumed Borel measurable, with $T(X_i)$ open in $\mathbb{R}^d$ for each $i \in \mathscr{I}$, and $T|_{X_i} : X_i \to T(X_i)$ a C^1 diffeomorphism which can be extended to a C^1 map on $\overline{X_i}$. We assume that T is *full branch*, i.e. $\overline{T(X_i)} = X$ for all $i \in \mathscr{I}$, and *expanding*, i.e. there exists $\beta > 1$ such that if $x, y \in X_i$ for some $i \in \mathscr{I}$, then $\|T(x) - T(y)\| \geq \beta \|x - y\|$. Each $T|_{X_i}$ has an *inverse branch* T_i, defined so that $T \circ T_i$ is the identity on the interior of X, and $T_i \circ T$ the identity on X_i, and satisfying $\sup_{x\in int(X)} \left\|T_i'(x)\right\|_{L(\mathbb{R}^d)} \leq \beta^{-1}$ for all $i \in \mathscr{I}$, where $\|\cdot\|_{L(\mathbb{R}^d)}$ denotes the induced operator norm on $L(\mathbb{R}^d) = L((\mathbb{R}^d, \|\cdot\|))$. We assume that $T : X \to X$ is *real analytic*, i.e. there is a bounded connected open set $D \subset \mathbb{C}^d$, with $X \subset D$, such that each T_i has a holomorphic extension to D.

For simplicity we shall take the hole H to be a union of some (but not all) elements of $\mathscr{X}$. In fact with some extra effort, and more cumbersome notation, the techniques described here extend to the case where H is a union of members of some refinement $\vee_{i=0}^{n-1} T^{-i} \mathscr{X}$ (a so-called Markov hole). Let $\mathscr{J} \subset \mathscr{I}$ be such that $\cup_{i\in\mathscr{J}} X_i = X \setminus H$. Transfer operators $\mathscr{L}_H$ for the open dynamical system $T|_{X\setminus H}$ then take the form

$$\mathscr{L}_H f = \sum_{i\in\mathscr{J}} w_i (f \circ T_i), \qquad (2.1)$$

where the *weight functions* w_i are related to the potential function φ by setting $w_i = \exp(\varphi \circ T_i)$ on X and assumed to admit a holomorphic extension to D which in turn extends continuously to $\overline{D}$. In the particular case $\varphi = -\log|T'|$, when w_i are the holomorphic extensions to D of $|T_i'|$ on X, the corresponding transfer operator is precisely the modified Perron-Frobenius operator $\mathscr{P}_H$. We shall always assume that D has the property that the closure of $\cup_{i\in\mathscr{J}} T_i(D)$ lies inside D itself, referring to such domains D as being *admissible* for the map T; this technical requirement, which we always assume without further comment, will ensure that $\mathscr{L}_H$ preserves suitable Banach spaces of functions holomorphic on D.

[1]When acting on these spaces, $\mathscr{P}_H$ has a strictly positive spectral radius δ, with $\delta > 0$ an eigenvalue such that $-\log\delta$ is the corresponding escape rate (see, e.g. [14] for one-dimensional maps); thus escape is at an exponential rate, rather than anything faster. Moreover, $\delta^{-n}\mathscr{P}_H^n 1 \to \varrho$, where ϱ is the density function for the Pianigiani-Yorke measure [15].

The structure of the article is as follows. We begin in Sect. 2.2 by considering transfer operators $\mathscr{L}_H$ acting on the Banach space[2] $U(D)$ of those holomorphic functions on D which extend continuously to $\overline{D}$ equipped with the usual supremum norm $\|w\|_{U(D)} = \sup_{z\in D} |w(z)|$. We show (Theorems 2.1 and 2.2) that in dimension $d = 1$, the eigenvalues $\lambda_n(\mathscr{L}_H)$ (arranged in order of decreasing modulus) converge to zero exponentially fast, deriving an explicit bound for $|\lambda_n(\mathscr{L}_H)|$. In higher dimensions $d \geq 2$, there are similar explicit bounds (see Theorem 2.3), though here the convergence to zero is[3] $O(\theta^{n^{1/d}})$ as $n \to \infty$, for some $\theta \in (0,1)$. In Sect. 2.3 we show that in fact the eigenvalues for $\mathscr{L}_H : U(D) \to U(D)$ are identical to those for $\mathscr{L}_H$ acting on a variety of Banach spaces $A(D)$ of holomorphic functions. This suggests the possibility of improving the bounds of Sect. 2.2 by judicious choice of $A(D)$, a strategy we pursue in Sect. 2.4 where $A(D)$ is chosen to be Hilbert Hardy space $H^2(D)$, yielding Theorems 2.6 and 2.7.

2.2 Eigenvalue Estimates via Weyl's Inequality

We begin with an explicit estimate on the eigenvalues of the modified Perron-Frobenius operator in dimension $d = 1$:

Theorem 2.1. *For an expanding interval map, the eigenvalues of the modified Perron-Frobenius operator $\mathscr{P}_H : U(D) \to U(D)$ satisfy*

$$|\lambda_n(\mathscr{P}_H)| \leq \theta^{n-1}\sqrt{n}\,\sup_{z\in D}\sum_{i\in\mathscr{I}} |T_i'(z)| \quad \textit{for all } n \geq 1\,, \tag{2.2}$$

provided each T_i' extends holomorphically to a disc $D \subset \mathbb{C}$, where $\theta < 1$ is such that $\cup_{i\in\mathscr{I}} T_i(D)$ is contained in the concentric disc whose radius is θ^2 times that of D.

The bounds in Theorem 2.1 are readily computed for specific maps T:

Example 2.1. As in [2], we consider the map

$$T(x) = \begin{cases} \frac{9x}{1-x} & \text{if } 0 \leq x \leq \frac{1}{10} \\ 10x - i & \text{if } \frac{i}{10} < x \leq \frac{i+1}{10},\ \text{for } 1 \leq i \leq 9 \end{cases}$$

---

[2]The study of transfer operators on this space $U(D)$ was inaugurated by Ruelle [18].

[3]Ruelle [18], following Grothendieck [11], stated the asymptotics were $O(\theta^n)$ as $n \to \infty$, independent of the dimension d, though Fried [10] corrected this to $O(\theta^{n^{1/d}})$. One novelty of our results, relative to Fried and Ruelle, is that the constant θ, as well as the implicit constant in the big-O asymptotics, is rendered explicit.

Note that the inverse branches $\{T_i\}_{0\le i\le 9}$ are given by

$$T_0(x) = \frac{x}{9+x}$$

and

$$T_i(x) = (x+i)/10 \quad \text{for } 1 \le i \le 9\,.$$

Choosing Markov hole $H = [1/5, 1]$ corresponds to setting $\mathscr{I} = \{0, 1\}$.

We claim that the eigenvalues of the modified Perron-Frobenius operator $\mathscr{P}_H : U(D) \to U(D)$ are bounded by

$$|\lambda_n(\mathscr{P}_H)| \le \frac{77}{320}\sqrt{5n}\left(\frac{1}{\sqrt{5}}\right)^n \quad \text{for all } n \ge 1\,. \tag{2.3}$$

In particular, note that the case $n = 1$ yields a bound on the escape rate γ (see, e.g. [6]) for this open dynamical system, namely, $\gamma \ge -\log 77/320$.

Let D be the disc of radius 1 centred at 0. Noting that $T_0(-1) = -1/8$, $T_1(-1) = 0$, $T_0(1) = 1/10$ and $T_1(1) = 1/5$, we see that $\cup_{i\in\mathscr{I}} T_i(D)$ is contained in the disc of radius $1/5$ centred at 0. This means we may set $\theta = 1/\sqrt{5}$ in Theorem 2.1. Note that $|T_0'(z)| + |T_1'(z)| = \frac{9}{|9+z|^2} + \frac{1}{10}$, and the supremum of this expression on D is the value $77/320$, attained (on the boundary of D) at $z = -1$. The bound (2.3) then follows from (2.2).

In fact Theorem 2.1 is a special case of the following one-dimensional result:

Theorem 2.2. *For an expanding interval map, the eigenvalues of the transfer operator $\mathscr{L}_H : U(D) \to U(D)$ satisfy*

$$|\lambda_n(\mathscr{L}_H)| \le \theta^{n-1}\sqrt{n}\,\sup_{z\in D}\sum_{i\in\mathscr{I}} |w_i(z)| \quad \textit{for all } n \ge 1\,, \tag{2.4}$$

provided each w_i and T_i extend holomorphically to the disc $D \subset \mathbb{C}$, where $\theta < 1$ is such that $\cup_{i\in\mathscr{I}} T_i(D)$ is contained in the concentric disc whose radius is θ^2 times that of D.

Proof. Let D' denote the concentric disc whose radius is $r = \theta^2$ times that of D. First, we observe that $\hat{\mathscr{L}}_H f := \sum_{i\in\mathscr{I}} w_i(f\circ T_i)$ defines a continuous operator $\hat{\mathscr{L}}_H : U(D') \to U(D)$. To see this, fix $f \in U(D')$ and note that $w_i(f\circ T_i) \in U(D)$ with $\|w_i(f\circ T_i)\|_{U(D)} \le \|w_i\|_{U(D)}\,\|f\|_{U(D')}$ for every $i \in \mathscr{I}$. But since we have $\|\hat{\mathscr{L}}_H f\|_{U(D)} \le \sum_{i\in\mathscr{I}} \|w_i\|_{U(D)}\,\|f\|_{U(D')}$, so $\hat{\mathscr{L}}_H f \in U(D)$ and $\hat{\mathscr{L}}_H$ is continuous. Now $\|\hat{\mathscr{L}}_H\| \le W =: \sup_{z\in D}\sum_{i\in\mathscr{I}} |w_i(z)|$, because for $f \in U(D')$ we have $|f(T_i(z))| \le \|f\|_{U(D')}$ for every $z \in D$, $i \in \mathscr{I}$; thus, by the maximum modulus principle, $\|\hat{\mathscr{L}}_H f\|_{U(D)} = \sup_{z\in D} |(\hat{\mathscr{L}}_H f)(z)| \le \sup_{z\in D}\sum_{i\in\mathscr{I}} |w_i(z)|\,|f(T_i(z))| \le W\,\|f\|_{U(D')}$.

Recall that if $L : B_1 \to B_2$ is a continuous operator between Banach spaces, then for $k \geq 1$, its *kth approximation number* $a_k(L)$ is defined by

$$a_k(L) = \inf\{ \|L - K\| \ : \ K : B_1 \to B_2 \text{ linear with } \operatorname{rank}(K) < k \} \ .$$

Note that in general $a_k(L_1 L_2) \leq \|L_1\| \, a_k(L_2)$ (see [16, 2.2]).

Now clearly $\mathscr{L}_H = \hat{\mathscr{L}}_H J$, where $J : U(D) \hookrightarrow U(D')$ denotes the canonical embedding, so

$$a_k(\mathscr{L}_H) \leq \|\hat{\mathscr{L}}_H\| a_k(J) \leq W a_k(J) \quad \text{for all } k \geq 1 \, . \tag{2.5}$$

Moreover, it can be shown that $\mathscr{L}_H$ is compact; in fact, it is of exponential class (see [3]) and in particular nuclear of any order.

Before proceeding recall that Weyl's inequality (see, e.g. [12]) asserts that $\prod_{k=1}^{n} |\lambda_k(\mathscr{L}_H)| \leq n^{n/2} \prod_{k=1}^{n} a_k(\mathscr{L}_H)$ for every $n \in \mathbb{N}$.[4]
Together with (2.5) this yields the inequality

$$|\lambda_n(\mathscr{L}_H)| \leq W n^{1/2} \prod_{k=1}^{n} a_k(J)^{1/n} \quad \text{for all } n \geq 1 \, , \tag{2.6}$$

because $|\lambda_n(\mathscr{L}_H)| \leq \prod_{k=1}^{n} |\lambda_k(\mathscr{L}_H)|^{1/n}$.

Using a result originally due to Babenko (see [1] or [17, Theorem VIII.2.1]) we see

$$a_l(J) \leq r^{l-1} \quad \text{for all } l \geq 1 \, ,$$

hence, $\prod_{l=1}^{n} a_l(J)^{1/n} \leq r^{\frac{1}{n}\sum_{l=1}^{n} l-1} = r^{(n-1)/2}$, so (2.6) becomes

$$|\lambda_n(\mathscr{L}_H)| \leq W n^{1/2} r^{(n-1)/2} \, ,$$

which is the desired bound (2.4). □

In higher dimension d the rate of eigenvalue decay is slower than exponential and can be shown to be $O(\theta^{n^{1/d}})$ as $n \to \infty$, for some $\theta \in (0, 1)$. The main new ingredient in the following result, proved in [5], is an estimate due to Farkov [8] on the approximation numbers of the embedding operator J in higher dimensions, namely $a_l(J) \leq r^{t_l}$, where $t_l := k$ for $\binom{k-1+d}{d} < l \leq \binom{k+d}{d}$.

Theorem 2.3. *In dimension $d \geq 1$, suppose the Euclidean ball $D \subset \mathbb{C}^d$ is such that $\cup_{i \in \mathscr{I}} T_i(D)$ is contained in the concentric ball whose radius is $r < 1$ times that of D. Setting $W := \sup_{z \in D} \sum_{i \in \mathscr{I}} |w_i(z)|$, the eigenvalues of $\mathscr{L}_H : U(D) \to U(D)$ can be bounded by*

$$|\lambda_n(\mathscr{L}_H)| < \frac{W}{r^d} n^{1/2} \, r^{\frac{d}{d+1}(d!)^{1/d} n^{1/d}} \quad \textit{for all } n \geq 1 \, . \tag{2.7}$$

[4]This is a Banach space version of Weyl's original inequality [19] in Hilbert space; the constant $n^{n/2}$ is optimal (see [12]).

2.3 The Common Spectrum

It turns out that a more oblique approach yields different, and sometimes better, bounds on the eigenvalues of $\mathscr{L}_H : U(D) \to U(D)$. This approach consists of varying the space upon which $\mathscr{L}_H$ acts. Clearly, in general $U(D)$ is not the only function space preserved by a transfer operator $\mathscr{L}_H$, and we would expect the spectrum of $\mathscr{L}_H$ to vary according to the space on which it acts. There is interest, however, in identifying a class of spaces $A(D)$ which are sufficiently closely related to $U(D)$ to ensure that the spectrum of $\mathscr{L}_H$ on these spaces is precisely the same as that of $\mathscr{L}_H : U(D) \to U(D)$. This motivates the following definition:

Definition 2.1. For a non-empty open connected set $D \subset \mathbb{C}^d$, a Banach space $A(D)$ of holomorphic functions $f : D \to \mathbb{C}$ is called *favourable* if it contains $U(D)$, with the natural embedding $U(D) \hookrightarrow A(D)$ having norm 1, and if $f \mapsto f(z)$ is continuous on $A(D)$ for each $z \in D$.

Transfer operators $\mathscr{L}_H$ can be shown (see [5]) to preserve all favourable spaces[5] $A(D)$, with the eigenvalues of $\mathscr{L}_H : A(D) \to A(D)$ related to a certain entire function:

Theorem 2.4. *The transfer operator $\mathscr{L}_H$ defined by (2.1) preserves every favourable space $A(D)$ of holomorphic functions on D. It has a well-defined spectral trace $\tau_{A(D)}(\mathscr{L}_H) = \sum_{n=1}^{\infty} \lambda_n(\mathscr{L}_H|_{A(D)})$ and spectral determinant* $\det_{A(D)}$, *related by*

$$\det_{A(D)}(I - z\mathscr{L}_H|_{A(D)}) = \exp\left(- \sum_{n=1}^{\infty} \frac{z^n}{n} \tau_{A(D)}(\mathscr{L}_H^n) \right), \tag{2.8}$$

for all $z \in \mathbb{C}$ in a suitable neighbourhood of 0, *and such that, counting multiplicities, the zeros of the entire function $z \mapsto \det(I - z\mathscr{L}_H|_{A(D)})$ are precisely the reciprocals of the eigenvalues of $\mathscr{L}_H : A(D) \to A(D)$.*

Motivated by the possibility that the trace and determinant do not in fact vary with the choice of favourable space $A(D)$, we follow Ruelle [18] in considering the following function:

Definition 2.2. For given weight functions w_i, $i \in \mathscr{I}$, the associated *dynamical determinant* is the entire function $\Delta : \mathbb{C} \to \mathbb{C}$, defined for all z of sufficiently small modulus by

$$\Delta(z) = \exp\left(- \sum_{n \in \mathbb{N}} \frac{z^n}{n} \sum_{\underline{i} \in \mathscr{I}^n} \frac{w_{\underline{i}}(z_{\underline{i}})}{\det(I - T'_{\underline{i}}(z_{\underline{i}}))} \right), \tag{2.9}$$

[5] As always, we are making the standing assumption that D is an admissible domain, i.e. that the closure of $\cup_{i \in \mathscr{I}} T_i(D)$ lies in D.

where for $\underline{i} = (i_1, \dots, i_n) \in \mathcal{J}^n$, we set $T_{\underline{i}} := T_{i_n} \circ \cdots \circ T_{i_1}$, and $w_{\underline{i}} := \prod_{k=1}^n w_{i_k} \circ T_{P_{k-1}\underline{i}}$, where $P_k : \mathcal{J}^n \to \mathcal{J}^k$ denotes the projection $P_k\underline{i} = (i_1, \dots, i_k)$ with the convention that $T_{P_0\underline{i}} = \text{id}$, and $z_{\underline{i}}$ denotes the (unique, by [7]) fixed point of $T_{\underline{i}}$ in D.

Theorem 2.5. *For every favourable space $A(D)$, the determinant of the transfer operator $\mathcal{L}_H : A(D) \to A(D)$ defined by (2.1) is precisely the dynamical determinant Δ, and its eigenvalue sequence is precisely the same as for $\mathcal{L} : U(D) \to U(D)$.*

Proof. The common trace formula

$$\tau_{A(D)}(\mathcal{L}_H^n) = \sum_{\underline{i} \in \mathcal{J}^n} \frac{w_{\underline{i}}(z_{\underline{i}})}{\det(I - T_{\underline{i}}'(z_{\underline{i}}))} \quad \text{for all } n \geq 1 \tag{2.10}$$

can be established (see [5]), valid for every favourable space $A(D)$ on which $\mathcal{L}_H$ acts, so that equality of determinants follows from comparison of (2.8) and (2.9). The equality of the eigenvalue sequences follows from the fact that the determinants are spectral. □

2.4 Hilbert Hardy Space

In view of Theorem 2.5, we are now at liberty to make particular choices of favourable spaces, in the hope of obtaining interesting new bounds on the eigenvalues of the transfer operator $\mathcal{L} : U(D) \to U(D)$.

For $p \in [1, \infty)$, the *Hardy space* $H^p(D)$ (see [13, Chap. 8.3]) is a favourable space, and we will be particularly interested in the Hilbert Hardy space $H^2(D)$.[6] The following eigenvalue bounds, valid in dimension 1, are obtained by choosing favourable space $A(D) = H^2(D)$ for $D \subset \mathbb{C}$ a disc:

Theorem 2.6. *With the hypotheses and notation of Theorem 2.2,*

$$|\lambda_n(\mathcal{L}_H)| \leq \frac{W}{\sqrt{1 - \theta^4}} \, \theta^{n-1} \quad \textit{for all } n \geq 1 \, . \tag{2.11}$$

[6]If D has C^2 boundary, then $H^2(D)$ can be identified with the $L^2(\partial D, \sigma)$-closure of $U(D)$, where σ denotes $(2d - 1)$-dimensional Lebesgue measure on the boundary ∂D, normalised so that $\sigma(\partial D) = 1$. The inner product in $H^2(D)$ is given (see [13, Chaps. 1.5 and 8]) by $(f, g) = \int_{\partial D} f^* \, \overline{g^*} \, d\sigma$, where, for $h \in H^2(D)$, the symbol h^* denotes the corresponding nontangential limit function in $L^2(\partial D, \sigma)$.

Proof. As in the proof of Theorem 2.3, let D' denote the concentric disc whose radius is $r = \theta^2$ times that of D, and let $J : H^2(D) \hookrightarrow H^\infty(D')$ denote canonical embedding. It can be shown that, for all $n \geq 1$,

$$|\lambda_n(\mathscr{L}_H)| \leq W \prod_{k=1}^{n} a_k(J)^{1/n}, \tag{2.12}$$

an inequality which is superior to (2.6), by virtue of the original Hilbert space version of Weyl's inequality, namely $\prod_{k=1}^n |\lambda_k(L)| \leq \prod_{k=1}^n a_k(L)$ (see [16, 3.5.1], [19]). An argument (see [4]) exploiting the interplay between the reproducing kernel of $H^2(D)$ and an orthonormal basis for $H^2(D)$ then allows the estimate

$$a_n(J) \leq \frac{r^{n-1}}{\sqrt{1-r^2}}, \tag{2.13}$$

and substituting into (2.12) yields the result. □

Example 2.2. Comparing (2.12) with (2.4), we see that Theorem 2.6 leads to improved eigenvalue bounds whenever $n > 1/(1-\theta^4)$. In Example 2.1 we can choose $\theta = 1/\sqrt{5}$; therefore, for all $n \geq 2 > 25/24$, the estimate

$$|\lambda_n(\mathscr{P}_H)| \leq \frac{77}{320}\frac{5}{\sqrt{24}}\left(\frac{1}{\sqrt{5}}\right)^{n-1}$$

derived from (2.12) is sharper than the previous bound (2.3) on the eigenvalues of the modified Perron-Frobenius operator.

A more elaborate version of the proof of Theorem 2.6 (see [4] for details) gives the following higher dimensional analogue, which for sufficiently large values of n yields estimates which are superior to those of Theorem 2.3:

Theorem 2.7. *With the hypotheses and notation of Theorem 2.3,*

$$|\lambda_n(\mathscr{L}_H)| < \frac{W\sqrt{d}}{r^d(1-r^2)^{d/2}}\, n^{(d-1)/(2d)}\, r^{\frac{d}{d+1}(d!)^{1/d} n^{1/d}} \quad \textit{for all } n \geq 1\,. \tag{2.14}$$

References

1. Babenko, K.I.: Best approximations to a class of analytic functions (Russian). Izv. Akad. Nauk SSSR Ser. Mat. **22**, 631–640 (1958)
2. Bahsoun, W., Bose, C.: Quasi-invariant measures, escape rates and the effect of the hole. Discrete Contin. Dyn. Syst. **27**, 1107–1121 (2010)
3. Bandtlow, O.F.: Resolvent estimates for operators belonging to exponential classes. Integr. Equ. Oper. Theory **61**, 21–43 (2008)

4. Bandtlow, O.F., Jenkinson, O.: Explicit a priori bounds on transfer operator eigenvalues. Comm. Math. Phys. **276**, 901–905 (2007)
5. Bandtlow, O.F., Jenkinson, O.: On the Ruelle eigenvalue sequence. Ergod. Theory Dyn. Syst. **28**, 1701–1711 (2008)
6. Demers, M., Young, L.-S.: Escape rates and conditionally invariant measures. Nonlinearity **19**, 377–397 (2006)
7. Earle, C.J., Hamilton, R.S.: A fixed point theorem for holomorphic mappings. In: Chern, S., Smale, S. (eds.) Global Analysis. Proceedings of Symposia in Pure Mathematics, vol. XVI, pp. 61–65. American Mathematical Society, Providence R.I. (1970)
8. Farkov, Yu. A.: The N-widths of Hardy-Sobolev spaces of several complex variables. J. Approx. Theory **75**, 183–197 (1993)
9. Ferguson, A., Pollicott, M.: Escape rates for Gibbs measures. Ergod. Theory Dyn. Syst. **32**, 961–988 (2012)
10. Fried, D.: Zeta functions of Ruelle and Selberg I. Ann. Sci. Ecole Norm. Sup. **9**, 491–517 (1986)
11. Grothendieck, A.: Produits tensoriels topologiques et espaces nucléaires. Mem. Am. Math. Soc. **16**, (1955)
12. Hinrichs, A.: Optimal Weyl inequality in Banach spaces. Proc. Am. Math. Soc. **134**, 731–735 (2006)
13. Krantz, S.G.: Function Theory of Several Complex Variables, 2nd edn. AMS Chelsea, New York (1992)
14. Liverani, C., Maume-Deschamps, V.: Lasota-Yorke maps with holes: conditionally invariant probability measures and invariant probability measures on the survivor set. Ann. Inst. H. Poincaré Probab. Stat. **39**, 385–412 (2003)
15. Pianigiani, G., Yorke, J.A.: Expanding maps on sets which are almost invariant: decay and chaos. Trans. Am. Math. Soc. **252**, 351–366 (1979)
16. Pietsch, A.: Eigenvalues and s-Numbers. Cambridge University Press, Cambridge (1987)
17. Pinkus, A.: n-widths in Approximation Theory. Springer, New York (1985)
18. Ruelle, D.: Zeta-functions for expanding maps and Anosov flows. Invent. Math. **34**, 231–242 (1976)
19. Weyl, H.: Inequalities between two kinds of eigenvalues of a linear transformation. Proc. Nat. Acad. Sci. USA **35**, 408–411 (1949)

Chapter 3
Periodic Points, Escape Rates and Escape Measures

Oscar F. Bandtlow, Oliver Jenkinson, and Mark Pollicott

Abstract For piecewise real analytic expanding Markov maps with Markov hole, it is shown that the escape rate and corresponding escape measure can be rapidly approximated using periodic points.

3.1 Introduction

For a dynamical system $T : X \to X$, a non-empty subset $H \subset X$ induces an *escape time function*:

$$e(x) = e_H(x) = \min\{n \geq 0 : T^n(x) \in H\},$$

the nomenclature motivated by interpreting H as a *hole* in phase space X, through which points may escape under iteration. The sequence of super-level sets $E_n = \{x \in X : e(x) > n\}$ decreases with n, and for a probability measure m on X, it is often the case that $m(E_n) \to 0$ as $n \to \infty$.

If T is a suitable hyperbolic map and m is, for example, Lebesgue measure, then the $m(E_n)$ approach zero at an exponential rate. In this case the exponential decay rate

$$\delta = \delta(T, H, m) = \lim_{n\to\infty} m(E_n)^{1/n}$$

is a quantity of interest; indeed

$$\varepsilon = \varepsilon(T, H, m) = -\log \delta(T, H, m)$$

O.F. Bandtlow (✉) • O. Jenkinson
School of Mathematical Sciences, Queen Mary University of London, Mile End Road, London, E1 4NS, UK
e-mail: ob@maths.qmul.ac.uk; omj@maths.qmul.ac.uk

M. Pollicott
Mathematics Institute, University of Warwick, Coventry, CV4 7AL, UK
e-mail: mpollic@maths.warwick.ac.uk

W. Bahsoun et al. (eds.), *Ergodic Theory, Open Dynamics, and Coherent Structures*, Springer Proceedings in Mathematics & Statistics 70, DOI 10.1007/978-1-4939-0419-8_3,

© Springer Science+Business Media New York 2014

is commonly referred to as the *escape rate* and has been widely studied (see, e.g. [3,4,8,9,12,13,21]). In certain special cases $\delta(T, H, m)$ can be found exactly,[1] though in general this is not feasible, so there is interest in developing methods for its efficient approximation.

The purpose of this note is to describe, in the context of analytic expanding maps T, a method for rapidly approximating $\delta = \delta(T, H, m)$. It relies on locating all periodic points of T, up to a certain period N, say. This yields (see Sect. 3.3 for further details) an approximation $\delta_N \approx \delta$, where the error satisfies

$$|\delta - \delta_N| \leq C\theta^{N^2}$$

for some constants $\theta \in (0, 1)$ and $C \geq 0$; in particular, the δ_N approximate δ *super-exponentially* fast.

For example, if the map $T : [0, 1] \to [0, 1]$ is defined, as in [4], by

$$T(x) = \begin{cases} \frac{9x}{1-x} & \text{if } 0 \leq x \leq \frac{1}{10} \\ 10x - i & \text{if } \frac{i}{10} < x \leq \frac{i+1}{10} \text{ for } 1 \leq i \leq 9, \end{cases}$$

and $H = [\frac{9}{10}, 1]$, we derive (see Sect. 3.5 for further details) the successive approximations

$$\delta_2 = 0.89937619148227610951885101534$$

$$\delta_3 = 0.90114292895376364489121035873 7$$

$$\delta_4 = 0.901139819292137417448614669069$$

$$\delta_5 = 0.901139820047631592907392158902$$

$$\delta_6 = 0.901139820047605710579196990120$$

$$\delta_7 = 0.901139820047605710706369756237$$

In fact these techniques also yield a means of rapidly approximating the corresponding *escape measure* μ, the T-invariant measure supported on the *survivor set* $E_\infty = e_H^{-1}(\infty)$, and maximising the quantity $h(m) - \int_{E_\infty} \log |T'| \, dm$ over all T-invariant probability measures m, where $h(\cdot)$ denotes metric entropy (see, e.g. [9, 13, 20]). For example, μ is completely determined by its sequence of nth moments $\mu(n) = \int x^n \, d\mu(x)$, which in general are not known exactly, but the periodic points of T can again be used (see Sect. 3.4 for the method and Sect. 3.5 for an example) to derive a sequence $\mu_N(n)$, where $|\mu(n) - \mu_N(n)| = O(\theta^{N^2})$ as $N \to \infty$.

---

[1]For example, if $T(x) = 3x \pmod 1$ on the interval $X = [0, 1]$, and H is the 'middle third' $(1/3, 2/3)$, then $m(E_n) = (2/3)^n$ for each n, so that $\delta(T, H, m) = 2/3$.

Using periodic points to calculate escape rates and related quantities is not a new idea. Indeed, there is a considerable body of work for rather general systems in the physics literature starting with [1,2] (see also [10,11,18,23] for later developments and applications). Restricting attention to analytic expanding maps, however, we are able to rigorously justify the approach and to provide precise estimates for the speed of convergence of the approximations.

This article is organised as follows. After some preliminaries on transfer operators and their determinants in Sect. 3.2, the method for approximating the escape rate is described in Sect. 3.3 and for the escape measure in Sect. 3.4. In the final Sect. 3.5, the speed of convergence of these methods is illustrated using the map T and hole H defined above.

3.2 Transfer Operators and Determinants

Suppose the unit interval[2] $I = [0,1]$ is partitioned as $I = I_1 \cup \cdots \cup I_d$, $d \geq 2$, where the I_i are closed intervals with pairwise disjoint interiors. We shall assume that $T : I \to I$ is such that $T|_{I_i}$ is real analytic, for each i, and *expanding* in the sense that $\min\{\,|T'(x)| : x \in I_i, 1 \leq i \leq d\,\} > 1$. We say that T is *Markov* if for each $1 \leq i \leq d$ the closure of $T(I_i)$ is a union of elements of the partition $\alpha = \{I_1, \ldots, I_d\}$, in which case α is referred to as the *Markov partition*. For each $n \geq 1$, define the usual refined partition $\alpha^{(n)} = \vee_{i=0}^{n-1} T^{-i}\alpha$. By a *Markov hole* we mean a union of members of $\alpha^{(n)}$, for some $n \geq 1$. The fact that T is expanding ensures that any subinterval $H \subset I$ can be approximated arbitrarily well by a Markov hole.[3]

Although the techniques described below apply, with slight modification, to general Markov holes H for Markov maps T, for simplicity of exposition we shall henceforth assume that for each $1 \leq i \leq d$, the closure of $T(I_i)$ equals I (the so-called *Bernoulli* case) and that the hole $H \subset I$ is a member of α.

We denote by $T_i : I \to I_i$ $(1 \leq i \leq d)$ the contractions which are inverse branches to T. By the implicit function theorem, the maps T_i are real analytic,

[2] For simplicity of exposition we restrict attention to one-dimensional dynamical systems, though in fact similar results apply to real analytic expanding Markov maps in higher dimensions. In dimension D the rate of convergence (of δ_N to δ, and of $\mu_N(n)$ to $\mu(n)$) can be shown to be $O(\theta^{N^{1+D^{-1}}})$ as $N \to \infty$, for some $0 < \theta < 1$; in particular it is super-exponential.

[3] This suggests the possibility of approximating the escape rate for non-Markov holes H, by using the methods of this paper for a sequence of Markov holes approximating H. More precisely, the escape rate can easily be seen to depend continuously on (the end points of) the hole, by a perturbation theorem of Keller and Liverani for the bounded variation semi-norm and L^1 (see [19]). Thus, for $\delta > 0$, provided n is sufficiently large, we can choose intervals $H_1 \subset H \subset H_2$ where H_1, H_2 are unions of elements of $\alpha^{(n)}$ and such that $\varepsilon(T, H_1, m) \leq \varepsilon(T, H, m) \leq \varepsilon(T, H_2, m)$ satisfy $0 \leq \varepsilon(T, H, m) - \varepsilon(T, H_1, m), \varepsilon(T, H_2, m) - \varepsilon(T, H, m) \leq \delta$. However, whereas the values $\varepsilon(T, H_1, m)$, $\varepsilon(T, H_2, m)$ can be approximated quickly, there is less explicit control of the dependence of n on δ.

since each $T|_{I_i}$ is real analytic. In particular, we can choose a bounded open neighbourhood $U \subset \mathbb{C}$ containing I such that

$$\overline{\cup_{i=1}^{d} T_i U} \subset U\,, \tag{3.1}$$

where here T_i denotes the relevant holomorphic extension to U.

Let $A^2(U)$ denote the Hilbert space of analytic functions $f : U \to \mathbb{C}$ which are square-integrable with respect to two-dimensional Lebesgue measure on U, equipped with the usual inner product.

We may now define a *transfer operator* $\mathscr{L}$ acting on $A^2(U)$ by

$$\mathscr{L} f(z) = \sum_{i=1}^{d} \epsilon_i T_i'(z) f(T_i z) \text{ where } f \in A^2(U)\,. \tag{3.2}$$

Here $\epsilon_i \in \{-1, 1\}$ denotes the sign of the derivative of T_i on I.

Using (3.1) it is not difficult to see that $\mathscr{L}$ maps $A^2(U)$ continuously into itself. In fact, on this space the transfer operator has strong spectral properties, which will be crucial for the results to follow. The spectral properties are conveniently described in terms of the theory of exponential classes developed in [5], which we briefly recall. Given positive real numbers a and γ, a bounded operator L on a Hilbert space is said to belong to the exponential class $E(a, \gamma)$ if

$$\sup_{n \in \mathbb{N}} s_n(L) \exp(a n^{\gamma}) < \infty\,,$$

where $s_n(L) = \inf\{\, \|L - K\| : \operatorname{rank}(K) < n \,\}$ denotes the nth *approximation number* of L. We now have the following result.

Proposition 3.1. *The transfer operator $\mathscr{L} : A^2(U) \to A^2(U)$ given in (3.2) belongs to the exponential class $E(a, 1)$ for some $a > 0$. In particular, $\mathscr{L}$ is trace class. Moreover, its eigenvalues decay at an exponential rate.*

Proof. The first assertion follows from [6, Theorem 5.9]. The second now follows since the approximation numbers of $\mathscr{L}$ are summable. The statement about the eigenvalue decay follows from [6, Lemma 5.11]. □

Given a hole $H \in \alpha$, without loss of generality assume that $H = I_d$. In order to analyse the corresponding escape rate, we consider the following modified operator:

Definition 3.1. Define $\mathscr{L}_H$ by

$$\mathscr{L}_H f(z) = \sum_{i=1}^{d-1} \epsilon_i T_i'(z) f(T_i z) \text{ where } f \in A^2(U) \text{ and } z \in U\,. \tag{3.3}$$

Equivalently, we can think of $\mathscr{L}_H$ as the original transfer operator $\mathscr{L}$ with the term corresponding to H removed. As a result, the modified transfer operator enjoys the same strong spectral properties as the original transfer operator.

Proposition 3.2. *The modified transfer operator* $\mathscr{L}_H : A^2(U) \to A^2(U)$ *given in (3.3) belongs to the exponential class* $E(a,1)$ *for some* $a > 0$. *In particular,* $\mathscr{L}_H$ *is trace class. Moreover, its eigenvalues decay at an exponential rate.*

Proof. See the proof of Proposition 3.1. □

Since $\mathscr{L}_H$ is trace class, it has a well-defined trace. Moreover, there is an explicit expression for the trace of any power of $\mathscr{L}_H$ in terms of fixed points of the iterates of the map:

Proposition 3.3. *For any* $n \in \mathbb{N}$, *we have*

$$\operatorname{tr}(\mathscr{L}_H^n) = \sum_{x \in \operatorname{Fix}_H(T^n)} \frac{\operatorname{sgn}((T^n)'(x))}{(T^n)'(x) - 1},$$

where $\operatorname{Fix}_H(T^n) = \{x \in [0,1] : T^n x = x, T^k x \notin H \text{ for } 0 \leq k < n\}$ *and the symbol* $\operatorname{sgn}(\xi) \in \{-1, 1\}$ *denotes the sign of* $\xi \in \mathbb{R}$.

Proof. This follows from [7, Theorem 4.2]. □

The traces can now be used to calculate the determinant of the operator $\mathscr{L}_H$.

Proposition 3.4. *The function* $z \mapsto \det(1 - z\mathscr{L}_H)$ *given for* z *of sufficiently small modulus by*

$$\det(I - z\mathscr{L}_H) = \exp\left(-\sum_{n=1}^{\infty} \frac{z^n}{n} \operatorname{tr}(\mathscr{L}_H^n)\right) \tag{3.4}$$

extends to an entire function, the zeros of which are exactly the reciprocals of the eigenvalues of $\mathscr{L}_H$ *(counting algebraic multiplicities).*

The Taylor coefficients c_n *of*

$$\det(I - z\mathscr{L}_H) = 1 + \sum_{n=1}^{\infty} c_n z^n \tag{3.5}$$

satisfy both the recurrence relation

$$c_n = -\frac{1}{n} \sum_{l=0}^{n-1} c_l \operatorname{tr}(\mathscr{L}_H^{n-l}) \text{ for } n \geq 1 \tag{3.6}$$

with $c_0 = 1$ *and Plemelj's formula*

$$c_n = \frac{(-1)^n}{n!} \det \begin{pmatrix} \operatorname{tr}(\mathcal{L}_H) & 1 & & & 0 \\ \operatorname{tr}(\mathcal{L}_H^2) & \operatorname{tr}(\mathcal{L}_H) & 2 & & \\ \vdots & \vdots & & \ddots & \\ \operatorname{tr}(\mathcal{L}_H^{n-1}) & \operatorname{tr}(\mathcal{L}_H^{n-2}) & \cdots & \operatorname{tr}(\mathcal{L}_H) & n-1 \\ \operatorname{tr}(\mathcal{L}_H^{n}) & \operatorname{tr}(\mathcal{L}_H^{n-1}) & \cdots & \operatorname{tr}(\mathcal{L}_H^{2}) & \operatorname{tr}(\mathcal{L}_H) \end{pmatrix}. \tag{3.7}$$

Moreover, we have

$$|c_n| = O(\theta^{n^2}) \quad \text{as } n \to \infty\,, \tag{3.8}$$

for some $0 < \theta < 1$.

Proof. For the recurrence formula and Plemelj's formula, see [22, Theorem 4.4.10]. The decay estimate for the Taylor coefficients is proved in [6, Theorem 6.1]. The remaining assertions follow from Lidskii's Trace Theorem (see, e.g. [14, Theorem 8.4, Chap. III]). □

Remark 3.1. Explicit estimates for θ, in terms of geometric properties of $T_i(U)$, can be found in [6, Theorem 6.1].

Proposition 3.5. *The following hold:*

(a) *The operator* $\mathcal{L}_H$ *has a simple eigenvalue* $\delta \in (0, 1]$, *strictly larger in modulus than all other eigenvalues, with corresponding eigenfunction* $\varrho \in A^2(U)$, *which is positive on* I.
(b) *There exists a probability measure* ν *supported on the survivor set* E_∞ *satisfying*

$$\int_{E_\infty} \mathcal{L}_H f \, d\nu = \delta \int_{E_\infty} f \, d\nu \text{ for all } f \in A^2(U)\,.$$

(c) *The probability measure* $\mu = \varrho\nu$ *supported on the survivor set* E_∞ *is* T*-invariant and coincides with the escape measure.*
(d) *The escape rate with respect to Lebesgue measure* m *satisfies*

$$\varepsilon(T, H, m) = -\log \delta\,.$$

Proof. The assertions in (a), (b) and (c) follow from results in [20]. To be precise, the existence of the eigenmeasure ν in (b) follows immediately from Theorem A in [20]. For (a) observe that (b) together with the compactness of $\mathcal{L}_H$ imply the existence of an eigenvector $\varrho \in A^2(U)$ corresponding to δ, which, by the positivity arguments used for the proof of Theorem A in [20], must have the stated properties. The same theorem also yields (c). Finally, (d) follows from the fact that

$$m(E_n) = \int_{I\setminus H} \mathcal{L}_H^n 1 \, dm$$

together with the spectral properties of $\mathcal{L}_H$ given in (a). □

3.3 Determining the Escape Rate

The results of Sect. 3.2 mean we can find the value $0 < \delta(T, H, m) \leq 1$ by considering the determinant:

Proposition 3.6. *The smallest zero (in modulus) of* $z \mapsto \det(I - z\mathscr{L}_H)$ *is simple, real and equal to* $\delta(T, H, m)^{-1}$.

Proof. By Proposition 3.5 the value $\delta(T, H, m)$ is a simple eigenvalue of the transfer operator $\mathscr{L}_H$ and also the largest in modulus. Combining this with Proposition 3.4, the assertions follow. □

Setting $\delta = \delta(T, H, m)$, the expansion (3.5) now gives

$$0 = 1 + \sum_{n=1}^{\infty} c_n \delta^{-n} = 1 + \sum_{n=1}^{N} c_n \delta^{-n} + O(\theta^{N^2})$$

leading naturally to the following definition:

Definition 3.2. For each $N \geq 1$ define δ_N to be the largest value (in modulus) such that

$$0 = 1 + \sum_{n=1}^{N} c_n \delta_N^{-n}.$$

This brings us to the first main result:

Theorem 3.1. *The values* δ_N *converge to* δ *at a super-exponential rate; more precisely,*

$$\delta_N = \delta + O(\theta^{N^2}) \quad \textit{as } N \to \infty.$$

Proof. By construction the sequence δ_N converges to δ. In order to estimate the speed of convergence, fix $N \geq 1$ and write

$$\Delta(z) = \det(1 - z\mathscr{L}_H) = 1 + \sum_{n=1}^{\infty} c_n z^n,$$

$$\Delta_N(z) = 1 + \sum_{n=1}^{N} c_n z^n.$$

By the mean value theorem, there is t_N on the line segment joining δ^{-1} and δ_N^{-1} such that

$$(\delta^{-1} - \delta_N^{-1})\, \Delta_N'(t_N) = \Delta_N(\delta^{-1}) - \Delta_N(\delta_N^{-1}) = \Delta_N(\delta^{-1}) = \Delta_N(\delta^{-1}) - \Delta(\delta^{-1}).$$

But since $\Delta'_N(t_N) \to \Delta'(\delta^{-1}) \neq 0$ by Proposition 3.6, it follows that $|\Delta'_N(t_N)|$ is bounded away from zero. Thus,

$$|\delta^{-1} - \delta_N^{-1}| \leq \frac{1}{|\Delta'_N(t_N)|} \sum_{n=N+1}^{\infty} |c_n| \delta^{-n} = O(\theta^{N^2}) \quad \text{as } N \to \infty$$

for some $0 < \theta < 1$. □

Remark 3.2. The implied constant in Theorem 3.1 can if necessary be explicitly estimated, using bounds on the Taylor coefficients c_n.

3.4 Determining the Escape Measure

In order to approximate the escape measure, we first introduce the following weighted transfer operator.

Definition 3.3. Let ϕ be a bounded holomorphic function on U. For t in a bounded neighbourhood V of $0 \in \mathbb{C}$, define the *weighted transfer operator* $\mathscr{L}_{H,t}$ by

$$\mathscr{L}_{H,t} f(z) = \sum_{i=1}^{d-1} \epsilon_i T_i'(z) e^{t\phi(z)} f(T_i z) \text{ where } f \in A^2(U) \text{ and } z \in U\,.$$

We now have analogues of the results from Sect. 3.2:

Proposition 3.7. *The operators $\mathscr{L}_{H,t} : A^2(U) \to A^2(U)$ satisfy:*

(a) For each $t \in V$, the operator $\mathscr{L}_{H,t}$ belongs to the exponential class $E(a,1)$ for some $a > 0$.
(b) The mapping $t \mapsto \mathscr{L}_{H,t}$ is holomorphic in the trace-class operator topology; in particular, the function $(z,t) \mapsto \det(1 - z\mathscr{L}_{H,t})$ is holomorphic on $\mathbb{C} \times V$.
(c) For any $t \in V$ and any $n \in \mathbb{N}$, we have

$$\operatorname{tr}(\mathscr{L}_{H,t}^n) = \sum_{x \in \operatorname{Fix}_H(T^n)} \frac{\operatorname{sgn}((T^n)'(x))\, e^{t\phi^{(n)}(x)}}{(T^n)'(x) - 1}$$

where, as before, $\operatorname{Fix}_H(T^n) = \{x \in [0,1] : T^n x = x, T^k x \notin H \text{ for } 0 \leq k < n\}$ and $\phi^{(n)} = \sum_{k=0}^{n-1} \phi \circ T^k$.
(d) The Taylor coefficients $c_{n,\phi}(t)$ of the determinant

$$\det(I - z\mathscr{L}_{H,t}) = 1 + \sum_{n=1}^{\infty} c_{n,\phi}(t) z^n$$

satisfy $\sup_{t \in V} |c_{n,\phi}(t)| = O(\theta^{n^2})$ as $n \to \infty$ for some $0 < \theta < 1$.

Proof. Assertion (a) follows from [6, Theorem 5.9], while assertion (b) follows from (a) and [14, Sect. 1.9, Chap. IV]. The formula for the traces in (c) is a consequence of [7, Theorem 4.2], and (d) follows from [6, Theorem 6.1]. □

Remark 3.3. Setting $t = 0$ we see that $\mathscr{L}_{H,0} = \mathscr{L}_H$; hence, $c_{n,\phi}(0) = c_n$ for all $n \geq 1$.

It turns out that the escape measure can be expressed as a quotient of the partial derivatives of $(z,t) \mapsto \det(I - z\mathscr{L}_{H,t})$. The proof of this relies on a formula for the derivative of a determinant which we briefly recall. Let $D \subset \mathbb{C}$ be an open neighbourhood of 0 and suppose that $D \ni s \mapsto L(s)$ is an operator-valued function which is holomorphic in the trace-class topology. If $\det(I - L(0)) \neq 0$, then

$$\frac{\mathrm{d}}{\mathrm{d}s}\det(I - L(s))|_{s=0} = -\det(I - L(0))\,\mathrm{tr}(\dot{L}(0)(I - L(0))^{-1}), \tag{3.9}$$

where $\dot{L}(0) = \frac{\mathrm{d}}{\mathrm{d}s}L(s)|_{s=0}$. For a proof see [22, 4.3.1.9 Proposition] or [14, Sect. 1.9, Chap. IV].

The calculation of the escape measure relies on the following result.

Proposition 3.8. *We have*

$$\int_{E_\infty} \phi \, d\mu = \delta \frac{\frac{\partial}{\partial t}\det(I - z\mathscr{L}_{H,t})|_{t=0,z=1/\delta}}{\frac{\partial}{\partial z}\det(I - z\mathscr{L}_{H,t})|_{t=0,z=1/\delta}}.$$

Proof. The proof is a simple application of formula (3.9), the only subtlety arising from the fact that both $\frac{\partial}{\partial t}\det(I - z\mathscr{L}_{H,t})$ and $\frac{\partial}{\partial z}\det(I - z\mathscr{L}_{H,t})$ vanish for $t = 0$ and $z = 1/\delta$. This problem, however, can be circumvented by choosing D to be a small punctured neighbourhood of $1/\delta$ such that $\det(1 - \zeta\mathscr{L}_{H,0}) \neq 0$ for $\zeta \in D$. We then apply formula (3.9) for $\zeta \in D$ and then take the limit $\zeta \to 1/\delta$.

We thus start by using (3.9) twice to obtain for any $\zeta \in D$

$$\frac{\partial}{\partial t}\det(I - \zeta\mathscr{L}_{H,t})|_{t=0} = -\det(I - \zeta\mathscr{L}_{H,0})\mathrm{tr}(\zeta\dot{\mathscr{L}}_{H,0}(I - \zeta\mathscr{L}_{H,0})^{-1})$$

where $\dot{\mathscr{L}}_{H,0} = \frac{\mathrm{d}}{\mathrm{d}t}\mathscr{L}_{H,t}|_{t=0}$, and

$$\frac{\partial}{\partial z}\det(I - z\mathscr{L}_{H,0})|_{z=\zeta} = -\det(I - \zeta\mathscr{L}_{H,0})\mathrm{tr}(\mathscr{L}_{H,0}(I - \zeta\mathscr{L}_{H,0})^{-1}).$$

We now observe that $\frac{\mathrm{d}}{\mathrm{d}t}\mathscr{L}_{H,t}|_{t=0} = M_\phi\mathscr{L}_{H,0}$ where $M_\phi : A^2(U) \to A^2(U)$ is the operator of multiplication by ϕ, that is, $M_\phi f = \phi f$ for $f \in A^2(U)$.

Before letting $\zeta \to 1/\delta$, we note that for $\zeta \in D$ we can write

$$\mathscr{L}_{H,0}(1 - \zeta\mathscr{L}_{H,0})^{-1} = \frac{\delta}{1 - \zeta\delta}\Pi + Q(\zeta),$$

where $\Pi f = \int_{E_\infty} f \, d\nu \; \varrho$ denotes the spectral projection associated to the eigenvalue δ and Q is a trace-class operator-valued holomorphic function on D. This follows from standard spectral theory (see, e.g. [22, Theorem 4.1.6]) together with the fact that δ is a simple eigenvalue of $\mathscr{L}_{H,0}$ by Proposition 3.5.

Now

$$\frac{\frac{\partial}{\partial t}\det(I - z\mathscr{L}_{H,t})|_{t=0,z=1/\delta}}{\frac{\partial}{\partial z}\det(I - z\mathscr{L}_{H,t})|_{t=0,z=1/\delta}} = \lim_{\zeta\to 1/\delta} \frac{\frac{\partial}{\partial t}\det(I - \zeta\mathscr{L}_{H,0})|_{t=0}}{\frac{\partial}{\partial z}\det(I - z\mathscr{L}_{H,0})|_{z=\zeta}}$$

$$= \lim_{\zeta\to 1/\delta} \zeta \frac{\delta \operatorname{tr}(M_\phi \Pi) + (1-\zeta\delta)\operatorname{tr}(M_\phi Q(\zeta))}{\delta\operatorname{tr}(\Pi) + (1-\zeta\delta)\operatorname{tr}(Q(\zeta))} = \frac{1}{\delta}\frac{\operatorname{tr}(M_\phi \Pi)}{\operatorname{tr}(\Pi)}$$

and the result follows by noting that $\operatorname{tr}(\Pi) = \int_{E_\infty} \varrho \, d\nu = 1$ and

$$\operatorname{tr}(M_\phi \Pi) = \int_{E_\infty} \phi \varrho \, d\nu = \int_{E_\infty} \phi \, d\mu \, .$$

□

Using Proposition 3.8 we can write

$$\int_{E_\infty} \phi \, d\mu = \delta \frac{\sum_{n=0}^{\infty} c'_{n,\phi}(0)\delta^{-n}}{\sum_{n=0}^{\infty} n\, c_{n,\phi}(0)\delta^{-(n-1)}} = \frac{\sum_{n=0}^{N} c'_{n,\phi}(0)\delta^{1-n}}{\sum_{n=0}^{N} n\, c_{n,\phi}(0)\delta^{1-n}} + O(\theta^{N^2}) \, , \quad (3.10)$$

for some $0 < \theta < 1$. Here we have used the fact that $c'_{n,\phi}(0) = O(\theta^{n^2})$ as $n \to \infty$ for some $0 < \theta < 1$, which follows from Proposition 3.7(d) and Cauchy's formula.

This leads naturally to the following definition:

Definition 3.4. For each $N \geq 1$, define $I_N(\phi)$ by

$$I_N(\phi) = \frac{\sum_{n=0}^{N} c'_{n,\phi}(0)\delta_N^{1-n}}{\sum_{n=0}^{N} n\, c_{n,\phi}(0)\delta_N^{1-n}} = \frac{\sum_{n=1}^{N} c'_{n,\phi}(0)\delta_N^{1-n}}{\sum_{n=1}^{N} n\, c_n \delta_N^{1-n}} \, .$$

This brings us to the second main result:

Theorem 3.2. *The values $I_N(\phi)$ converge to $\int_{E_\infty} \phi \, d\mu$ at a super-exponential rate; more precisely,*

$$I_N(\phi) = \int_{E_\infty} \phi \, d\mu + O(\theta^{N^2}) \quad \text{as } N \to \infty$$

for some $0 < \theta < 1$.

Proof. This follows from (3.10) and Theorem 3.1. □

Remark 3.4. Similar approximating formulae, in the context of invariant measures equivalent to Lebesgue measure, have been derived in [15–17] using a slightly different approach.

Importantly, it is possible to efficiently calculate each $c'_{n,\phi}(0)$ using periodic points:

Proposition 3.9. *Setting*

$$b_{\phi,n} = \frac{1}{n} \sum_{x \in \mathrm{Fix}_H(T^n)} \frac{\mathrm{sgn}((T^n)'(x))\phi^{(n)}(x)}{(T^n)'(x) - 1}, \tag{3.11}$$

we have

$$c'_{\phi,n}(0) = -\sum_{i=1}^{n} b_{\phi,i} c_{n-i} \quad \textit{for all } n \geq 1. \tag{3.12}$$

Proof. Let z belong to a sufficiently small disc centred at the origin. Then we have

$$\begin{aligned} \frac{\partial}{\partial t} \det(1 - z\mathscr{L}_{H,t})|_{t=0} &= -\det(1 - z\mathscr{L}_{H,0}) \sum_{m=1}^{\infty} \frac{z^m}{m} \frac{\partial}{\partial t} \mathrm{tr}(\mathscr{L}_{H,t}^n)|_{t=0} \\ &= -(1 + \sum_{n=1}^{\infty} c_n z^n) \sum_{m=1}^{\infty} b_{\phi,m} z^m . \end{aligned} \tag{3.13}$$

On the other hand,

$$\frac{\partial}{\partial t} \det(1 - z\mathscr{L}_{H,t})|_{t=0} = 1 + \sum_{n=1}^{\infty} c'_n(0) z^n , \tag{3.14}$$

and the result now follows by comparing coefficients in (3.13) and (3.14). □

3.5 An Example

As in [4], we consider the map

$$T(x) = \begin{cases} \frac{9x}{1-x} & \text{if } 0 \leq x \leq \frac{1}{10} \\ 10x - i & \text{if } \frac{i}{10} < x \leq \frac{i+1}{10} \text{ for } 1 \leq i \leq 9 \end{cases}$$

and $H = [\frac{9}{10}, 1]$.

Note that the inverse branches $\{T_i\}_{0\leq i\leq 9}$ are given by

$$T_0(x) = \frac{x}{9+x}$$

and

$$T_i(x) = (x+i)/10 \quad \text{for } 1 \leq i \leq 9 .$$

Writing

$$a_n = \frac{1}{n}\text{tr}(\mathscr{L}_H^n) = \frac{1}{n}\sum_{x\in \text{Fix}_H(T^n)} \frac{1}{(T^n)'(x)-1},$$

these a_n can be computed by locating the members of $\text{Fix}_H(T^n)$, all of which are quadratic numbers.

For example, there are 9 members of $\text{Fix}_H(T)$, denoted $x_0, x_1, \ldots, x_8$, say. For each $1 \leq i \leq 8$, we see that

$$\frac{1}{T'(x_i)-1} = \frac{1}{10-1} = \frac{1}{9},$$

whereas

$$\frac{1}{T'(x_0)-1} = \frac{1}{9-1} = \frac{1}{8}.$$

Therefore,

$$a_1 = \sum_{i=0}^{8} \frac{1}{T'(x_i)-1} = \frac{1}{8} + \frac{8}{9} = \frac{73}{72}.$$

The computation of a_2 is only slightly more involved. For the fixed point 0, we have

$$\frac{1}{(T^2)'(0)-1} = \frac{1}{81-1} = \frac{1}{80},$$

whereas for those 64 period-2 points $x_{ij} = (T_i \circ T_j)(x_{ij})$ with $1 \leq i, j \leq 8$, we have

$$\frac{1}{(T^2)'(x_{ij})-1} = \frac{1}{100-1} = \frac{1}{99}.$$

It remains to consider the 8 period-2 points of the form $x_{0i} = (T_0 \circ T_i)(x_{0i})$ and the 8 period-2 points of the form $x_{i0} = (T_i \circ T_0)(x_{i0})$, for $1 \leq i \leq 8$. In fact since $(T^2)'(x_{0i}) = (T^2)'(x_{i0})$, it suffices to consider the points x_{0i}, and a calculation gives

$$(T_0 \circ T_i)(x) = \frac{x+i}{x+90+i} \quad , \quad (T_0 \circ T_i)'(x) = \frac{90}{(x+90+i)^2},$$

$$x_{0i} = 5\left(\sqrt{\left(9+\frac{i-1}{10}\right)^2 + \frac{i}{25}} - 9 - \frac{i-1}{10}\right),$$

from which we compute

$$\begin{aligned} a_2 &= \frac{1}{2}\left(\frac{1}{80} + \frac{64}{99} + 2\sum_{i=1}^{8} \frac{1}{(T^2)'(x_{0i}) - 1}\right) \\ &= 0.41099534583625112158865416285 8\ldots \end{aligned}$$

Subsequent values a_n can be computed similarly, for example:

$$a_3 = 0.244247986872392594300895837121\ldots$$

$$a_4 = 0.164881484924536515073990416986\ldots$$

$$a_5 = 0.118849630250109944686793773181\ldots$$

$$a_6 = 0.089248843422890449580723889612\ldots$$

$$a_7 = 0.068936195289851448498303594869\ldots$$

3.5.1 The Escape Rate

We are now in a position to compute the power series coefficients c_i of the determinant $\det(I - z\mathscr{L}_H) = 1 + \sum_{i=1}^{\infty} c_i z^i$. Specifically, the formulae of Proposition 3.4 give

$$c_1 = -a_1$$

$$c_2 = -a_2 + \frac{a_1^2}{2}$$

$$c_3 = -a_3 + a_1 a_2 - \frac{a_1^3}{6}$$

$$c_4 = -a_4 + \frac{a_2^2}{2} + a_1 a_3 - \frac{a_1^2 a_2}{2} + \frac{a_1^4}{24}$$

$$c_5 = -a_5 + a_1a_4 + a_2a_3 - \frac{a_1^2a_3}{2} - \frac{a_1a_2^2}{2} + \frac{a_1^3a_2}{6} - \frac{a_1^5}{120}$$

$$c_6 = -a_6 + \frac{a_3^2}{2} + a_1a_5 + a_2a_4 - \frac{a_1^2a_4}{2} - a_1a_2a_3 - \frac{a_2^3}{6} + \frac{a_1^3a_3}{6} + \frac{a_1^2a_2^2}{4} - \frac{a_1^4a_2}{24} + \frac{a_1^6}{720}$$

$$c_7 = -a_7 + a_1a_6 + a_2a_5 + a_3a_4 - \frac{a_1^2a_5}{2} - a_1a_2a_4 - \frac{a_1a_3^2}{2} - \frac{a_2^2a_3}{2}$$
$$+ \frac{a_1^3a_4}{6} + \frac{a_1^2a_2a_3}{2} + \frac{a_1a_2^3}{6} - \frac{a_1^4a_3}{24} - \frac{a_1^3a_2^2}{12} + \frac{a_1^5a_2}{120} - \frac{a_1^7}{5040}.$$

Substituting the above numerical values[4] of a_n into the formulae for the c_i then gives

$$c_1 = -\frac{73}{72} = -1.01388888888888888888888888888\ldots$$
$$c_2 = 0.10298999366992171791751867664 8\ldots$$
$$c_3 = -0.00125238060300195381957803905 7\ldots$$
$$c_4 = 1.99475450153693261420976047639 3\ldots \times 10^{-6}$$
$$c_5 = -4.36711791065831134367103560290 0\ldots \times 10^{-10}$$
$$c_6 = 1.34821551235686339969318798546 5\cdots \times 10^{-14}$$
$$c_7 = -5.96955940686956115988494761374 1\ldots \times 10^{-20}.$$

These values of c_i allow us to form, for $1 \leq N \leq 7$, the degree-N polynomial approximation

$$\Delta_N(z) = 1 + \sum_{i=1}^{N} c_i z^i$$

to the determinant. The smallest root z_N of Δ_N can then be computed as follows:

$$z_1 = 72/73 = 0.98630136986301369863013 6\ldots$$
$$z_2 = 1.11188177924955318420101201507 6\ldots$$

[4]Of course we use higher precision for the a_n, ensuring that the values c_i are correct to the precision given.

$$z_3 = 1.109701877327063363180409111227\ldots$$

$$z_4 = 1.109705706696569182143392132129\ldots$$

$$z_5 = 1.109705705766218331774455583303\ldots$$

$$z_6 = 1.109705705766250204483482219528\ldots$$

$$z_7 = 1.109705705766250204326875729570\ldots$$

and inverting these gives the sequence of approximations $\delta_N = z_N^{-1}$ to $\delta(T, H, m)$ already listed in Sect. 3.1.

3.5.2 The Escape Measure

The escape measure μ is completely determined by its set of *moments* $\int_{E_\infty} x^n \, d\mu(x)$, $n \geq 0$. Each nth moment can be rapidly approximated by setting $\phi(x) = x^n$ then using the approach described in Sect. 3.4. Here we shall illustrate this in the case $n = 1$: the first moment $\mu(1) = \int_{E_\infty} x \, d\mu(x)$ is often called the *barycentre*, or *resultant*, of the measure μ.

Since $\phi(x) = x$ is fixed, we write $b_n = b_{\phi,n}$ (see (3.11)), so that

$$b_n = \frac{1}{n} \sum_{x \in \mathrm{Fix}_H(T^n)} \frac{\sum_{i=0}^{n-1} T^i x}{(T^n)'(x) - 1}.$$

We find that

$$b_1 = 4/9 = 0.444444444444444444444444444444\ldots$$

$$b_2 = 0.363146979940866817710676390686\ldots$$

$$b_3 = 0.323945697078082902031586942946\ldots$$

$$b_4 = 0.291597918113354097600085433302\ldots$$

$$b_5 = 0.262738636423342281952526356399\ldots$$

$$b_6 = 0.236761095523224368789249278048\ldots$$

$$b_7 = 0.213354539113042148099894783840\ldots$$

Recall that the coefficients $d_i = c'_{\phi,i}(0)$ (where $\phi(x) = x$) are given by formula (3.12). It follows that, for example, the first four[5] d_i are given by

$$d_1 = -b_1$$

$$d_2 = a_1 b_1 - b_2$$

$$d_3 = -b_3 + b_1 a_2 + a_1 b_2 - \frac{a_1^2 b_1}{2}$$

$$d_4 = a_2 b_2 + b_1 a_3 + a_1 b_3 - b_4 - a_1 b_1 a_2 - \frac{a_1^2 b_2}{2} + \frac{a_1^3 b_1}{6}$$

Substituting the numerical values of a_n, b_n into the formulae for the d_i gives us

$$\begin{aligned}
d_1 &= -4/9 = -0.44444444444444444444\ldots \\
d_2 &= 0.08747030400975046623994089326 4\ldots \\
d_3 &= -0.00152833960244703092715945867\ldots \\
d_4 &= 3.13319345309491769809247791617 0\ldots \times 10^{-6} \\
d_5 &= -8.40390182408161094002529348420\ldots \times 10^{-10} \\
d_6 &= 3.09098501937266448635392181469 8\ldots \times 10^{-14} \\
d_7 &= -1.60253894897971331452691425140\ldots \times 10^{-19}
\end{aligned}$$

The approximations

$$\mu_N(1) = \frac{\sum_{n=1}^{N} d_n z_N^{n-1}}{\sum_{n=1}^{N} n c_n z_N^{n-1}}$$

to the integral $\mu(1) = \int_{E_\infty} x \, d\mu(x)$ are then

$$\mu_2(1) = 0.44235438367466453221492914515 6\ldots$$

$$\mu_3(1) = 0.44213597759819689311366774805 5\ldots$$

$$\mu_4(1) = 0.44213667629780872206512523192 2\ldots$$

$$\mu_5(1) = 0.44213667605386536904818124984 5\ldots$$

---

[5]In the calculation that follows, we use d_i for $1 \leq i \leq 7$, though the algebraic formulae for d_i in terms of a_n, b_n are a little long to conveniently give here (e.g. the analogous expression for d_7 consists of a sum of 30 terms).

$$\mu_6(1) = 0.442136676053875847256104872452\ldots$$

$$\mu_7(1) = 0.442136676053875847197526214497\ldots$$

Acknowledgements We would like to thank Carl Dettmann, Wolfram Just and Gregor Tanner for bibliographic assistance.

References

1. Artuso, R., Aurell, E., Cvitanović, P.: Recycling of strange sets: I. Cycle expansions. Nonlinearity **3**, 325–359 (1990)
2. Artuso, R., Aurell, E., Cvitanović, P.: Recycling of strange sets: II. Applications. Nonlinearity **3**, 361–386 (1990)
3. Bahsoun, W.: Rigorous numerical approximation of escape rates. Nonlinearity **19**, 2529–2542 (2006)
4. Bahsoun, W., Bose, C.: Quasi-invariant measures, escape rates and the effect of the hole. Discrete Contin. Dyn. Syst. **27**, 1107–1121 (2010)
5. Bandtlow, O.F.: Resolvent estimates for operators belonging to exponential classes. Integr. Equ. Oper. Theory **61**, 21–43 (2008)
6. Bandtlow, O.F., Jenkinson, O.: Explicit eigenvalue estimates for transfer operators acting on spaces of holomorphic functions. Adv. Math. **218**, 902–925 (2008)
7. Bandtlow, O.F., Jenkinson, O.: On the Ruelle eigenvalue sequence. Ergod. Theory Dyn. Syst. **28**, 1701–1711 (2008)
8. Bunimovich, L.A., Yurchenko, A.: Where to place a hole to achieve fastest escape rate. Israel J. Math. **182**, 229–252 (2011)
9. Collet, P., Martínez, S., Schmitt, B.: The Yorke-Pianigiani measure and the asymptotic law on the limit Cantor set of expanding systems. Nonlinearity **7**, 1437–1434 (1994)
10. Cvitanović, P., Eckhardt, B., Rosenqvist, P.E., Russberg, G., Scherer, P.: Dynamical averaging in terms of periodic orbits. Physica D **83**, 109–123 (1995)
11. Cvitanović, P., Dettmann, C.P., Mainieri, R., Vattay, G.: Trace formulas for stochastic evolution operators: weak noise perturbation theory. J. Stat. Phys. **93**, 981–999 (1998)
12. Demers, M., Young, L.-S.: Escape rates and conditionally invariant measures. Nonlinearity **19**, 377–397 (2006)
13. Ferguson, A., Pollicott, M.: Escape rates for Gibbs measures. Ergod. Theory Dyn. Syst. **32**, 961–988 (2012)
14. Gohberg, I.C., Krein, M.G.: Introduction to the Theory of Linear Non-Selfadjoint Operators. American Mathematical Society, Providence (1969)
15. Jenkinson, O., Pollicott, M.: Computing invariant densities and metric entropy. Comm. Math. Phys. **211**, 687–703 (2000)
16. Jenkinson, O., Pollicott, M.: Ergodic properties of the Bolyai-Rényi expansion. Indag. Math. N.S. **11**, 399–418 (2000)
17. Jenkinson, O., Pollicott, M.: Orthonormal expansions of invariant densities for expanding maps. Adv. Math. **192**, 1–34 (2005)
18. Just, W.: On the thermodynamic approach towards time-series analysis. J. Phys. A **27**, 3029–3049 (1994)
19. Keller, G., Liverani, C.: Stability of the spectrum for transfer operators. Ann. Scuola Norm. Sup. Pisa Cl. Sci. **28**(4), 141–152 (1999)
20. Liverani, C., Maume-Deschamps, V.: Lasota-Yorke maps with holes: conditionally invariant probability measures and invariant probability measures on the survivor set. Ann. Inst. H. Poincaré Probab. Stat. **39**, 385–412 (2003)

21. Pianigiani, G., Yorke, J.A.: Expanding maps on sets which are almost invariant: decay and chaos. Trans. Am. Math. Soc. **252**, 351–366 (1979)
22. Pietsch, A.: Eigenvalues and s-Numbers. Cambridge University Press, Cambridge (1987)
23. Tanner, G., Richter, K., Rost, J.M.: The theory of two electron atoms: between ground state and complete fragmentation. Rev. Mod. Phys. **72**, 497–544 (2000)

Chapter 4
A Multi-time Step Method to Compute Optical Flow with Scientific Priors for Observations of a Fluidic System

Ranil Basnayake and Erik M. Bollt

Abstract Optical flow is a classical problem in computer vision, but the concepts must be adapted for applications to other fields such as fluid mechanics and dynamical systems. Our approaches are based on an inverse problem formalism, considering imposed scientific priors in the form of a cost function that rewards an assumed infinitesimal generator commensurate with assumed physics of the observed density evolution. This leads to a practical and principled approach to analyze an observed dynamical system. Additionally we present here for the first time a new multi-frame version of the functional coupling of multiple images. Following the calculus of variations, this yields a coupled set of Euler–Lagrange PDEs which serve as an assimilation method that inputs video frames as driving terms. The solution of the PDE which follows is the vector field, as designed. Data from an oceanographic system will be highlighted. It is also shown here how these flow fields can be used to analyze mixing and mass transport in the fluid system being imaged.

4.1 Introduction

The goal of an optical flow method is to compute vector fields, considering the apparent motion between time-adjacent images of the same scene. Horn and Schunck in [10] introduced the original optical flow algorithm to detect the motion field of a moving object making two assumptions. The first assumption, a brightness constraint, states that the image brightness of a point on the brightness pattern is constant over time for small motions. This includes a notion of rigid body motion. Then an energy functional is obtained measuring the errors of the brightness

R. Basnayake • E.M. Bollt (✉)
Department of Mathematics, Clarkson University, 8 Clarkson Avenue Box 5815, Potsdam, NY 13699, USA
e-mail: basnayrk@clarkson.edu; bolltem@clarkson.edu

W. Bahsoun et al. (eds.), *Ergodic Theory, Open Dynamics, and Coherent Structures*, Springer Proceedings in Mathematics & Statistics 70, DOI 10.1007/978-1-4939-0419-8_4,

© Springer Science+Business Media New York 2014

constraint over the image domain so that the velocity components u and v along the x and y directions, respectively, are obtained by minimizing the energy functional. However, in general, minimizing this energy functional is an ill-posed problem. Hence the above functional is regularized by making the second assumption that the expected flow is smooth, called the smoothness constraint. While the Horn and Schunck derivation was made in terms of local considerations only, the same PDE could be derived as a conservation law. The functional obtained from the brightness constraint is called the data term and the smoothness constraint is called the regularization term. The total energy functional includes a regularization term to the data term with weighting factor α, which is called the regularization parameter.

After regularizing the energy functional, the flow components u and v can be reconstructed by minimizing the derived energy functional. We achieve this by choosing a suitable regularization parameter α [9, 12, 13] to balance the desire to match the data fidelity but also to compromise with some form of regularity. We minimize the energy functional using a calculus of variations [6] approach. We apply Euler–Lagrange equations, first-order necessary conditions to have an extremum and a functional as explained in Sect. 4.2.1, to the optical flow energy functional. For this problem, we have two coupled Euler–Lagrange equations to be solved for u and v. Fortunately, the resulting PDE system allows for known and relatively simpler numerical techniques such as the Gauss–Seidel method, the gradient descent method, or the LU factorization with Gaussian elimination.

The above general optimization framework of optical flow computations allows reconstruction of both flow components u and v for a given image data. However, the only measured data is the image intensity. Since there is only one measurement, it is more stable to reconstruct just one unknown function in such a way that the flow components u and v can be obtained from this computed function. One way we suggested in [14] to help with the stability of the optimization problem while at the same time representing the fluid nature of our problems of interests is to reconstruct the stream/potential function ψ and then compute u and v by taking the Hamiltonian/usual gradient on the stream/potential function.

Our main goal here is to analyze unsteady fluid flow dynamics inferred using a sequence of image data of the system, taken by a movie camera or even from a satellite. Our previous methods required that there were only small changes of scene between each image, but otherwise the optimization approach just described yields a spatially regularized vector field in as far as the regularity term in the cost function is emphasized. However, if the scene in the movie data changes significantly between frames of the movie due to a relatively fast changing non-autonomous system, then there can be undesirable irregularity in the inferred vector field. This motivates us to develop a new approach to emphasize temporal regularity, a new approach of computing optical flow, using multiple images rather than just two. We will call this approach a *multi-time step optical* flow computation, which we introduce and develop here for the first time. For the computation of one stream function at a time, the multi-time step method and the stream function method are the same. However, when we compute more than one stream function for consecutive time points of the system simultaneously, we wish to adjust the functional to emphasize that the stream functions of two consecutive time-adjacent motions have similar behavior. This new assumption would incorporate an additional term into the energy functional with a weighting factor β. We will demonstrate our method with a benchmark data set representing a gyre and also the sea surface temperature (SST) data from the coast of Oregon, USA, as shown in Sect. 4.6.

Finally, we will demonstrate a use of the computed vector field to analyze mixing and mass transport in the fluid system being imaged. Several methods such as determining Lagrangian coherent structures (LCS) [2, 7, 18] and coherent pairs [5] are available to achieve this goal. In this endeavor we compute finite time Lyapunov exponents (FTLE) at each and every point in the system to determine LCSs. In the computation of FTLEs, we consider two nearby points at time t_0 and measure the separation of the trajectories over the time period $[t_0, T]$. If the separation is relatively high, the set of corresponding points are barriers for mixing and mass transport in the fluid system. These separation barriers on the FTLE fields are the LCSs. The LCS for the SST data set is computed using the vector fields obtained from the multi-time step method and shown in Fig. 4.9.

4.2 Classical Optical Flow Method

According to the original Horn and Schunck formulation of optical flow [10], the image brightness $I(x, y, t)$ at a point (x, y) is assumed to be locally conserved over the time if the motion is small. This implies that

$$I(x, y, t) = I(x + u, y + v, t + 1), \tag{4.1}$$

where u and v are the velocity components along the x and y directions, respectively, and here we consider a unit time interval. Note that $I(x, y, t)$ is the assumed data function. $I : \Omega \times \mathbb{R}^+ \to \mathbb{R}$ represents the gray scale color intensity of an image at time t, where (x, y) is in the scene domain Ω. Generally, actual data from a digital movie camera is pixelated at discrete spatial locations $\{x_i, y_j\}_{i,j=1}^{p,q}$ at times t_k, $\{I_{i,j,k}\}_{i,j,k=1}^{p,q,T}$ at times t_k as a complete data set, where $t_1 = 0$ and t_k is the time after $k - 1$ units. In other words, the image one was taken at $t = 0$ and the image k was taken after $k - 1$ time units from image 1 have been taken. Here p and q are the number of rows and the number of columns of the input images. The Taylor series expansion on Eq. (4.1) is

$$I_t + I_x u + I_y v = \delta, \tag{4.2}$$

where δ is the errors accumulated from the higher-order terms and I_t, I_x, and I_y are partial derivatives of I with respect to t, x, and y, respectively. The data term can be obtained by integrating the errors of the brightness constraint over the image domain as given by

$$E_b(u, v) = \int_{\Omega} \left(I_t + I_x u + I_y v\right)^2 \, d\Omega. \tag{4.3}$$

To develop u and v, as suggested by the brightness constraint objective in Eq. (4.2), we state the functional in Eq. (4.3). However, minimizing $E_b(u,v)$ in Eq. (4.3) is an ill-posed problem. To avoid ill-posedness of this problem, the data term must be regularized. We approach the regularization by assuming the expected flow is smooth. This implies that the partial derivatives of the velocity components u and v exist and hence suggests the regularization term should become

$$R(u,v) = \int_{\Omega} \left(u_x^2 + u_y^2 + v_x^2 + v_y^2\right) d\Omega. \tag{4.4}$$

Now the problem can be reformulated in terms of an energy functional obtained by combining the data term and the regularization term with weighting the second term by nonnegative regularization parameter α. Then the total energy functional to be minimized is given by

$$E(u,v) = \int_{\Omega} (I_t + I_x u + I_y v)^2 \, d\Omega + \alpha \int_{\Omega} \left(u_x^2 + u_y^2 + v_x^2 + v_y^2\right) d\Omega. \tag{4.5}$$

4.2.1 Euler–Lagrange Equations

To minimize the functional in Eq. (4.5), we use a calculus of variations approach, and in this subsection, we will demonstrate methods to minimize our functionals as in [6]. First consider minimization of a simple functional and necessary conditions to have a minimum to be solved for u and v.

Suppose we are given a functional $J(u)$ and to be determined is the optimizer $u(x) = \hat{u}(x)$ over the interval $a \le x \le b$

$$J(u) = \int_a^b F(x,u,u_x)\, dx. \tag{4.6}$$

Here $F(x,u,u_x)$ is a function with continuous first and second partial derivatives with respect to all of its arguments. Also, let $u(x)$ be a continuously differentiable function on $[a,b]$ which satisfies the boundary conditions

$$u(a) = A \qquad \text{and} \qquad u(b) = B. \tag{4.7}$$

Optimization necessitates that the first variations are stationary. Analogous to the first derivative of a function, here we obtain the first variation of the given functional. First, we give an increment $h(x)$ to the function $u(x)$ with the boundary conditions

$$h(a) = h(b) = 0. \tag{4.8}$$

The corresponding increment ΔJ in Eq. (4.6) with respect to the increment of $h(x)$ is

$$\Delta J = \int_a^b [F(x, u + h, u_x + h_x) - F(x, u, u_x)]\, dx. \tag{4.9}$$

The first variation of the functional Eq. (4.9) with respect to the argument u is obtained by taking the first-order terms of the Taylor series expansion on the integrand of Eq. (4.9) as

$$\delta J = \int_a^b [F_u(x, u, u_x)h + F_{u_x}(x, u, u_x)h_x]\, dx. \tag{4.10}$$

Theorem 4.1 ([6]). *A necessary condition for the differentiable functional $J(u)$ to have an extremum for $u = \hat{u}$ is that its variation vanish for $u = \hat{u}$, i.e., that*

$$\delta J(h) = 0 \tag{4.11}$$

for $u = \hat{u}$ and all admissible h.

From the above theorem, the necessary condition, called the Euler–Lagrange equation, for the functional in Eq. (4.6) to have an extremum can be obtained as

$$\frac{\partial F}{\partial u} - \frac{d}{dx}\left(\frac{\partial F}{\partial u_x}\right) = 0. \tag{4.12}$$

This statement can be expanded to allow for many variables, and we are interested to expand the results to a functional of the form

$$J(u, v) = \int_\Omega F(x, y, u, v, u_x, u_y, v_x, v_y)\, d\Omega, \tag{4.13}$$

allowing for vector fields $\langle u(x, y), v(x, y)\rangle$ in the plane $(x, y) \in \mathbb{R}^2$. Since there are two argument functions u and v, we may have two coupled Euler–Lagrange equations. The Euler–Lagrange equations for the functional in Eq. (4.13) follow as

$$\begin{aligned} \frac{\partial F}{\partial u} - \frac{\partial}{\partial x}\left(\frac{\partial F}{\partial u_x}\right) - \frac{\partial}{\partial y}\left(\frac{\partial F}{\partial u_y}\right) &= 0 \\ \frac{\partial F}{\partial v} - \frac{\partial}{\partial x}\left(\frac{\partial F}{\partial v_x}\right) - \frac{\partial}{\partial y}\left(\frac{\partial F}{\partial v_y}\right) &= 0. \end{aligned} \tag{4.14}$$

The above Euler–Lagrange equations are specialized below for functionals such as of the form in Eq. (4.5), and hence we can determine the velocity components by solving the resulting Euler–Lagrange PDE equations for u and v.

Another way to compute the Euler–Lagrange equations for a functional $J(u)$ in Eq. (4.6) is to compute the Gateaux derivative $J(u)$ with respect to u and set it to zero. The Gateaux derivative of the functional $J(u)$ is obtained as

$$DJ(u) = \frac{d}{d\tau} J(u + \tau h)|_{\tau=0} \tag{4.15}$$

for all admissible h.

4.2.2 Solution to the Optical Flow Problem

In general, when we minimize the optical flow problem, assuming existence and uniqueness of the solution, we determine the minimum by solving the Euler–Lagrange equations for each argument variable. The issues of existence and uniqueness of solutions of the PDEs in Eq. (4.14) follow the theory of convex optimization including discussion of convexity, coercivity, and lower semicontinuity of the specific functional $J(u, v)$. This goes beyond the needs of the discussion here, but we refer to the excellent textbook [19]. The following theorem from [19] explains sufficient conditions to have a unique minimum for the functional $J(u, v)$.

Theorem 4.2 ([19]). *Assume that $J : H \to \mathbb{R}$ is weakly lower semicontinuous and coercive and that C is a closed, convex subset of H. Then J has a minimizer over C. If, furthermore, J is also strictly convex, then the minimizer is unique.*

In the minimization process, first, we apply the equations in Eq. (4.14) to the functional in Eq. (4.5) to obtain the Euler–Lagrange equations, the gradients of the energy functional with respect to u and v, as

$$u_t = I_x(I_t + I_x u + I_y v) + \alpha \Delta^2 u, \qquad \text{and} \qquad v_t = I_y(I_t + I_x u + I_y v) + \alpha \Delta^2 v. \tag{4.16}$$

When we numerically solve the above system for u and v, one way of reaching the solution is to use numerical iterative methods such as the gradient descent method or the Gauss–Seidel method. For instance, the gradient decent algorithm is an iterative method which updates u and v for given initial conditions u_0 and v_0 as

$$\begin{aligned} u^{(k+1)} &= u^{(k)} - \delta t[I_x(I_t + I_x u^{(k)} + I_y v^{(k)}) + \alpha \Delta^2 u^{(k)}] \\ v^{(k+1)} &= v^{(k)} - \delta t[I_y(I_t + I_x u^{(k)} + I_y v^{(k)}) + \alpha \Delta^2 v^{(k)}] \end{aligned} \tag{4.17}$$

Here k represents the iteration number and δt is the time step. Recall that $u^{(k)}$ and $v^{(k)}$ must be discretely represented on the grid $\{x_i, y_j\}_{i,j=1}^{p,q}$ and derivatives must be numerically approximated by finite differences.

On the other hand, we can linearize the Laplacian terms in the Euler–Lagrange equations by introducing operators to compute second partial derivatives of u and v and then use direct methods to solve the resulting linear system. We will explain how to use the operator matrices to simplify a system in Sect. 4.3.

4.3 Stream Function Method

The Horn and Schunck optical flow method was developed to capture rigid body motion. Later research [14] applies this method to capture vector fields in fluid systems but changes must better respect expected fluid behavior. When we deal with a fluid system, dealing with the stream function of the motion is an accurate and precise way to develop velocity components. Note that in order to use the stream function, we assume the fluid is incompressible. Therefore, we convert the optical flow method into a stream function formulation. If the stream function is $\psi(x, y)$, then the velocity components are obtained as

$$\langle u, v\rangle = \langle -\psi_y, \psi_x\rangle. \tag{4.18}$$

Substituting $-\psi_y$ and ψ_x into Eq. (4.5), the energy functional in terms of the stream function ψ becomes

$$E(\psi) = \int_\Omega \left(I_t - I_x\psi_y + I_y\psi_x\right)^2 d\Omega + \alpha \int_\Omega (\psi_{xx}^2 + \psi_{yy}^2 + \psi_{xy}^2 + \psi_{yx}^2) d\Omega \tag{4.19}$$

One of the other advantages of the stream function method is we can impose the regularity directly to the stream function rather than to the flow components. For instance, if the resulting flow is known to be sparse, then the appropriate regularization term [17] in the usual optical flow computation "uv- method" is

$$R(u, v) = \int_\Omega (\mid u \mid + \mid v \mid)\, d\Omega. \tag{4.20}$$

Minimization of this regularization term emphasizes u to be zero in some places or v to be zero in some places which does not imply the flow to be zero. However, in the stream/potential function method, we can apply the scientific prior directly to the flow by regularizing the energy functional by the choice

$$R(\psi) = \int_\Omega \mid \nabla\psi \mid d\Omega. \tag{4.21}$$

This would measure the sparsity of the flow rather than components of the flow.

Next we review the development of Euler–Lagrange equations for the stream function method. Taking the Gateaux derivative as in Eq. (4.15), the gradients of the data fidelity

$$\int_\Omega \left(I_t - I_x\psi_y + I_y\psi_x\right)^2 d\Omega, \quad \text{and the regularization term}$$

$$\int_\Omega \left(\psi_{xx}^2 + \psi_{yy}^2 + \psi_{xy}^2 + \psi_{yx}^2\right) d\Omega, \quad \text{are obtained as}$$

$$A^*(I_t + A)\psi \quad \text{and} \quad (B + B^*)\psi,$$

respectively. Here A^* is the adjoint of operator A, and A and B are obtained as

$$A = -I_x D_y + I_y D_x \quad \text{and}$$

$$B = D_{xx} D^*_{xx} + D_{yy} D^*_{yy} + D_{xy} D^*_{xy} + D_{yx} D^*_{yx},$$

following standard considerations of calculus variations [19]. In the operators A and B, the operator $D_{\bullet\bullet}$ is a matrix operator of size $m \times m$ to compute the partial derivatives of a given vector of size $m \times 1$ with respect to indices $\bullet\bullet$. Here we stack the given matrix of size $p \times q$ into $m \times 1$ vector and $m = pq$. To develop the operator matrices $D_{\bullet\bullet}$, we use finite difference approximations with suitable boundary conditions. Finally, choosing a suitable regularization parameter α, the stream function ψ can be determined by solving the following Euler–Lagrange equation:

$$[A^* A + \alpha(B + B^*)]\psi = -A^* I_t. \tag{4.22}$$

Now the system can be solved for ψ by taking the LU decomposition on $A^* A + \alpha(B + B^*)$ and then applying Gaussian elimination. The velocity components u and v can be derived from Eq. (4.18).

4.4 Multi-time Step Method

Now we introduce the new multi-time step method of computing optical flow for a sequence of several time-dependent images to compute n time-dependent vector fields simultaneously. Furthermore, we can use higher-order finite difference approximations to compute I_t rather than the forward difference approximation necessary when just two images are available. In this endeavor our multi-time step optical flow algorithm is obtained based on the stream/potential optical flow algorithm, and the energy functional for $n = 1$ is given in Eq. (4.19); the solution can be obtained by solving the system Eq. (4.22). When we consider $n = 2$ or more, we include the additional term in the data fidelity by introducing regularity in the time direction by assuming two consecutive stream functions have similar behavior. Suppose we are given T time-adjacent images as a movie of a dynamical system. Then evolving the window is slow enough that considerations of continuously evolving frame views allow inference of the underlying dynamical systems, though the flow is unsteady. In other words assume $I(x, y, t)$ is continuous with respect to t throughout the scene. In fact we cope with this assumption by including a new term with a weighting factor. For instance, if there are only two stream functions ψ_1 and ψ_2, the additional minimizing integral would be

$$\int_\Omega (\psi_1 - \psi_2)^2 \, d\Omega \tag{4.23}$$

added to the chosen data fidelity and regularization terms already designed for assumed scientific priors. In this case we use three images at a time to compute the flow in two different time instances unless we use higher-order finite difference approximations.

We include into the new functional a parameter $\beta > 0$ in the previous energy functional, and the resulting energy functional is given by

$$\begin{aligned} E(\psi_1, \psi_2) = \int_\Omega (I_{1t} - I_{1x}\psi_{1y} + I_{1y}\psi_{1x})^2 + (I_{2t} - I_{2x}\psi_{2y} + I_{2y}\psi_{2x})^2 \, d\Omega \\ + \beta \int_\Omega (\psi_1 - \psi_2)^2 \, d\Omega \\ + \alpha \int_\Omega (\psi_{1xx}^2 + \psi_{1yy}^2 + \psi_{1xy}^2 + \psi_{1yx}^2) d\Omega \\ + \alpha \int_\Omega (\psi_{2xx}^2 + \psi_{2yy}^2 + \psi_{2xy}^2 + \psi_{2yx}^2) d\Omega. \end{aligned} \tag{4.24}$$

Thus we have data fidelity in two time instances at once, the term to emphasize time regularity and the spatial regularity.

Now taking the Gateaux derivative as in Eq. (4.15) of the functional in Eq. (4.24), the Euler–Lagrange equations corresponding to ψ_1 and ψ_2 are the system of PDEs as

$$\begin{aligned} [A_1^* A_1 + \beta(\psi_1 - \psi_2) + \alpha(B + B^*)]\psi_1 &= -A_1^* I_{1t} \\ [A_2^* A_2 - \beta(\psi_1 - \psi_2) + \alpha(B + B^*)]\psi_2 &= -A_2^* I_{2t}, \end{aligned} \tag{4.25}$$

where

$$A_1 = \left(-I_{1x} D_y + I_{1y} D_x\right),$$

$$A_2 = \left(-I_{2x} D_y + I_{2y} D_x\right) \quad \text{and}$$

$$B = D_{xx} D_{xx}^* + D_{yy} D_{yy}^* + D_{xy} D_{xy}^* + D_{yx} D_{yx}^*.$$

We can generalize the energy functional for n stream functions at n successive time instances as shown in the following:

$$\begin{aligned} E(\psi_1, \psi_2, \ldots, \psi_n) = \sum_{k=1}^{n} \int_\Omega \left(I_{kt} - I_{kx}\psi_{ky} + I_{ky}\psi_{kx}\right)^2 d\Omega \\ + \beta \sum_{k=1}^{n-1} \int_\Omega (\psi_k - \psi_{k-1})^2 \, d\Omega \\ + \alpha \sum_{k=1}^{n} \int_\Omega (\psi_{kxx}^2 + \psi_{kyy}^2 + \psi_{kxy}^2 + \psi_{kyx}^2) d\Omega. \end{aligned} \tag{4.26}$$

Similarly to $n = 2$, the Euler–Lagrange equations can be obtained for any number n, using the Gateaux derivative as in Eq. (4.15) of the functional in Eq. (4.26) with respect to $\psi_1, \psi_2, \ldots$, and ψ_n. The system of Euler–Lagrange equations for any integer n is a system of n partial differential equations and is obtained as

$$A_1^*(I_{1t} + A_1\psi_1)+\beta(\psi_1-\psi_2)+\alpha(B + B^*)\psi_1 = 0$$

$$A_k^*(I_{kt}+A_k\psi_k)+\beta(-\psi_{k-1}+2\psi_k-\psi_{k+1})+\alpha(B + B^*)\psi_k = 0, \quad \text{for } k = 2, 3, \ldots n-1$$

$$A_n^*(I_{nt}+A_n\psi_n)+\beta(\psi_{n-1}-\psi_n)+\alpha(B + B^*)\psi_n = 0,$$

where

$$A_k = \left(-I_{kx}D_y + I_{ky}D_x\right) \quad \forall k \text{ and}$$

$$B = D_{xx}D_{xx}^* + D_{yy}D_{yy}^* + D_{xy}D_{xy}^* + D_{yx}D_{yx}^*.$$

The above set of Euler–Lagrange equations can be reformulated as a linear system as

$$[K + \alpha L]z = b \tag{4.27}$$

where

$$K = \begin{bmatrix} A_1^*A_1 + \beta & -\beta & & & \\ -\beta & A_2^*A_2 + 2\beta & -\beta & & \\ \ddots & \ddots & \ddots & & \\ & & -\beta & A_{n-1}^*A_{n-1} + 2\beta & -\beta \\ & & & \beta & A_n^*A_n - \beta \end{bmatrix},$$

$$L = \begin{bmatrix} B+B^* & & & & \\ & B+B^* & & & \\ & & \ddots & & \\ & & & B+B^* & \\ & & & & B+B^* \end{bmatrix}, \quad z = \begin{bmatrix} \psi_1 \\ \psi_2 \\ \vdots \\ \psi_{n-1} \\ \psi_n \end{bmatrix} \text{ and } b = \begin{bmatrix} A_1^*I_{1t} \\ A_2^*I_{2t} \\ \vdots \\ A_{n-1}^*I_{(n-1)t} \\ A_n^*I_{nt} \end{bmatrix}.$$

All the entries of the matrices K and L are block matrices of the size $m \times m$, where $m = pq$ and $p \times q$ are the dimensions of the image domain. The entries of the vectors z and b are vectors of the size $m \times 1$.

To solve the above system, first, we take the LU decomposition of the matrix $K + \alpha L$ and then use Gaussian elimination. Solution of the above system yields $\psi_1(x, y) \equiv \psi_1(x, y, t_1)$, $\psi_2(x, y) \equiv \psi_2(x, y, t_2), \ldots, \psi_n(x, y) \equiv \psi_n(x, y, t_n)$, and taking the Hamiltonian gradient in Eq. (4.18) on each separately, the vector fields for each time instance, $t_1, t_2, \ldots, t_n$ can be obtained. That is, the velocity components $u_1, u_2, \ldots, u_n$ and $v_1, v_2, \ldots, v_n$ are computed from $\psi_1, \psi_2, \ldots, \psi_n$ using $\langle u_k, v_k\rangle = \langle -\psi_{ky}, \psi_{kx}\rangle$ for $k = 1, 2, \ldots n$. To improve the accuracy of the results, instead of first-order finite differences for the time derivative of I, higher-order finite difference approximations can be used when more than two images are available.

4.5 Scientific Priors

Toward our discussion of designing scientific prior information into our functional, we first review some well-known inverse problem theory. Recall that in linear algebra, solving a system $Au = z$ when A and u are given is called a forward problem. When A and z are available and the system is solved for u, it is called an inverse problem. For an inverse problem, a unique solution can be obtained by

$$u = A^{-1}z, \quad \text{if } A^{-1} \text{ exists.}$$

If A is not invertible, then the system may have infinitely many solutions or no solutions. In either case, it is standard to reformulate the problem as a minimization problem in such a way that from infinitely many solutions, we can select or emphasize a desired solution. Then the new problem can be written as

$$\arg\min_{u} \|Au - z\|^2. \tag{4.28}$$

In the modified problem, we find u in such a way to minimize the distance between Au and z according to the Euclidean norm. This alternative approach of solving the original problem ensures at least one solution to the original problem. We achieve the unique solution by introducing a new constraint to the reformulate problem. For instance, if we add $\|u\|_2$, to the reformulated problem, then such a minimal solution emphasizes the Euclidean perspective that the good solution should be on a radial closest to the origin. We call this notion the scientific prior of this simple linear problem. The procedure called imposing a scientific prior on the solution is done by adding a new term, a regularization term $R(\psi)$, to the modified problem with a nonnegative weighting parameter α. On the other hand, the scientific prior can be applied to the problem at the beginning of the construction of the model, by choosing different operators for A according to the prior knowledge of the data.

Analogously, building scientific prior information into functionals allows our inverse problem solutions for vector fields to emphasize expected physics. Now we will discuss various operators and regularizations to emphasize expected scientific prior information.

In the nominal optical flow algorithm, Horn and Schunck assumed conservation of image brightness intensity locally for rigid body motion. For the entire domain Ω with time step t, the conservation of image brightness I is emphasized by

$$E(\psi) = \int_{\Omega} \left(I_t - I_x\psi_y + I_y\psi_x\right)^2 \, d\Omega, \tag{4.29}$$

where ψ is the corresponding stream function for the motion. However, later researchers were interested in fluid motion and assumed that the image brightness I behaves according to the continuity equation over time in order to allow for divergent flow fields. Therefore, Corpetti in [4] proposed the data fidelity term

$$E(\psi) = \int_{\Omega} \left(I_t - I_x\psi_y + I_y\psi_x - I\psi_{yx} + I\psi_{xy}\right)^2 \, d\Omega. \tag{4.30}$$

Also Weickert in [20] improved the data fidelity by imposing the constancy of spatial brightness gradient, instead of brightness constancy, with the following:

$$E(\psi) = \int_{\Omega} \left(I_{xt} - I_{xx}\psi_y + I_{xy}\psi_x\right)^2 + \left(I_{yt} - I_{yx}\psi_y + I_{yy}\psi_x\right)^2 \, d\Omega. \tag{4.31}$$

Further, researchers in [3] have combined the data fidelities Eqs. (4.29) and (4.31) together with a nonnegative parameter β. The resulting data fidelity is obtained as

$$\begin{aligned} E(\psi) = \int_{\Omega} \left(I_t - I_x\psi_y + I_y\psi_x\right)^2 \, d\Omega + \beta \int_{\Omega} \left(I_{xt} - I_{xx}\psi_y + I_{xy}\psi_x\right)^2 \, d\Omega \\ + \beta \int_{\Omega} \left(I_{yt} - I_{yx}\psi_y + I_{yy}\psi_x\right)^2 \, d\Omega. \end{aligned} \tag{4.32}$$

In this data fidelity, either image brightness is constant or the gradient of the image brightness is constant, by adjusting β to emphasize either expected outcome or underlying physics. Through this kind of approach, the notion of modeling, the physics behind creating the scenes evolution in the movie data, is designed into the functional.

In addition to the data fidelities, it is important to regularize the energy functional, as understood by Andrey Tikhonov as shown in [11], to extract important information from ill-posed problem, both for functional analysis and optimization theoretic reasons of well posedness, and to further emphasize expected physics. For instance, if the resulting flow is expected to be sparse, then the regularization term that emphasizes sparsity is the total variation of the stream function. The regularization term can be written as

$$R_1(\psi) = \int_{\Omega} \mid \nabla\psi \mid d\Omega. \tag{4.33}$$

If, however, the flow field is expected to be smooth, then the appropriate regularization is the Horn and Schunck-type regularization

$$R_2(\psi) = \int_{\Omega} (\psi_{xx}^2 + \psi_{yy}^2 + \psi_{xy}^2 + \psi_{yx}^2) d\Omega. \tag{4.34}$$

On the other hand, if we impose the regularity on flow components, we would use

$$R_3(\psi) = \int_{\Omega} (\psi_x^2 + \psi_y^2) d\Omega. \tag{4.35}$$

According to the above explanation, we can develop different data terms and regularization terms using the known physics of the data and the expected flow. We can construct various multi-time step algorithms according to the prior knowledge of systems being imaged. Hence applying a suitable algorithm, the motion field of the observed system can be determined and then used to analyze the dynamics of the system.

4.6 Results from Multi-time Step Method

In this section, we will demonstrate the importance of using the data term in Eq. (4.29) and the regularization term in Eq. (4.34). Also we present the improvement of the accuracy of our algorithm with larger n. For the above purpose, we introduce a simple benchmark data set and an oceanic data set. Then we compare and benchmark some data terms and regularization terms introduced in Sect. 4.5.

4.6.1 Synthetic Data

Since we are interested in applying our algorithm to capture fluid motions, we constructed a benchmark data set, called the gyre, which is stereotypical of fluid motion. The stream function

$$\psi(x, y) = \sin(\pi x) \sin(\pi y) \tag{4.36}$$

on the domain $[0, 1] \times [0, 1]$ was considered, and the vector field governed by Eq. (4.36) is

$$\langle u, v \rangle = \langle -\pi \sin(\pi x) \cos(\pi y), \pi \cos(\pi x) \sin(\pi y) \rangle. \tag{4.37}$$

Evolving an initial condition over the domain $[0, 1] \times [0, 1]$ using the autonomous vector field in Eq. (4.37) according to the continuity equation

$$\frac{dI}{dt} = -(I_x u + I_y v + I u_x + I v_y), \tag{4.38}$$

a sequence of images is obtained. Two images are selected after transients and are shown in Fig. 4.1, images (a) and (b), and the vector field in Eq. (4.37) is shown in image (c).

When we compare the reconstructed vector fields with the true vector field, we need a figure of merit for comparison. Therefore, we computed angular error between the computed flow and the true flow and then the mean over the domain to obtain mean angular error (MAE) as explained in [15]. In this computation we include a third unit component to both computed and true velocity components to avoid the incorrect measurements near the points where flow magnitude is close to zero and sufficiently large.

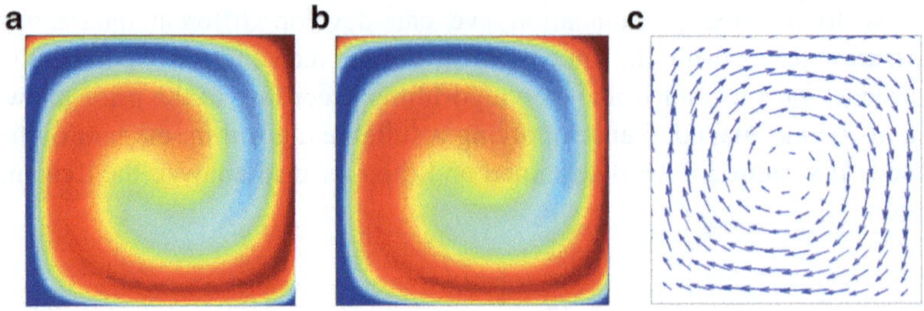

Fig. 4.1 Gyre data and true flow—images (**a**) and (**b**) show two later time instances of an initial density that has been evolved according to Eq. (4.38) with velocity components given by Eq. (4.37). The true flow field is shown in (**c**)

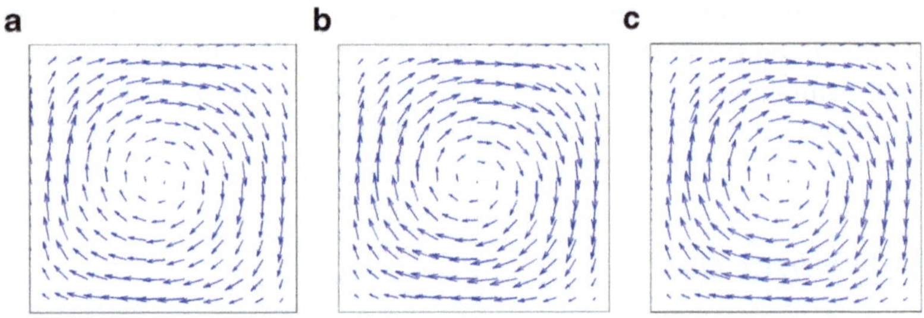

Fig. 4.2 Gyre flow with data term in Eq. (4.29)—images (**a**), (**b**), and (**c**) show the computed flow fields for the images (a) and (b) shown in Fig. 4.1 from the data term in Eq. (4.29) with regularization terms in Eqs. (4.33)–(4.35), respectively

As explained in Sect. 4.5, there are various ways to develop an energy functional to reconstruct vector fields. However, we choose the data term in Eq. (4.29) and regularization term in Eq. (4.34), introduced in the original Horn and Schunck method, to develop our multi-time step method. We introduce two other possible regularization terms in Eqs. (4.33) and (4.35) to cope with optical flow functionals. We combine three regularization terms in Eqs. (4.33)–(4.35) with the data term in Eq. (4.29), and the flow reconstructions are shown in Fig. 4.2. Note that in this presentation, we use the lagged diffusivity fixed point iteration method as in [1] to reconstruct the vector fields, when the regularization term in Eq. (4.33) is combined with any data term. Furthermore, we use multi-time step method with $n = 1$ to compare the data terms and the regularization terms.

The computed MAE for the energy functionals constructed from the data term in Eq. (4.29) with the regularization terms in Eqs. (4.33)–(4.35) are 2.4247°, 0.9837°, and 2.2790°, respectively. Among the three regularization terms, the regularization term in Eq. (4.34) gives least MAE. Further, we can use the data term in Eq. (4.30) instead of Eq. (4.29) with the regularization terms in Eqs. (4.33)–(4.35) to develop another three different energy functionals, and three different reconstructions are shown in Fig. 4.3.

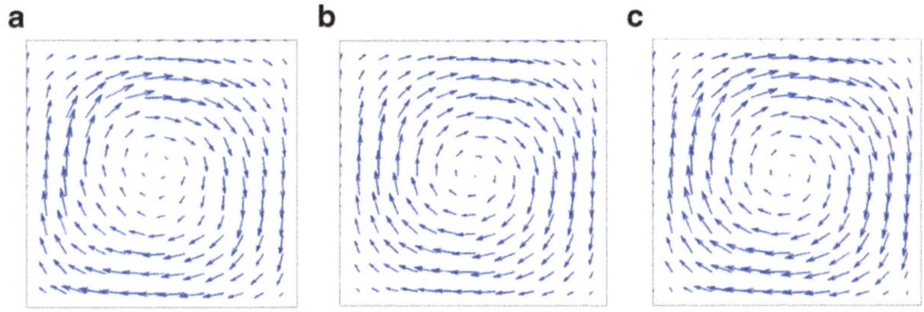

Fig. 4.3 Gyre flow with data term in Eq. (4.30)—images (**a**), (**b**), and (**c**) show the computed flow fields for the images (a) and (b) shown in Fig. 4.1 from the data term in Eq. (4.30) with regularization terms in Eqs. (4.33)–(4.35), respectively

The MAEs obtained from minimizing energy functionals of the data term in Eq. (4.30) with the regularization terms in Eqs. (4.33)–(4.35) are 3.3909°, 0.9895°, and 2.4787°, respectively. In this case also, the energy functional obtained from data term in Eq. (4.30) with the regularization terms in Eq. (4.34) gives the least MAE even though it is slightly bigger than the combination of the data term in Eq. (4.29) with regularization terms in Eq. (4.34). This verifies our selection for $n = 1$, and hence for larger n also, we use the data term in Eq. (4.29) with regularization terms in Eq. (4.34).

The next step is to compare the results by varying the step size n, the number of stream functions ψ computed at a time using multiple images. We applied our multi-time step method on an image sequence of gyre data set of which (a) and (b) of Fig. 4.1 show the fourth and fifth images of that, respectively. For instance, if we apply a first-order finite difference approximation to estimate I_t with the step size $n = 1$, then we compute one stream function ψ_1 using two images, image 1 and image 2. The velocity components $(u_1, v_1) = (-\psi_{1y}, \psi_{1x})$ represent the motion field between image 1 and image 2. If however we choose the step size $n = 2$, then we compute two stream functions ψ_1 and ψ_2 using three images. Then $(u_1, v_1) = (-\psi_{1y}, \psi_{1x})$ is the motion field between image 1 and image 2 and $(u_2, v_2) = (-\psi_{2y}, \psi_{2x})$ is the motion field between image 2 and image 3, respectively. Continuing in this manner, we can increase the step size n. Note that, if there is a sequence of nine images, we can compute eight vector fields for each consecutive image pair. When the step size is $n = 1$, eight separate computations are necessary, but when $n = 2$, only four computations are necessary and so on. Again we emphasize that the advantage of choosing larger n is that the time regularity is emphasized as seen clearly in the computations. Addition of terms of the form Eq. (4.23) penalizes large changes of ψ between successive frames. As explained, the results from $n = 1, 2, 3$, and 4 for the gyre image sequence are shown in images (a), (b), (c), and (d) in Fig. 4.4, respectively.

The MAEs for $n = 1, 2, 3$, and 4 are 0.9837°, 0.9826°, 0.9807°, and 0.9883°, respectively. In this example, the accuracy of the algorithm improves until $n = 3$. In all the above reconstructions, the regularization parameter α was selected so that it minimizes the MAE and the parameter β was fixed at $\beta = 0.01$.

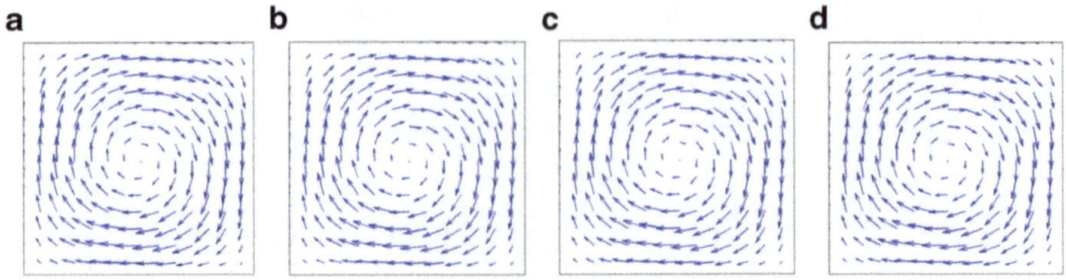

Fig. 4.4 Gyre flow from multi-time step method—images (**a**), (**b**), (**c**), and (**d**) show vector fields computed on the image (b) in Fig. 4.1 from multi-time step method with $n = 1, 2, 3$, and 4, respectively. The true flow field is shown in (**a**). While all estimated vector fields are visually similar, the mean angular errors are improving up to $n = 3$

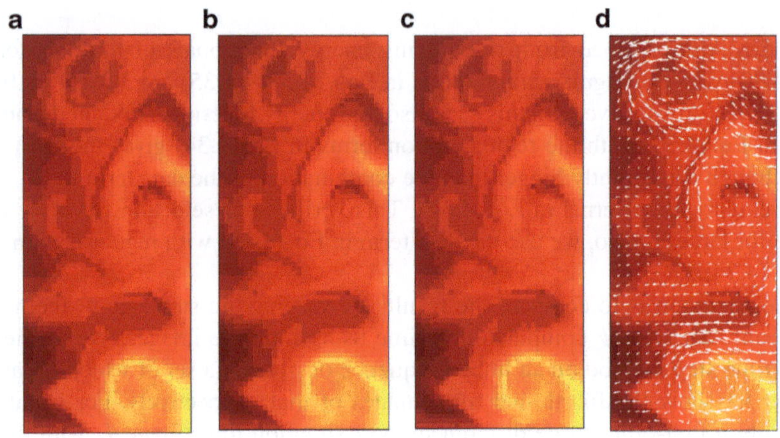

Fig. 4.5 SST data and true flow—three consecutive images of the SST data set are shown in (**a**), (**b**), and (**c**), respectively, and the flow on image (**b**) is shown in the image (**d**)

4.6.2 An Oceanographic Data Set

Now we apply the algorithm to a natural scenario which shows the SST off the coast of Oregon, USA. This data set was generated from a 3-D ocean model, using data obtained from geostationary operational environmental satellite (GOES) [16]. In Fig. 4.5, images (a), (b), and (c) show SST data of three consecutive hours on August 1, 2002. The image (d) represents the true vector field of the mixing temperature corresponding to the image (b).

Since we have a time-dependent sequence of images of the SST data, we can apply the multi-time step method to compute the vector fields. When the step size is n, we compute n consecutive vector fields at a time and required $n + 4$ images, if we apply fourth-order finite difference approximations to compute I_t. The Fig. 4.6 represents the computed flow fields on the image (b) shown in Fig. 4.5 with the step size $n = 1, 2, 3$, and 4 in images (a), (b), (c), and (d), respectively. In each case, the

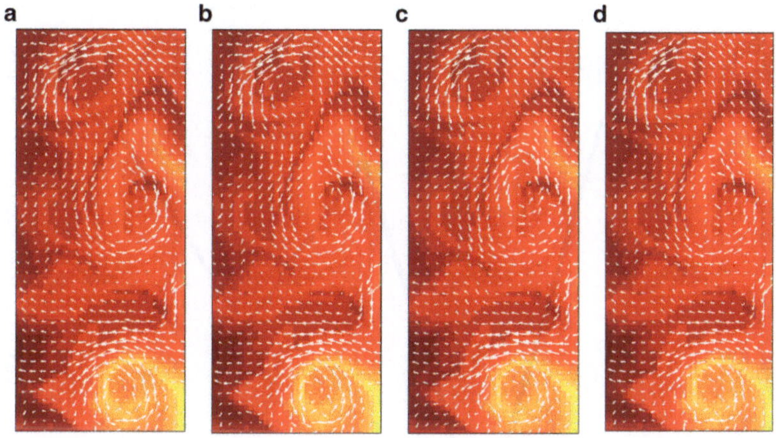

Fig. 4.6 SST flow—the computed flow fields for the data showed in Fig. 4.5 with n equals to 1, 2, and 3 are shown in (**a**), (**b**), (**c**), and (**d**), respectively. While all these are roughly similar and so not immediately different to visual inspection, there are visible differences that are apparent in closer inspection

algorithm captures the gyres accurately, and it is clearly visible that when $n = 3$, the algorithm captures the laminar flow as appears just above the bottom gyre. Except for the above laminar flow, all the other vectors represent a similar behavior and the differences are not visible. Now we can compare the two vector fields consisting of large numbers of vectors by comparing single numbers and in the comparison of the step sizes; we use the percentage MAE.

Figure 4.7 shows the graph of percentage MAE of the computed flow relative to the true flow versus the step size used to compute the flow fields. According to the graph, the percentage MAE fluctuates and the minimum is achieved when $n = 3$ as we can see in the flow fields. Note that, for all the reconstructions in Fig. 4.5, the regularization parameter α was selected so that it minimizes the MAE and the parameter β was fixed at $\beta = 0.01$.

4.7 Mixing and Transport Barriers

As an application, we will now discuss a transport analysis inferred directly from observed spatiotemporal movie data. Toward the identification of mixing and transport barriers, we compute FTLE. These are scalar values for each point in the domain D as explained in [7, 8] and obtained LCS. In this case, for a given point $\mathbf{x} = \langle x(t), y(t)\rangle$, a flow map ϕ_T is obtained by evolving $\mathbf{x}$ over a time period $[t, t + T]$ using the velocity components $\mathbf{v} = \langle u(x, y, t), v(x, y, t)\rangle$.

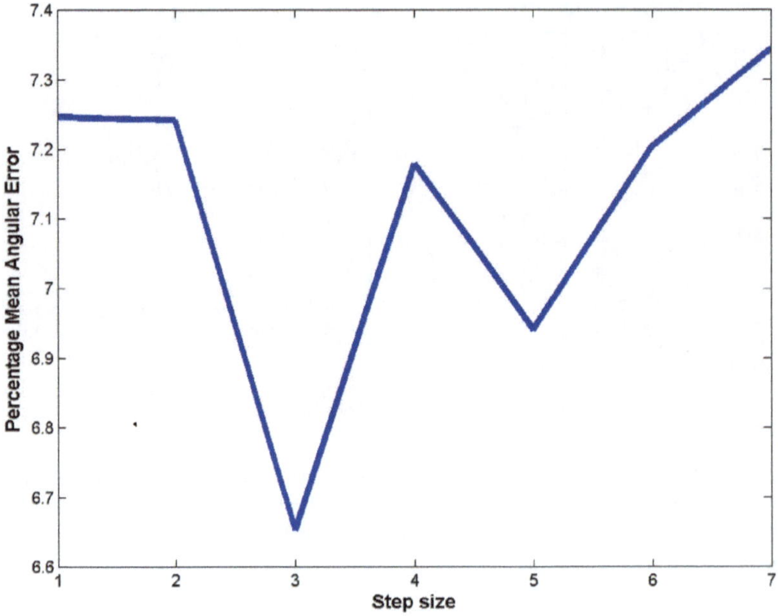

Fig. 4.7 Percentage MAE vs. step size—the graph shows the percentage of the mean angular error for the computed flow by changing the step size n on the image (b) shown in Fig. 4.5. For the parameters as we specified, the $n = 3$, multi-time step method is overall best

Then the Jacobian matrix of the flow map ϕ_T is obtained $J = \frac{d\phi_T(\mathbf{x})}{d\mathbf{x}}$, and therefore, the finite time strain tensor of $\mathbf{v} = \langle u(x,y,t), v(x,y,t)\rangle$ along the trajectory $\mathbf{x} = \langle x(t), y(t)\rangle$ is obtained as

$$M = \frac{d\phi_T(\mathbf{x})^*}{d\mathbf{x}}\frac{d\phi_T(\mathbf{x})}{d\mathbf{x}}, \tag{4.39}$$

where A^* is the adjoint of A. Then the FTLE value at a point $\mathbf{x}$ over time T is given by

$$\sigma^T = \frac{1}{|T|} ln\sqrt{\lambda_{max}(M)}. \tag{4.40}$$

When the FTLEs are computed for the entire domain of the system, the set of points in the domain corresponding to relatively high FTLE values are suggested to act as pseudo barriers for mixing and transport of the system, although it is known that some can be simplified due to shear behavior. For instance, if we consider the double gyre with the stream function $\psi(x,y,t) = C\sin(\pi f(x,t))\sin(\pi y)$, where $f(x,t) = \epsilon\sin(\omega t)x^2 + (1 - 2\epsilon\sin(\omega t))x$ over the domain $D = [0,2] \times [0,1]$, the corresponding vector field can be obtained as

$$\langle u, v\rangle = \langle -\pi C\sin(\pi f(x,t))\cos(\pi y), \pi C\cos(\pi f(x,t))\sin(\pi y)\frac{\partial f}{\partial x}\rangle, \tag{4.41}$$

a b

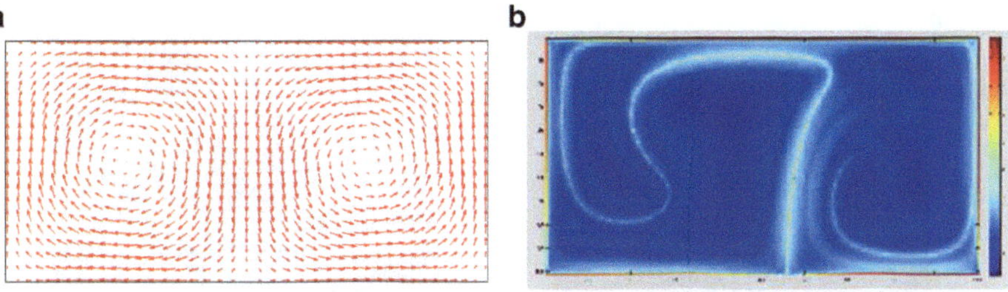

Fig. 4.8 Double gyre flow and FTLE field—the flow field for the non-autonomous double gyre at time $t = 0$ and the computed FTLE field for $T = 10$ are shown in images (**a**) and (**b**), respectively

where the constants are $C = 0.1, \omega = \frac{2\pi}{10}$ and $\epsilon = 0.25$. The vector field that represents $\langle u, v \rangle$ at $t = 0$ and the computed FTLE field using $\langle u, v \rangle$ with $T = 10$ are shown in Fig. 4.8. The red color represents relatively high FTLE values and the blue color represents the relatively small FTLE values. In [18], the authors have proven that flux across the LCS is close to zero, and hence it is difficult for fluids in the system to cross the LCS. Therefore, these LCSs act as barriers to the mixing and transport.

The next example represents the FTLE field for the SST data set using the computed flow from the multi-time step method with $n = 3$. In Fig. 4.9, the flow field obtained for the SST data on August 2, 2002, is shown in image (a) and the FTLE field obtained from those computed vector fields is shown in (b). When we compute FTLE fields, we need more than three time-adjacent vector fields. Since we have a sequence of images, we apply multi-time step method on the image sequence to obtain a sequence of non-autonomous vector fields. If we use fourth-order finite difference approximations to estimate I_t, we need first 7 images of the sequence to compute three time-adjacent vector fields on images 3, 4, and 5, respectively. To compute the vector fields on images 6, 7, and 8, we apply multi-time step method again on images 4 to 10, and we can continue the procedure until we reach the end of the image sequence. In this case, we evolve 10 time steps forward in time to get the FTLE field. The blue color represents relatively low FTLE values and the red color represents relatively high FTLE values. The red color ridges, LCSs, act as the mixing barriers to the heat.

4.8 Conclusion

We have developed a new optical flow algorithm to extract vector fields of an observed system from spatiotemporal (movie) data. When a sequence of images is available for an observed system, we can emphasize regularity not only spatially but also in time. In spatial regularization, we emphasize known physics of both the data and the expected flow, and in regularization in time, we emphasize the similar

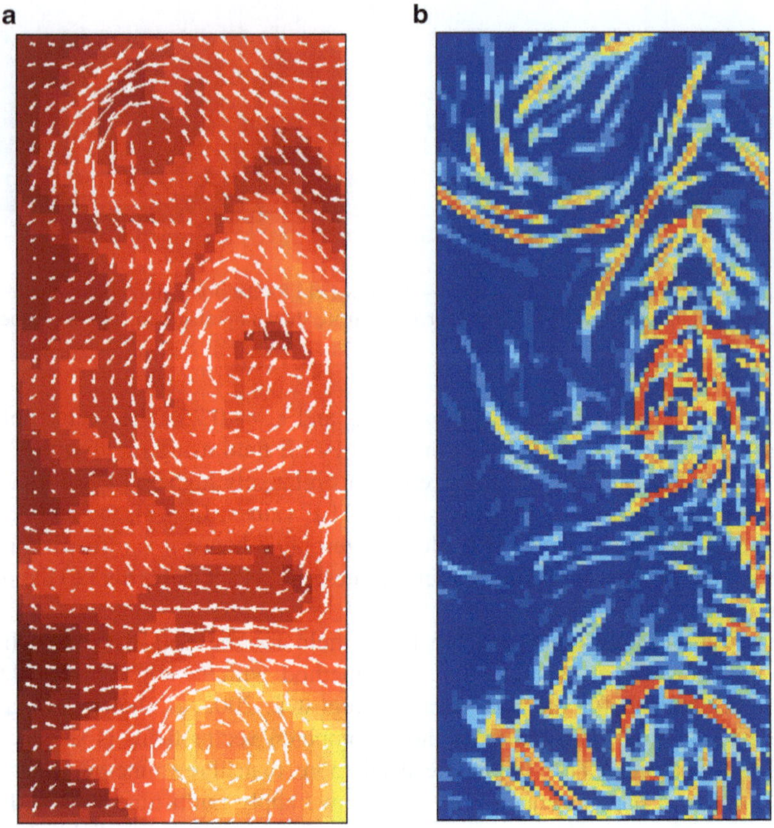

Fig. 4.9 SST flow and FTLE field—the computed flow field for the SST data and the FTLE field are shown in images (**a**) and (**b**), respectively

behavior of the flow fields. From these computed vector fields, we can analyze dynamics of the system by computing LCS. However, according to the results shown in Sect. 4.6.2, when the number of frames n simultaneously used in our new multi-time step method increases, the accuracy of the algorithm improves. Furthermore, we use the same regularization parameter to compare the results, and it does not need to be the same. Our future goal is to improve the algorithm to select the best step size n and the suitable regularization parameter.

References

1. Basnayake, R., Luttman, A., Bollt, E.: A lagged diffusivity method for computing total variation regularized fluid flow. Contemp. Math. **586**, 59–66 (2013)
2. Bollt, E., Luttman, A., Kramer, S., Basnayake, R.: Measurable dynamics analysis of transport in the Gulf of Mexico during the oil spill. Int. J. Bifurcat. Chaos **22**(03), 1230012 (2012)

3. Brox, T., Bruhn, A., Papenberg, N., Weickert, J.: High accuracy optical flow estimation based on a theory for warping. In: Pajdla, T., Matas, J. (eds.) Proceedings of 8th European Conference on Computer Vision, vol. 4, pp. 25–36 (2004)
4. Corpetti, T., Mémin, E., Pérez, P.: Adaptation of standard optic flow methods to fluid motion. In: 9th International Symposium on Flow Visualization, pp. 1–10 (2000)
5. Froyland, G., Santitissadeekorn, N., Monahan, A.: Transport in time-dependent dynamical systems: finite-time coherent sets. Preprint arXiv:1008.1613 (2010)
6. Gelfand, I., Fomin, S.: Calculus of Variations. Dover Publications, Mineola (2000)
7. Haller, G.: Lagrangian coherent structures from approximate velocity data. Phys. Fluid. **14**, 1851 (2002)
8. Haller, G., Poje, A.: Finite time transport in aperiodic flows. Phys. D Nonlinear Phenom. **119**(3–4), 352–380 (1998)
9. Hansen, P.C.: The L-curve and its use in the numerical treatment of inverse problems. In: Johnston, P. (ed.) Computational Inverse Problems in Electrocardiology. Advances in Computational Bioengineering, pp. 119–142. WIT Press, Southampton (2000)
10. Horn, B.K.P., Schunck, B.G.: Determining optical flow. Artif. Intell. **17**, 185–203 (1981)
11. Kabanikhin, S.I.: Inverse and Ill-Posed Problems: Theory and Applications, vol. 55. De Gruyter, Germany (2011)
12. Krawczyk-StańDo, D., Rudnicki, M.: Regularization parameter selection in discrete ill-posed problems: the use of the U-curve. Int. J. Appl. Math. Comput. Sci. **17**(2), 157–164 (2007)
13. Krawczyk-Stańdo, D., Rudnicki, M.: The use of L-curve and U-curve in inverse electromagnetic modelling. Intell. Comput. Tech. Appl. Electromagn. **119**, 73–82 (2008)
14. Luttman, A., Bollt, E., Basnayake, R., Kramer, S.: A stream function approach to optical flow with applications to fluid transport dynamics. Proc. Appl. Math. Mech. **11**(1), 855–856 (2011)
15. McCane, B., Novins, K., Crannitch, D., Galvin, B.: On benchmarking optical flow. Comput. Vis. Image Understand. **84**, 126–143 (2001). doi:10.1006/cviu.2001.0930
16. Office of Satellite Operation. http://www.oso.noaa.gov/goes/ (2011)
17. Schmidt, M., Fung, G., Rosales, R.: Fast optimization methods for l1 regularization: a comparative study and two new approaches. Machine Learning: ECML 2007, pp. 286–297 (2007)
18. Shadden, S., Lekien, F., Marsden, J.: Definition and properties of Lagrangian coherent structures from finite-time Lyapunov exponents in two-dimensional aperiodic flows. Phys. D Nonlinear Phenom. **212**(3), 271–304 (2005)
19. Vogel, C.R.: Computational Methods for Inverse Problems. Frontiers in Mathematics. SIAM, Philadelphia (2002)
20. Weickert, J., Bruhn, A., Papenberg, N., Brox, T.: Variational optic flow computation: from continuous models to algorithms. In: Alvarez, L. (ed.) International Workshop on Computer Vision and International Workshop on Computer Vision and Image Analysis, IWCVIAÃ'03, Las Palmas de Gran Canaria (2003)

Chapter 5
Numerical Approximation of Conditionally Invariant Measures via Maximum Entropy

Christopher Bose and Rua Murray

Abstract It is well known that open dynamical systems can admit an uncountable number of (absolutely continuous) conditionally invariant measures (ACCIMs) for each prescribed escape rate. We propose and illustrate a convex optimisation-based selection scheme (essentially maximum entropy) for gaining numerical access to some of these measures. The work is similar to the maximum entropy (MAXENT) approach for calculating absolutely continuous invariant measures of nonsingular dynamical systems but contains some interesting new twists, including the following: (i) the natural escape rate is not known in advance, which can destroy convex structure in the problem; (ii) exploitation of convex duality to solve each approximation step induces important (but dynamically relevant and not at first apparent) localisation of support; and (iii) significant potential for application to the approximation of other dynamically interesting objects (e.g. invariant manifolds).

5.1 Introduction

The study of classical dynamical systems concerns the existence and stability of invariant sets under the action of a transformation $T : X \to X$. Depending on the setting, X may be a measure space, a topological space (with or without a metric structure), a differentiable manifold, a Banach space, and so on. In each case, orbits

C. Bose
Department of Mathematics and Statistics, University of Victoria, PO Box 3060 STN CSC, Victoria, BC, Canada V8W3R4
e-mail: cbose@uvic.ca

R. Murray (✉)
Department of Mathematics and Statistics, University of Canterbury, Private Bag 4800, Christchurch 8140, New Zealand
e-mail: rua.murray@canterbury.ac.nz

W. Bahsoun et al. (eds.), *Ergodic Theory, Open Dynamics, and Coherent Structures*, Springer Proceedings in Mathematics & Statistics 70, DOI 10.1007/978-1-4939-0419-8_5,
© Springer Science+Business Media New York 2014

defined by iterative application of T remain in X. For an **open dynamical system**, T is defined only on a subset $A \subsetneq X$, and there are $x \in A$ for which $T(x) \notin A$. Such x are said to *escape*.

Open dynamical systems may be studied in their own right (the paper of Demers and Young [12] gives a summary of important questions) or may be used to study metastable states in closed dynamical systems. In the latter case, a subset $A \subset X$ is *metastable* if $T(A) \setminus A$ is in some sense *small relative to* A. Work on making this precise dates at least to 1979, when Pianigiani and Yorke [22] introduced **conditionally invariant measures** (CIMs) (see Sect. 5.1.2 below) and used them to study metastability in expanding interval maps.[1] More recently, Homburg and Young [19] made productive use of CIM to analyse intermittent behaviour near saddle-node and boundary crisis bifurcations in unimodal families. Many authors have continued to obtain results connecting escape rates and metastable behaviour of closed systems; see, for example, [1, 2, 13, 17, 18, 20].

One of the interesting challenges is to find CIMs which models the escape statistics of orbits distributed according to some 'natural' initial measure m on A. In closed dynamical systems, there may exist a unique ergodic invariant measure μ which is absolutely continuous (AC) with respect to m. Via Birkhoff's ergodic theorem, such measures μ describes the orbit distribution of large[2] sets of initial conditions. By contrast, an open system may support uncountably many AC conditionally invariant measures (ACCIMs) [12, Theorem 3.1], so ascribing dynamical significance on the basis of absolute continuity alone does not make sense. Recently, progress has been made in a variety of settings, identifying ACCIMs whose densities arise as eigenfunctions of certain quasicompact *conditional transfer operators* acting on suitable Banach spaces. Such ACCIMs may be considered 'natural' (see [12] for discussion), giving a well-defined *escape rate* from A. See, for example, [6] for dynamics on Markov towers, [9, 10] for interval maps modelled by Young towers, [7, 8] for expanding circle maps and subshifts of finite type, and [21] for interval maps with BV potentials. Extending these techniques to higher-dimensional settings such as billiards and Lorentz gas is an area of much current interest [11].

This chapter develops a new class of computational methods for the explicit approximation of conditionally invariant probability measures on A. Our ideas use convex optimisation: the criteria for conditional invariance are expressed as a sequence of *moment conditions* over L^1 (integration against a suitable set of L^∞ test functions), and the principle of maximum entropy (MAXENT) is used to select (convergent) sequences of measures that are approximately conditionally invariant, in other words, *approximate CIMs*. The entropy maximisation is solved via standard convex duality techniques, although attainment in the dual problem may necessitate a nonobvious (but dynamically meaningful) reduction of the domain on which the

---

[1] The motivation in [22, p. 353] went beyond interval maps, including preturbulent phenomena in the now famous Lorenz equations, and metastable structures in atmospheric and other fluid flows and complex systems.

[2] In the sense of positive m-measure

maximisation is done. The required steps are achievable for piecewise constant test functions (similar in spirit to Ulam's method [15] but with a completely different mathematical foundation). The chapter is structured as follows: first, we introduce notation for our study of open systems and formulate the ACCIM problem (and its uncountable multiplicity of solutions) via *conditional transfer operators*; next, the MAXENT problem is set up and analysed; the Ulam-style test functions are introduced in Sect. 5.3, and the domain reduction and some numerical examples are given to illustrate the method; we finish with some concluding remarks.

5.1.1 Nonsingular Open Dynamical Systems

Let (X, m) be a measure space. We consider the dynamics generated by a transformation on a **subset of** X **which fails to be forward invariant**; such a dynamical system is called **open** and may or may not support any recurrent behaviour. Let $A \subsetneq X$ be measurable and let $T : A \to X$ be a measurable transformation where:

- $H_0 := T(A) \setminus A$ is a measurable subset of X (called the *hole*).
- $m(A \cap T^{-1}H_0) > 0$.
- $m(E) > 0$ whenever $m(T^{-1}E) > 0$ and E is a measurable subset of X.
- T is locally finite-to-one (for each $x \in A$, $T^{-1}(x) = \{x_{-1} \in A : T(x_{-1}) = x\}$ is either empty or finite).

Definition 5.1. Let $m|_A$ denote the restriction of the measure m to A. We call[3] $(T, A, m|_A)$ satisfying the above conditions a **nonsingular open dynamical system.**

Notice that $T(x)$ is defined only for $x \in A$, and the 'hole' H_0 can be used to define a *survival time* for each $x \in A$:

$$\tau(x) := \begin{cases} n & \text{if } x, T(x), \ldots T^n(x) \in A \text{ and } T^{n+1}(x) \in H_0 \\ \infty & \text{if } T^k(x) \in A \ \forall k \in \mathbb{Z}_+. \end{cases}$$

When $\tau(x) = n < \infty$, $T^n(x) \in H_1 := A \cap T^{-1}(H_0)$ and such orbits of T terminate at time $\tau(x)+1$. Only those x for which $\tau(x) = \infty$ can exhibit recurrent behaviour.

For all that follows, it is convenient to decompose A into invariant and transient parts. Define:

- The n **step survivor set** as

$$A_n := \{x \in A : \tau(x) \geq n\} = \{x : x, T(x), \ldots, T^n(x) \in A\} = \cap_{k=0}^{n} T^{-k} A$$

[3]Clearly $m \circ T^{-1} \ll m$ so that $T : (A, m|_A) \to (X, m)$ is a nonsingular transformation, but $T : (A, m|_A) \to (X, m|_A)$ fails to be nonsingular, as $m|_A \circ T^{-1}(H_0) = m(A \cap T^{-1}(H_0)) > 0$ while $m|_A(H_0) = 0$.

- $A_\infty := \cap_{n\geq 0} A_n$.
- $H_n := A_{n-1} \setminus A_n = \{x : \tau(x) = n-1\}$ for $1 < n < \infty$.

Notice that if $x \in H_n$, then $T^k(x) \in H_{n-k}$ for $0 < k \leq n$. The orbit of x 'falls into the hole' at time n (escapes) and is lost to the system thereafter. As well as escape from A, we need to account for the possibility that backward orbits may not be defined ($T : A_1 \to A$ may not be *onto*). Since some $x \in A$ may have no preimages in A, define the following subsets of A:

- $K_0 := \{x : A \cap T^{-1}x = \emptyset\}$
- $K_n := \{x : \emptyset \neq (A \cap T^{-n}(x)) \subset K_0\} = \{x : \min\{k : A \cap T^{-k}x = \emptyset\} = n+1\}$
- $K_\infty := \{x_0 :$ there is no sequence $\{x_{-n}\}_{n=1}^\infty$ such that $T(x_{-n}) = x_{-(n-1)}$, $n > 0\}$
- $H_\infty := \cup_{n>0}(H_n \setminus K_\infty)$

Points in K_∞ are 'backward transient', while points in H_∞ are 'forward transient'. Lemma 5.1 contains some facts about the action of T on the various sets H_n, K_n. The reader may easily verify that:

- $A_0 = A$ and $T(A_n) \subseteq A_{n-1}$.
- $H_n \cap H_m = \emptyset$ if $n \neq m$, $H_n \subseteq A_{n-1}$ and $H_n \cap A_n = \emptyset$.
- $T(H_n) \subseteq H_{n-1}$.
- $A \cap T^{-1}(K_n) \subseteq \cup_{m<n} K_m$ and $K_{n+1} \subseteq T(K_n)$.
- $\cup_{n=0}^\infty K_n \subseteq K_\infty$, and the union on the left may be finite or infinite (or even the empty set if T is onto A).

Any of the containments above may be strict. In order to avoid unduly messy formulas, from this point on, we will generally assume the range of the map T^{-1} is restricted to A.

Lemma 5.1. *Let $(T, A, m|_A)$ be a nonsingular open dynamical system. If $\Omega := A_\infty \setminus K_\infty$, then A admits the disjoint decomposition $A = K_\infty \cup \Omega \cup H_\infty$ and:*

(a) $T^{-1}(\cup_{n\geq 0} K_n) \subseteq \cup_{n\geq 0} K_n \pmod{m|_A}$.
(b) $T(\Omega) = \Omega$.
(c) $T : (H_n \setminus K_\infty) \to (H_{n-1} \setminus K_\infty)$ *is onto and nonsingular (with respect to the obvious restrictions of m).*
(d) $K_\infty = \cup_{n=0}^\infty K_n$.

Proof. (a) Note that $T^{-1}K_n \subseteq \cup_{m<n} K_m$ (for each $n > 0$) and $T^{-1}K_0 = \emptyset$.
(b) If $x \in \Omega$, then $x \in A_\infty$ so $T^n(x) \in A_\infty$. Thus, Ω is the set of points whose future orbit is contained in A and has at least one backward orbit in A.
(c) Let $x \in H_{n-1} \setminus K_\infty$. Then there is a sequence $\{x_{-k}\}_{k=1}^\infty$ such that $T(x_{-k}) = x_{-(k-1)}$ and $T(x_{-1}) = x$. Clearly $x_{-1} \in H_n \setminus K_\infty$.
(d) First, suppose that $x \notin \cup_{n\geq 0} K_n$. Then $x \notin K_0$ so $\emptyset \neq T^{-1}x$. If $T^{-1}x \subseteq \cup_{n\geq 0} K_n$, then there are $N_1, \ldots, N_j$ such that $T^{-1}x \subseteq K_{N_1} \cup \cdots \cup K_{N_j}$. Putting

$N = 1 + \max\{N_1, \ldots, N_j\}$, one has $x \in K_N$, a contradiction. Thus, there is at least one $x_{-1} \in T^{-1}x$ such that $x_{-1} \notin \cup_{n\geq 0} K_n$. The proof is completed by induction. □

Example 5.1. Let $X = \mathbb{R}^2$, $A = [0,1]^2$ and $T(x,y) = (2x, 1/2y)$. Then $H_n = (2^{-n}, 2^{-(n-1)}] \times [0,1]$, and $A_\infty = \{0\} \times [0,1]$. On the other hand, $K_n = [0,1] \times (2^{-(n+1)}, 2^{-n}]$, so $K_\infty = [0,1] \times (0,1]$. The 'recurrent set' $A_\infty \setminus K_\infty = \{(0,0)\}$ is a fixed point (so genuinely recurrent), and $A_\infty \cap K_\infty = \{0\} \times (0,1]$ is part of the stable manifold to $(0,0)$. Notice that $H_\infty = (0,1] \times \{0\}$ is part of the unstable manifold to $(0,0)$.

5.1.2 Escape, Conditionally Invariant Measures and Their Supports

We now make precise the notion of escape rates and establish some important connections with CIMs.

Definition 5.2. The **escape rate** of a probability measure m_0 on A is

$$\rho_{m_0} := \lim_{n\to\infty} -\frac{1}{n} \log m_0(A_n) = \lim_{n\to\infty} -\frac{1}{n} \log m_0\{x : \tau(x) \geq n\}$$

(when such a limit exists). The open system $(T, A, m|_A)$ will satisfy the **escape hypothesis** iff

$$m(A_\infty) = 0. \tag{5.1}$$

Clearly, if there is an escape rate $\rho_m > 0$, then (5.1) holds.

Definition 5.3. A probability measure μ on A is a CIM iff there is $\alpha \in (0,1)$ such that

$$\mu(T^{-1}E) = \alpha\, \mu(E) \qquad \forall \quad \text{measurable } E \subseteq A.$$

Note that if μ is a CIM, then

$$\mu\{\tau \geq n\} = \mu(A_n) = \mu(A \cap T^{-1}A_{n-1}) = \alpha\, \mu(A_{n-1}) = \cdots = \alpha^n\, \mu(A) = \alpha^n.$$

Thus, $\rho_\mu = -\log\alpha$ and $\mu\{x : \tau(x) \geq n\} = \mu(A_n) = e^{-\rho_\mu n}$, so that initial conditions distributed according to μ display geometric escape. Provided $H_\infty \neq \emptyset$, Lemma 5.1(c) implies the existence of at least one backward semi-orbit $\{x_{-k}\}_{k\geq 0}$ (with $T(x_{-k}) = x_{-(k-1)}$). Demers and Young [12] point out that a CIM can be obtained as $(1-\alpha)\sum_{k=0}^{\infty} \alpha^k \delta_{x_{-k}}$. However, such CIMs describe only a single orbit, and it remains an interesting challenge to find a CIM which models the escape statistics of the 'natural' initial measure $m|_A$.

The domain decomposition of Lemma 5.1 and the following Lemma 5.2 reveal that that A decomposes into three pieces:

(i) A backward transient part K_∞ which cannot support any CIM but includes any local basins of attraction (we will later identify numerically certain parts of K_∞ and exclude them for computational reasons). The intuition behind this fact is that the lack of preimages of points in K_∞ means there is no way to 'replenish' mass which is lost to the hole.
(ii) An envelope $\Omega = A_\infty \setminus K_\infty$ for the 'recurrent' piece which can support invariant measures, but not CIMs.
(iii) A transient part H_∞ which is the place to look for CIM (and includes any local unstable manifolds).

Lemma 5.2. *Let $(T, A, m|_A)$ be a nonsingular open dynamical system and let Ω, K_∞ and H_∞ be as defined previously. Then:*

(a) If μ is an invariant or CIM on A, then $\mu(K_n) = 0$ for all n (and $\mu(K_\infty) = 0$).
(b) If μ is an invariant measure, then $\mu(H_\infty) = 0$.
(c) If μ is a CIM, then $\mu(\Omega) = 0$.

Proof. (a) Suppose that $\mu \circ T^{-1} = \alpha\,\mu$ for some $\alpha \in (0, 1]$. Then

$$\alpha^{n+1}\,\mu(K_n) = \mu \circ T^{-(n+1)}(K_n) = \mu \circ T^{-1}(T^{-n} K_n) \leq \mu(T^{-1} K_0) = \mu(\emptyset) = 0.$$

By part (d) of Lemma 5.1, $\mu(K_\infty) = \mu(\cup_n K_n) \leq \sum_n \mu(K_n) = 0$.
(b) If μ is an invariant measure and $\mu(H_n) > 0$, then by the Poincaré recurrence theorem almost every $x \in H_n$ recurs to H_n infinitely often. But if $x \in H_n$, then $\{k > n \,:\, T^k x \in H_n\} = \emptyset$, so $\mu(H_n) = 0$. It follows that $\mu(\cup H_n) = 0$ and hence $\mu(H_\infty) = 0$.
(c) By Lemma 5.1(b), $\Omega \subseteq T^{-1}(T(\Omega)) \subseteq T^{-1}\Omega$ so that

$$\mu(\Omega) \leq \mu \circ T^{-1}(\Omega) = \alpha\,\mu(\Omega) < 1\,\mu(\Omega).$$

Hence $\mu(\Omega) = 0$. □

Example 5.1 revisited. Let $X = \mathbb{R}^2$, $A = [0, 1]^2$ and $T(x, y) = (2x, 1/2y)$. Since $\Omega = \{0\}$, $K_\infty = [0, 1] \times (0, 1]$ and $H_\infty = (0, 1] \times \{0\}$, the only invariant measure is concentrated on the fixed point at 0 and all CIMs are concentrated on H_∞ (the unstable manifold to $(0, 0)$).

Remark 5.1. As suggested already, a discrete variant of the set K_∞ arises naturally in the numerical methods described below. When T is countable-to-one, it can occur that $K_\infty \neq \cup_n K_n =: K'_\infty$, but this does not alter the result of Lemma 5.2(a).

5.1.3 Conditional Transfer Operators and the Multiplicity of ACCIMs

We complete the introduction by characterising CIMs as eigenvectors of certain conditional transfer operators. This provides a concrete mathematical setting for the approximation algorithms and gives a useful technical tool for establishing the existence of absolutely continuous CIM.

For each $k \geq 0$, put $m_k = m|_{A_k}$ (so that $m_0 = m|_A$). Then $T : (A_{k+1}, m_{k+1}) \to (A_k, m_k)$ is a nonsingular transformation, so that $m_{k+1} \circ T^{-1} \ll m_k$ and a *conditional Frobenius–Perron operator*

$$\mathbf{L}_k : L^1(A_{k+1}; m_{k+1}) \to L^1(A_k; m_k)$$

can be defined in the usual manner:

$$\mathbf{L}_k f = \frac{d}{dm_k}([f\, m_{k+1}] \circ T^{-1}).$$

Dual to $\mathbf{L}_k$ is the *(conditional) Koopman operator* $U_k : L^\infty(A_k; m_k) \to L^\infty(A_{k+1}; m_{k+1})$ with the action

$$U_k \psi = \psi \circ T.$$

The relation

$$\int_{A_k} (\mathbf{L}_k\, \varphi)\, \psi\, dm_k = \int_{A_{k+1}} \varphi\, U_k \psi\, dm_{k+1} \tag{5.2}$$

is automatic for $\varphi \in L^1(A_{k+1}; m_{k+1}), \psi \in L^\infty(A_k; m_k)$. In particular, for any $\varphi \in L^1(A; m_0)$ and $\psi \in L^\infty(A; m_0)$,

$$\int_{A_0} \mathbf{L}_0(\varphi\, \mathbf{1}_{A_1})\, \psi\, dm = \int_{A_1} \varphi\, U_0 \psi\, dm. \tag{5.3}$$

Lemma 5.3. *Let $(T, A, m|_A)$ be a nonsingular open dynamical system and let $\mu \ll m$ be a measure such that $\mu(A_0) = 1$. Then μ is a CIM with escape rate $-\log \alpha$ if and only if $\mathbf{L}_0(\mathbf{1}_{A_1} \frac{d\mu}{dm}) = \alpha \frac{d\mu}{dm}$.*

Proof. Let $\varphi = \frac{d\mu}{dm}$. Then for $E \subseteq A_0$, one has $T^{-1}E \subseteq A_1$ so that using (5.3)

$$\int_E \mathbf{L}_0(\mathbf{1}_{A_1}\varphi)\, dm_0 = \int_{A_1} \varphi\, U_0 \mathbf{1}_E\, dm_1 = \int \varphi\, \mathbf{1}_{T^{-1}E}\, dm = \mu(T^{-1}E),$$

since $\alpha \int_E \varphi\, dm = \alpha\, \mu(E) = \mu(T^{-1}E)$. □

Lemma 5.3 characterises absolutely continuous conditionally invariant measures (ACCIMs) as those whose density functions solve a conditional transfer operator equation: $\mathbf{L}_0(\mathbf{1}_{A_1}\varphi) = \alpha\,\varphi$. However, in contrast to the typical situation for nonsingular dynamical systems, this equation may have an uncountable number of solutions for each α if no additional regularity is specified; see [12, Theorem 3.1] and discussion therein. We now give a version of this result.

Theorem 5.1. *Let (T, A, m) be a nonsingular open dynamical system. If there is $\kappa > 0$ such that $\mathbf{L}_0\mathbf{1}_{A_1} \geq \kappa\,\mathbf{1}_{H_\infty}$ and $m(H_\infty) > 0$, then for every $\alpha \in (0,1)$ there is a CIM which is AC with respect to m and has escape rate $-\log\alpha$.*

Proof. There is at least one N for which $m(H_N \setminus K_\infty) > 0$. By an inductive application of Lemma 5.1(c), $m(H_1 \setminus K_\infty) > 0$. Now let $\mu_1 \ll m|_{H_1\setminus K_\infty}$ be a finite measure and put $\varphi_1 = \frac{d\mu_1}{dm}$. Note that $\mathbf{1}_{A_1}\varphi_1 = 0$. Next, we construct (inductively) a sequence of integrable functions φ_k, supported on $H_k \setminus K_\infty$ such that each $\mathbf{L}_0(\mathbf{1}_{A_1}\varphi_{k+1}) = \mathbf{L}_k\varphi_{k+1} = \varphi_k$. Let $\varphi_k \in L^1(H_k \setminus K_\infty; m_k)$ be given. Assume that φ_k is bounded (the general case follows from the bounded case by an approximation argument). On $H_{k+1} \setminus K_\infty$ put

$$\varphi_{k+1} := \frac{\varphi_k \circ T}{U_k \mathbf{L}_k \mathbf{1}_{H_{k+1}\setminus K_\infty}}$$

(note that the denominator is bounded below by $\kappa\,\mathbf{1}_{H_{k+1}\setminus K_\infty}$). Let $\mu_j = \varphi_j\, m_j$ for $j = k, k+1$ and $E \subseteq H_k \setminus K_\infty$. Then

$$\begin{aligned} \mu_{k+1} \circ T^{-1}E &= \int_{H_{k+1}\setminus K_\infty} \varphi_{k+1}\, U_k \mathbf{1}_E \, dm \\ &= \int_{A_{k+1}} U_k(\varphi_k\, \mathbf{1}_E / \mathbf{L}_k \mathbf{1}_{H_{k+1}\setminus K_\infty})\, \mathbf{1}_{H_{k+1}\setminus K_\infty}\, dm \\ &= \int_{A_k} \varphi_k\, \mathbf{1}_E \, dm = \mu_k(E). \end{aligned}$$

Thus, $\varphi_k = \frac{d}{dm_k}\mu_k = \frac{d}{dm_k}\mu_{k+1} \circ T^{-1} = \mathbf{L}_k \frac{d\mu_{k+1}}{dm_{k+1}} = \mathbf{L}_k\varphi_{k+1}$. Using $E = H_k \setminus K_\infty$ and Lemma 5.1(c), we see that $\int \varphi_k\, dm = \int \varphi_{k+1}\, dm$. Finally, put $\varphi = \frac{1-\alpha}{\mu_1(A_0)} \sum_{k=1}^{\infty} \alpha^{k-1}\varphi_k$. Then, $\int_{A_0} \varphi\, dm = 1$ and $\mathbf{L}_0(\mathbf{1}_{A_1}\varphi) = \alpha\,\varphi$. The theorem follows from Lemma 5.3. □

Remark 5.2. The proof given above is essentially the one from [12]; the different conditions are to account for the fact that we have not imposed any topological (or smoothness) restrictions on T. Note that each choice of finite AC measure on $H_1 \setminus K_\infty$ gives a different ACCIM.

5.2 Convex Optimisation for the ACCIM Problem

We now describe a selection principle for ACCIMs based on the Shannon–Boltzmann entropy. The first idea is to encode the criteria for being a CIM into a sequence of moment conditions and to search for *approximate* CIMs which locally resemble the measure m. This leads to the optimisation problems $(P_{n,\alpha})$, where the entropy-maximising density is sought, subject to meeting the first n moment conditions for conditional invariance (with escape rate $-\log\alpha$). Then, in Sect. 5.2.2, we recall some standard results from convex optimisation which allow the MAXENT problem $(P_{n,\alpha})$ to be recast in dual form. Theorem 5.2 identifies a condition which is both necessary and sufficient for solvability of the dual problem. Section 5.2.3 introduces a *domain reduction* technique which ensures that the conditions of Theorem 5.2 are met, revealing an interesting connection between the structure of the moment conditions and the backward transient sets K_∞. The main result is Theorem 5.3: an explicit formula for the solution of $(P_{n,\alpha})$.

5.2.1 Moment Formulation of the ACCIM Problem

By Lemma 5.3, if μ is an ACCIM and $\varphi = \frac{d\mu}{dm}$, then

$$\mathbf{L}_0(\mathbf{1}_{A_1}\varphi) = \alpha\,\varphi, \qquad \alpha = \int_{A_1}\varphi\,dm = \mu(A_1).$$

This is equivalent to

$$\int_{A_0}[\mathbf{L}_0(\mathbf{1}_{A_1}\varphi) - \alpha\,\varphi]\,\psi\,dm = 0 \qquad \forall\psi\in L^\infty(A;m), \qquad \int(\mathbf{1}_{A_1}\,\varphi)\,dm = \alpha$$

and hence, using (5.3),

$$\int_{A_0}[\mathbf{1}_{A_1}\psi\circ T - \alpha\,\psi]\,\varphi\,dm = 0 \qquad \forall\psi\in L^\infty(A;m), \qquad \int_{A_1}\varphi\,dm = \alpha.$$

To obtain a computationally tractable representation of these conditions, observe that it suffices to verify for all ψ in a weak* dense subset of $L^\infty(A;m_0)$.

Definition 5.4. Let $\{\psi_j\}_{j=1}^\infty \subset L^\infty(A;m_0)$ be a sequence whose span is weak* dense and put $\psi_0 = \mathbf{1}_A$. Fix $\alpha\in(0,1]$. Then

$$\mathscr{F}_n := \Big\{0\le\varphi\in L^1(A;m_0)\ :\ \int_{A_1}\varphi\,dm = \alpha, \int_A\varphi\,\psi_0\,dm = 1, \qquad \text{and}$$
$$\int_A\big[\mathbf{1}_{A_1}\psi_j\circ T - \alpha\,\psi_j\big]\,\varphi\,dm = 0,\ j = 1,\dots,n\Big\}. \tag{5.4}$$

are **approximately conditionally invariant densities** with escape rate $-\log\alpha$.

Notice that each $\mathscr{F}_{n+1} \subset \mathscr{F}_n$. If a sequence $\{f_n\}$ is chosen such that each $f_n \in \mathscr{F}_n$ and $f_n \xrightarrow{weak} f_\infty$, then $f_\infty \in \cap_{n>0}\mathscr{F}_n$. Such an f_∞ is the density of a CIM. Using arguments similar to those leading up to Theorem 5.2 in [4], one has weak (and indeed L^1) convergence of such a sequence when selecting f_n to solve

$$\text{maximise } H(f) \quad s.t. \quad f \in \mathscr{F}_n \qquad (P_{n,\alpha})$$

where H is a suitably chosen functional. We use the Shannon–Boltzmann entropy

$$H(f) := -\int_A f(x) \log f(x)\, dm(x)$$

(where $t \log t$ is set to 0 when $t = 0$ and ∞ when $t < 0$). If T admits an ACCIM μ for which $H(\frac{d\mu}{dm}) > -\infty$, then each problem $(P_{n,\alpha})$ has a unique solution f_n, and $\lim f_n$ exists both weakly and in L^1 (proofs can be adapted from [4]).

Each *primal problem* $(P_{n,\alpha})$ is concave, admitting a solution $f_{n,\alpha}$ depending on both n and α. As we illustrate with numerical examples (Sect. 5.3.4), the role of α is interesting, being a parameter that is tunable to produce a range of escape rates[4]: for α near 0, escape is rapid (with mass of the ACCIM tending to concentrate on the first few preimages of the hole); for α near 1, escape is slow with mass concentrated nearer to Ω.

In order to identify the entropy-maximising ACCIM, we propose a nested approach: at the outer level, for each fixed n, optimise $H(f_{n,\alpha})$ (over α); as an 'inner' step, each $f_{n,\alpha}$ is computed to solve $(P_{n,\alpha})$.

Remark 5.3. The optimisation problem $(P_{n,\alpha})$ can be reformulated to remove α as a variable. One simply replaces the jth moment condition in (5.4) with

$$\int_{A_0} \left[\mathbf{1}_{A_1}\psi_j \circ T - (\textstyle\int_{A_1} \varphi\, dm)\, \psi_j\right] \varphi\, dm = 0$$

for each ψ_j. This destroys the linearity of the constraint and potentially the convexity of the optimisation problem.

5.2.2 *Convex Duality for Problem $(P_{n,\alpha})$*

Problems like $(P_{n,\alpha})$ are never solved directly. Instead, a 'Lagrange multipliers' approach converts the problem to an equivalent finite-dimensional unconstrained optimisation. For the benefit of readers not familiar with this type of argument, we outline the steps leading to this 'dual formulation'. Let n, α and $\{\psi_k\}_{k=1}^n$ be fixed.

[4]The flexibility to tune α without impact on numerical effort is reminiscent of the use of Ulam's method to calculate the topological pressure of piecewise smooth dynamical systems by varying an inverse temperature parameter [16].

To simplify matters we assume that the test functions form a partition of unity over A, so $\psi_0 = \mathbf{1}_A = \sum_{k=1}^{n} \psi_k$ and

$$0 = \int_{A_0} [\mathbf{1}_{A_1} \mathbf{1}_{A_0} \circ T - \alpha\, \mathbf{1}_{A_0}]\, \varphi\, dm = \int_{A_1} \varphi\, dm - \alpha \int_{A_0} \varphi\, dm$$

follows from the corresponding conditions for $\psi_1, \dots, \psi_n$. The normalisation $\int_{A_0} \varphi\, dm = 1$ is thus a consequence of $\int_{A_1} \varphi\, dm = \alpha$, so only one of those conditions is needed.

Definition 5.5. Define $\mathbb{M} : L^1(A; m_0) \to \mathbb{R}^{n+1}$ by

$$(\mathbb{M}\varphi)_0 = \int_{A_1} \varphi\, dm \qquad \text{and} \quad (\mathbb{M}\varphi)_j = \int_A \big[\mathbf{1}_{A_1} \psi_j \circ T - \alpha\, \psi_j\big]\, \varphi\, dm$$

for $j = 1, \dots, n$. Let $\mathbb{M}^* : \mathbb{R}^{n+1} \to L^\infty(A; m_0)$ be defined by

$$\mathbb{M}^* \lambda = \lambda_0\, \mathbf{1}_{A_1} + \sum_{j=1}^{n} \lambda_j (\mathbf{1}_{A_1} \psi_j \circ T - \alpha\, \psi_j).$$

Let $\mathbf{e} = [1, 0, \dots, 0]^T \in \mathbb{R}^{n+1}$, put $Q(\lambda) := \alpha\, \lambda^T \mathbf{e} - \int_A \exp(\mathbb{M}^* \lambda - 1)\, dm$ and define a *dual problem*:

$$\text{maximise } Q(\lambda) \qquad \text{s.t. } \lambda \in \mathbb{R}^{n+1}. \qquad (D_{n,\alpha})$$

We now outline how $(D_{n,\alpha})$ is related to $(P_{n,\alpha})$. First, note that

$$f \in \mathscr{F}_n \Leftrightarrow \mathbb{M} f = \alpha\, \mathbf{e} \quad \text{and} \quad \lambda^T(\mathbb{M} f) = \int_A \mathbb{M}^* \lambda\, f\, dm \quad \forall f \in L^1(A; m).$$

For every $\lambda \in \mathbb{R}^{n+1}$,

$$\begin{aligned} \sup_{f \in \mathscr{F}_n} H(f) &= \sup_{\{f\,:\, \mathbb{M} f = \alpha\, \mathbf{e}\}} H(f) \\ &= \sup_{\{f\,:\, \mathbb{M} f = \alpha\, \mathbf{e}\}} [H(f) + \lambda^T(\mathbb{M} f - \alpha\, \mathbf{e})] \end{aligned}$$

This last expression can be estimated by

$$\begin{aligned} \sup_{\{f\,:\, \mathbb{M} f = \alpha\, \mathbf{e}\}} [H(f) + \lambda^T(\mathbb{M} f - \alpha\, \mathbf{e})] &\le \sup_{f \in L^1(A;m)} [H(f) + \lambda^T(\mathbb{M} f - \alpha\, \mathbf{e})] \\ &= -\alpha\, \lambda^T \mathbf{e} + \sup_{f \in L^1(A_0;m)} \left[\int_A \mathbb{M}^* \lambda\, f\, dm - (-H(f))\right] \\ &= -\alpha\, \lambda^T \mathbf{e} + H^*(\mathbb{M}^* \lambda) \\ &= -\alpha\, \lambda^T \mathbf{e} + \int_A \exp(\mathbb{M}^* \lambda - 1)\, dm = -Q(\lambda) \end{aligned}$$

where H^* is the *Fenchel conjugate* of the convex functional $-H$, and the second to last equality is a nontrivial result in convex analysis (see Rockafellar [23] and Borwein and Lewis [3]). Observe that $-Q(\lambda)$ is an upper bound on $H(f)$ for all $f \in \mathscr{F}_n$ and $\lambda \in \mathbb{R}^{n+1}$ so that the (negative of the) solution to $(D_{n,\alpha})$ provides an upper bound on the solution to $(P_{n,\alpha})$. This is called the *principle of weak duality*. In fact, $(D_{n,\alpha})$ is a differentiable, unconstrained, concave maximisation problem, and our method involves solving it.

Theorem 5.2 (Dual attainment). *Let α, n be fixed.*

(a) λ^* solves $(D_{n,\alpha})$ if and only if $f_n := \exp(\mathbb{M}^*\lambda^* - 1) \in \mathscr{F}_n$ and $H(f_n) = -Q(\lambda^*)$.
(b) The problem $(D_{n,\alpha})$ attains its maximum if and only if

$$0 \neq \lambda \in \{\ker \mathbb{M}^* \oplus span(\mathbf{e})\}^{\perp} \quad \Rightarrow \quad [\mathbb{M}^*\lambda]^+ \neq 0 \; m\text{-a.e.}. \tag{5.5}$$

Proof. (a) This is a standard result in dual optimisation theory and is a consequence of the fact that λ^* solves $(D_{n,\alpha})$ iff $\alpha\,[\mathbf{e}]_j - [\mathbb{M}\exp(\mathbb{M}^*\lambda^* - 1)]_j = \frac{\partial Q}{\partial \lambda_j}|_{\lambda^*} = 0$ for $j = 0, \ldots, n$.
(b) Sufficiency of (5.5) is established by minor modifications to the proof of Theorem 3.3 in [5]. For necessity, suppose that $\lambda^T\mathbf{e} = 0$, $0 \neq \lambda \in \{\ker \mathbb{M}^*\}^{\perp}$ and $M^*\lambda \leq 0$. Then there are $\kappa > 0$ and $E \subseteq A$ such that $m(E) > 0$ and $M^*\lambda \leq -\kappa\,\mathbf{1}_E$. Then, for any $\lambda^\dagger \in \mathbb{R}^{n+1}$ and $t > 0$,

$$Q(\lambda^\dagger + t\,\lambda) \geq Q(\lambda^\dagger) + (1 - e^{-\kappa t}) \int_E \exp(\mathbb{M}^*\lambda^\dagger - 1)\,dm > Q(\lambda^\dagger).$$

Hence Q cannot attain its maximum.

□

5.2.3 Domain Reduction and Dual Optimality Conditions

The condition (5.5) incorporates some important facts about ACCIMs. First, by Theorem 5.1, there exists an ACCIM. It follows from this that $\mathscr{F}_n \neq \emptyset$ and $\alpha\,\mathbf{e} \in \text{Range}(\mathbb{M}) = \{\ker \mathbb{M}^*\}^{\perp}$ (this is the reason for separating out the direction $\mathbf{e}$). Second, the support of each ACCIM must be disjoint from subsets of A associated with 'bad functions'. (This is made precise in Lemma 5.4 below.) A function ψ will be called a *bad function* if $\mathbf{1}_{A_1}\,\psi \circ T - \alpha\,\psi \leq 0$ (but not equal to 0 m-a.e.). If $\lambda \in \mathbb{R}^{n+1}$ is such that $[\lambda]_0 = 0$ and $\mathbb{M}^*\lambda \leq 0$ (but nonzero), then $\psi = \sum_{j=1}^{n}[\lambda]_j\,\psi_j$ is a *bad function*. The condition (5.5) for solvability of $(D_{n,\alpha})$ is equivalent to there being no bad functions in span$\{\psi_j\}_{j=1}^{n}$. We are going to show that bad functions may exist (Example 5.2), but they are irrelevant to the ACCIMs (their supports are disjoint from H_∞; see Lemmas 5.2(c) and 5.4) and can be excised from the problems $(P_{n,\alpha})$ and $(D_{n,\alpha})$ (Lemma 5.5). We call this latter procedure *domain reduction*.

Example 5.2. If $x \in \cup_{n\geq 0} K_n$, let $N(x) := \min\{k : T^{-k}(x) \cap A_0 = \emptyset\}$. Note that $N(x) + 1 \leq N(T(x))$ (where $N(y) = \infty$ if $y \notin \cup_{n\geq 0} K_n$). Define $\psi(x) = (\alpha/2)^{N(x)}$. Then $-(\alpha/2)\psi = (\alpha/2)\psi - \alpha\psi \geq \psi \circ T - \alpha\psi$. Hence $\mathbf{1}_{A_1} \psi \circ T - \alpha\psi < 0$ on $\cup_{n\geq 0} K_n$.

Lemma 5.4. *Let* $\alpha \in (0,1)$ *and suppose that* $\psi \in L^\infty(A; m)$ *satisfies* $\mathbf{1}_{A_1} \psi \circ T \leq \alpha\,\psi$. *Then* $\psi|_{\cup_{k>0} H_k} \geq 0$ *and* $\psi|_{A\setminus K_\infty} \leq 0$. *In particular,* $m(H_\infty \cap supp(\psi)) = 0$.

Proof. First, let $x \in H_1$. Then $\mathbf{1}_{A_1}(x) = 0$ so $0 = \mathbf{1}_{A_1} \psi \circ T(x) \leq \alpha\,\psi(x)$, so $\psi|_{H_1} \geq 0$. Now suppose that $x \in H_k$. Then $T^{k-1}(x) \in H_1$ so that

$$0 \leq \psi(T^{k-1}(x)) \leq \alpha\,\psi(T^{k-2}(x)) \leq \cdots \leq \alpha^{k-1}\psi(x).$$

Thus, $\psi|_{H_k} \geq 0$. On the other hand, if $x \notin K_\infty$, then for each $k > 0$ there is at least one x_{-k} such that $T^k(x_{-k}) = x$. Then $\psi(x) = \psi \circ T^k(x_{-k}) \leq \alpha^k\,\psi(x_{-k}) \leq \alpha^k \|\psi\|_\infty$. Letting $k \to \infty$, $\psi(x) \leq 0$. □

To apply Theorem 5.2 when $K_\infty \neq \emptyset$, we need to ensure that the chosen test functions $\{\psi_j\}_{j=1}^n$ are *unable to detect* bad functions. To do this, we exploit a **basis-specific domain reduction**: remove from the domain A the support of any function $h = \mathbb{M}^*\lambda$ where $h \leq 0$ and $\lambda \in \text{Range}(\mathbb{M})/\text{span}\{\mathbf{e}\}$. Let $\hat{A}$ denote this reduced domain.

Lemma 5.5. *In the notation of this section, suppose that* $\hat{A}$ *is measurable and* $f \in \mathscr{F}_n$. *Then* $f = f\,\mathbf{1}_{\hat{A}}$ *m-a.e.*

Proof. Suppose that $m(\text{supp}(f) \setminus \hat{A}) > 0$ and let λ be such that $\lambda^T\mathbf{e} = 0$, $\mathbb{M}^*\lambda \leq 0$ and $\text{supp}(\mathbb{M}^*\lambda) \cap \text{supp}(f) \subseteq A_0 \setminus \hat{A}$ has positive measure. Then, $\mathbb{M}f = \alpha\,\mathbf{e}$ so that

$$0 = \lambda^T(\mathbb{M}f) = \int_{A_0} \mathbb{M}^*\lambda\, f\, dm < 0,$$

an obvious contradiction. □

In view of Lemma 5.5, m can be replaced with $\hat{m} = m|_{\hat{A}}$ in the definition of the problem $(P_{n,\alpha})$ *without any change to the set* $\mathscr{F}_n$. The value of the problem is also unchanged, since there is no contribution to $H(f)$ from those places where f takes the value 0. The duality theory is now applied to the measure space $(A_0, \hat{m})$, and the corresponding dual problem is

$$\text{maximise } \hat{Q}(\lambda) := \alpha\,\lambda^T\mathbf{e} - \int_{\hat{A}} \exp(\mathbb{M}^*\lambda - 1)\, dm \qquad \text{s.t. } \lambda \in \mathbb{R}^{n+1}. \qquad (\hat{D}_{n,\alpha})$$

Notice that if $\mathbb{M}^*\lambda \leq 0$ m-almost everywhere, then the domain reduction ensures that $\mathbb{M}^*\lambda = 0 \hat{m}$-a.e. Thus, all potentially problematic λ have been pushed into $\ker \mathbb{M}^*$ (modulo $\hat{m}$). In particular, condition (5.5) is satisfied for the reduced domain. The previous results can be collected in our main theorem.

Theorem 5.3. *Let α, n be fixed and suppose that $\hat{A}$ is measurable. Then $(\hat{D}_{n,\alpha})$ attains its maximum at finite λ^* and $f_n = \mathbf{1}_{\hat{A}} \exp(\mathbb{M}^*\lambda^* - 1)$ solves $(P_{n,\alpha})$.*

We note that $\mathbb{M}^*$ may have nontrivial kernel (modulo $\hat{m}$), so the optimising λ^* can be non-unique. We also make the following observations:

- The reduced domain $\hat{A}$ depends on n, possibly α, and may be *very difficult to determine for general test functions.*
- Assuming the escape hypothesis (5.1), we have $A \setminus \hat{A} \subseteq K_\infty$ (mod m) [$m(A_\infty) = 0$ by (5.1) which together with Lemma 5.4 shows that $\text{supp}(\psi) \subseteq K_\infty$ (mod m) for any bad function ψ; the observation follows].
- If $\hat{A}$ is overestimated, then condition (5.5) fails and the dual optimisation problem does not have a solution for finite λ. Nevertheless it would be a simple matter to set up the dual formulation $(D_{n,\alpha})$ and seek a numerical 'solution' of this infeasible optimization problem without first verifying the optimality condition in (5.5); such a numerical approach is bound to be both unstable and misleading. See Borwein and Lewis [3] for further discussion of this and related issues.

Notwithstanding these warnings, in Sect. 5.3, we show how to compute $\hat{A}$ for piecewise constant test functions based on a measurable partition of A.

5.3 A MAXENT Procedure for Approximating ACCIMs

Under the conditions of Theorem 5.1, there are many ACCIMs for each escape rate. If at least one of these has a density with finite Shannon–Boltzmann entropy, then the solutions of a sequence of problems $(P_{n,\alpha})$ will converge (in L^1) as $n \to \infty$ to the density of an ACCIM. This, in principle, allows one to select an 'entropy-maximising' ACCIM; the entropy maximisation spreads mass as uniformly as possible, given the condition of being a CIM. Solutions to each problem $(P_{n,\alpha})$ can be calculated via convex duality, provided there are no 'bad functions' ($\mathbb{M}^*\lambda$ which fail the condition (5.5) in Theorem 5.2). This condition can be ensured by a basis-dependent domain reduction (Lemma 5.5 and Theorem 5.3), leading to a domain-reduced dual problem $(\hat{D}_{n,\alpha})$. We now make a specific choice of test functions, reminiscent of Ulam's method [14, 15, 24]. We identify the reduced domain $\hat{A}$ (Lemma 5.6), derive the relevant optimality equations (Lemma 5.7) and present a convergent fixed point method for their solution.

5.3.1 Piecewise Constant Test Functions and Domain Reduction

Let $\{\psi_j\}$ be obtained from a sequence of increasingly fine partitions of A. In particular, let $\mathscr{B}_n$ be a partition of A into measurable subsets $\{B_1, \ldots, B_n\}$ and put $\psi_j = \mathbf{1}_{B_j}$. Notice that $\mathbf{1}_A = \sum_{j=1}^n \psi_j$ so the partition of unity assumption is satisfied (cf. Sect. 5.2.2). To derive and solve the optimality equations for $(\hat{D}_{n,\alpha})$, notice that $\mathbb{M}^*\lambda$ is a piecewise constant function, on elements of $\mathscr{B}_n \vee \{T^{-1}\mathscr{B}_n, H_1\}$:

$$\begin{aligned}\mathbb{M}^*\lambda &= \mathbf{1}_{A_1} \sum_{j,k=1}^{n} (\lambda_0 + \lambda_j - \alpha\,\lambda_k)\mathbf{1}_{B_j} \circ T\mathbf{1}_{B_k} + \mathbf{1}_{H_1} \sum_{k=1}^{n} (-\alpha\,\lambda_k)\,\mathbf{1}_{B_k} \\ &= \sum_{j,k=1}^{n} (\lambda_0 + \lambda_j - \alpha\,\lambda_k)\mathbf{1}_{B_k \cap T^{-1}B_j} - \alpha \sum_{k=1}^{n} \lambda_k\,\mathbf{1}_{H_1 \cap B_k}\end{aligned} \tag{5.6}$$

(note that $\mathbf{1}_{A_1} = \mathbf{1}_{A \cap T^{-1}A} = \sum_{jk} \mathbf{1}_{B_k \cap T^{-1}B_j}$).

Definition 5.6. For the partition $\mathscr{B}_n$, form a matrix C and vector $\mathbf{c}$ by putting

$$C_{kj} = m(B_k \cap T^{-1}B_j) \qquad \text{and} \qquad c_k = m(H_1 \cap B_k) \qquad j,k = 1, \ldots, n.$$

A set B_j is **reachable** from B_k if there is $n > 0$ such that $(C^n)_{kj} > 0$; write $k \rightsquigarrow j$.

Remark 5.4. The entries of the matrix C are the same data needed to compute the (sub)stochastic transition matrices used by Ulam's method.

Lemma 5.6. *Suppose that (T, A, m) is a nonsingular open dynamical system and that $m(A_\infty) = 0$. Fix α, n and let $\hat{A}$ be the reduced domain when $\mathbb{M}^*$ is constructed from the partition $\mathscr{B}_n$. Then $\hat{A}$ is the union of those B_k where either $k \rightsquigarrow k$ or there is at least one i for which $i \rightsquigarrow i \rightsquigarrow k$; in particular, $\hat{A}$ is measurable.*

Proof. Let $\lambda^T\,\mathbf{e} = 0$ and suppose that $\mathbb{M}^*\lambda \leq 0$. From (5.6), we immediately have

$$\lambda_j \leq \alpha\lambda_k \quad \text{when } C_{kj} > 0 \qquad \text{and} \qquad \lambda_k \geq 0 \quad \text{when } c_k > 0.$$

Since C is a non-negative matrix, $i \rightsquigarrow k$ iff there is a string $i = i_0, i_1 \ldots, i_n = k$ such that each $C_{i_l i_{l+1}} > 0$. Thus, by induction, if $i \rightsquigarrow k$, then there is an $n > 0$ such that $\lambda_k \leq \alpha^n \lambda_i$. First, if $c_k > 0$ and $i \rightsquigarrow k$, we infer that $\lambda_i \geq 0$. Next, since $m(A_\infty) = 0$, for every B_i, there is an n for which $m(B_i \cap H_n) > 0$. Then, since T is nonsingular, there is B_l such that $C_{il} > 0$ and $m(B_l \cap H_{n-1}) > 0$. By induction, there is a k for which $i \rightsquigarrow k$ and $c_k > 0$. Hence, $\lambda_i \geq 0$ for all i. Now, if $k \rightsquigarrow k$, again use the inequality $\lambda_k \leq \alpha^n \lambda_k$ to infer that $\lambda_k \leq 0$ and hence $\lambda_k = 0$. Similarly, if $i \rightsquigarrow i \rightsquigarrow k$, $\lambda_k \leq \alpha^n\,\lambda_i = 0$, so also $\lambda_k = 0$. Suppose that k is one of the indices identified in the statement of the lemma. Then (5.6) implies that $\mathbf{1}_{B_k}\mathbb{M}^*\lambda = \sum_j \lambda_j \mathbf{1}_{B_k \cap T^{-1}B_j} \geq 0$; since $M^*\lambda \leq 0$, $B_k \cap \operatorname{supp}(\mathbb{M}^*\lambda) = \emptyset$. To complete the proof, let $\mathscr{K}$ denote those $\hat{k}$ which fail the condition in the statement. For each such $\hat{k}$, let $N(\hat{k}) = \max\{N : (C^N)_{i\hat{k}} > 0\ \exists i\}$; $N(\hat{k})$ may be 0. (Note that if $(C^N)_{ik} > 0$ for $N > n$, then there is a sequence $i = i_0, i_1, \ldots, i_n = k$ for which $C_{i_l i_{l+1}} > 0$; this list must contain at least one repeat, implying $k \notin \mathscr{K}$.) Note that if $C_{i\hat{k}} > 0$, then $N(i) + 1 \leq N(\hat{k})$. Finally, for each $\hat{k} \in \mathscr{K}$, put $\lambda_{\hat{k}} = (\alpha/2)^{N(\hat{k})}$, with $\lambda_k = 0$ for $k \notin \mathscr{K}$. Then, $C_{i\hat{k}} > 0$ implies $\lambda_i(\alpha/2) \geq \lambda_{\hat{k}}$. Hence $\lambda_{\hat{k}} - \alpha\,\lambda_i \leq -\lambda_{\hat{k}} < 0$. It follows that $\operatorname{supp}(\mathbb{M}^*\lambda) = \cup_{\hat{k}\in\mathscr{K}} B_{\hat{k}}$. □

Remark 5.5. The set $\hat{A}$ identified by the lemma is the union of all B_k which are reachable from the strongly connected components of the directed graph implied by the nonzero elements of the matrix C. This can be found quickly and easily.

Now, form the matrix $\hat{C}$ and vector $\hat{\mathbf{c}}$ by retaining those entries where B_k is identified as belonging to $\hat{A}$ and setting the rest to 0. These ingredients can be used to obtain explicit formulae for the optimality conditions for $(\hat{D}_{n,\alpha})$. Using (5.6),

$$\hat{Q}(\lambda) = \alpha\lambda_0 - \sum_{jk} \exp(\lambda_0 - 1 + \lambda_j - \alpha\,\lambda_k)\hat{C}_{kj} - \sum_k \exp(-1 - \alpha\,\lambda_k)\,\hat{c}_k.$$

Because $\hat{Q}$ is differentiable and concave, the maximising λ^* is found by solving the first-order conditions $\frac{\partial \hat{Q}}{\partial \lambda_i} = 0$. The following lemma writes these conditions in a more convenient form.

Lemma 5.7. *Assume the conditions of Lemma 5.6 and let $\hat{A}$ be as given there. Let $\hat{C}, \hat{\mathbf{c}}$ be obtained similarly to Definition 5.6, but using $\hat{m} = m|_{\hat{A}}$ in place of m. If $\{x_i\}_{i=1}^n$ are positive numbers solving*

$$x_i^{1+\alpha} = \alpha\,\frac{\sum_j \hat{C}_{ij} x_j + \hat{c}_i}{\sum_k \hat{C}_{ki} x_k^{-\alpha}}$$

and λ_0^* satisfies $e^{\alpha\lambda_0^* - 1}\sum_j \hat{C}_{ij} x_j x_i^{-\alpha} = \alpha$, then $\lambda_i^* := \log(x_i) - \lambda_0^*$ give the solution to $(\hat{D}_{n,\alpha})$.

Proof. By differentiation, the optimality equations for $(\hat{D}_{n,\alpha})$ are

$$\begin{aligned}
0 &= \alpha - \textstyle\sum_{jk} \exp(\lambda_0 - 1 + \lambda_j - \alpha\,\lambda_k)\hat{C}_{kj} && (i = 0)\\
0 &= \alpha \textstyle\sum_j \exp(\lambda_0 - 1 + \lambda_j - \alpha\,\lambda_i)\hat{C}_{ij} \quad - \textstyle\sum_k \exp(\lambda_0 - 1 + \lambda_i - \alpha\,\lambda_k)\hat{C}_{ki} \\
&\qquad + \alpha\,\exp(-1 - \alpha\,\lambda_i)\,\hat{c}_i && (1 \leq i \leq n).
\end{aligned}$$

The $i = 0$ equation is a normalisation. By putting $x_i = e^{\lambda_i + \lambda_0}$ for $1 \leq i \leq n$, the latter equations are equivalent to

$$0 = \alpha \sum_j \hat{C}_{ij} x_j x_i^{-\alpha} - \sum_k \hat{C}_{ki} x_i x_k^{-\alpha} + \alpha\,\hat{c}_i x_i^{-\alpha}.$$

Multiplying by x_i^{α} and rearranging gives the equations in the statement of the lemma. □

5.3.2 Iterative Solution of the Optimality Equations

We now summarise the numerical method:

1. Specify α ($= e^{-\rho}$ where ρ is the preferred escape rate).
2. Fix a measurable partition $\mathscr{B}_n = \{B_j\}_{j=1}^n$ of A.
3. Obtain the matrix C and vector $\mathbf{c}$ of partition overlap masses (as specified in Definition 5.6).
4. Use Lemma 5.6 to identify $\hat{A}$ and thus form the dual problem $(\hat{D}_{n,\alpha})$.
5. Solve the optimality equations via Lemma 5.7. This can be accomplished with a fixed point iteration: set $\mathbf{x}_0 = [1, \ldots, 1]^T$ and iterate

$$\mathbf{x}_{t+1} = \Psi(\mathbf{x}_t) \qquad \text{where} \qquad [\Psi(\mathbf{x})]_i = \left(\alpha \, \frac{\sum_j \hat{C}_{ij} x_j + \hat{c}_i}{\sum_k \hat{C}_{ki} x_k^{-\alpha}}\right)^{1/(1+\alpha)}$$

   until desired accuracy is achieved.
6. Recover the optimal λ^* via Lemma 5.7 and solution $f_{n,\alpha}$ to $(P_{n,\alpha})$ from Theorem 5.3.
7. (Optional) Calculate $H(f_{n,\alpha})$.

5.3.3 Sketch Proof of Convergence of the Fixed Point Iteration

Assume the escape hypothesis (5.1).

Without loss of generality, assume that all sums in the definition of Ψ are nonempty.[5] Because $(\hat{D}_{n,\alpha})$ actually has a solution, there is $\mathbf{y}^*$ for which $\Psi(\mathbf{y}^*) = \mathbf{y}^*$. For any $\mathbf{x} \in \mathbb{R}^n_+$, let

$$V(\mathbf{x}) = \min\left\{R \,:\, \frac{1}{R} \leq \frac{x_i}{y_i^*} \leq R, 1 \leq i \leq n\right\}.$$

Clearly $V(\mathbf{x}) \geq 1$ and $V(\mathbf{x}) = 1$ iff $\mathbf{x} = \mathbf{y}^*$. Moreover,

$$[\Psi(\mathbf{x})]_i \leq \left(\alpha \, \frac{V(\mathbf{x}) \sum_j \hat{C}_{ij} y_j^* + \hat{c}_i}{V(\mathbf{x})^{-\alpha} \sum_k \hat{C}_{ki} (y_k^*)^{-\alpha}}\right)^{1/(1+\alpha)} \leq V(\mathbf{x}) \, [\Psi(\mathbf{y}^*)]_i = V(\mathbf{x}) \, y_i^*. \quad (5.7)$$

Together with a similar inequality involving $1/V$, one has $V \circ \Psi \leq V$. Thus, $\{V \circ \Psi^t(\mathbf{x}_0)\}$ is a decreasing sequence, bounded below by 1. Because $V(\mathbf{x}_0) < \infty$, all

[5] Note that $\hat{C}_{ki} = 0 \forall k$ only if $B_i \cap \hat{A} = \emptyset$. In this case also each $\hat{C}_{ij} = \hat{c}_i = 0$ and the value of $\mathbb{M}^* \lambda$ on B_i is irrelevant to the solution of $(P_{n,\alpha})$ (by Lemma 5.5). The function Ψ can be defined to be 1 on such coordinates.

$\{\mathbf{x}_t\}$ are confined to a closed, bounded rectangle in $\mathbb{R}^n$; let $\mathbf{x}_*$ be a limit point of $\{\mathbf{x}_t\}$. Then $V \circ \Psi(\mathbf{x}_*) = V(\mathbf{x}_*)$.

Suppose that i is such that[6] $[\Psi(\mathbf{x}_*)]_i = V(\mathbf{x}_*) y_i^*$. An inductive argument [using the equality form of (5.7)] shows that $[\mathbf{x}_*]_k = V(\mathbf{x}_*) y_k^*$ and $\hat{c}_k = V(\mathbf{x}_*)\hat{c}_k$ whenever $i \rightsquigarrow k$. Since there is at least one k with $\hat{c}_k > 0$ reachable from i, $V(\mathbf{x}_*) = 1$. Thus, $\mathbf{x}_* = \mathbf{y}^*$ and $\mathbf{x}_t \to \mathbf{y}^*$.

5.3.4 Examples

We present two simple examples to demonstrate the effectiveness of the method; each implementation takes only a few dozen lines of MATLAB code.

Example 5.3 (Tent map with slope 3). Let $X = \mathbb{R}$, $A = [0, 1]$ and put

$$T(x) = \begin{cases} 3\,x & x < 0.5 \\ 3\,(1 - x) & x > 0.5 \end{cases}$$

Then, $A_1 = [0, 1/3] \cup [2/3, 1]$ and $H_1 = (1/3, 2/3)$. The 'natural' ACCIM is Lebesgue measure with density $f_* = 1$ and corresponding value of $\alpha = 2/3$. In this case, $K_n = \emptyset = K_\infty$ (for all n) and the survivor set $\Omega = A_\infty$ is the usual middle third Cantor set. At a selection of values of $\alpha \in (0, 1)$, we applied the MAXENT method using the partition-based test functions $\{\psi_j = \mathbf{1}_{[(j-1)/1000, j/1000)}\}_{j=1}^{1000}$. The results are depicted in Fig. 5.1. As expected, for small values of α, escape is rapid and the ACCIMs are strongly concentrated on the hole H_1 and its first few preimages. For α near 1, escape is slow and the ACCIMs are more strongly concentrated around the repelling Cantor set A_∞; see Fig. 5.2. The MAXENT method can be tuned to produce a 'most uniform' approximate ACCIM, and the maximal entropy solution is in fact the constant density function, appearing at $\alpha = 2/3$.

Example 5.4 (A linear saddle). Let $A = [-1, 1]^2$ and m Lebesgue measure on $X = \mathbb{R}^2$; put

$$T(x, y) = (2\,x, 0.8\,y).$$

Then $K_n = [-1, 1] \times \pm(0.8^{(n+1)}, 0.8^n]$, $A_\infty = \{0\} \times [-1, 1]$ and $H_\infty = [-1, 1] \times \{0\} \setminus (0, 0)$. This linear map has a saddle-type fixed point at $(0, 0)$. The only invariant measure is the delta measure at 0. All CIMs are supported on the local unstable manifold to the origin; in this case, the segment of the x-axis contained in A. Indeed, $m(H_\infty) = 0$ and there are no ACCIMs. There are, however, many CIMs which are AC with respect to the one-dimensional Lebesgue measure on the x-axis, and these

[6] A similar argument works if i is such that $[\Psi(\mathbf{x}_*)]_i = y_i^* / V(\mathbf{x}_*)$.

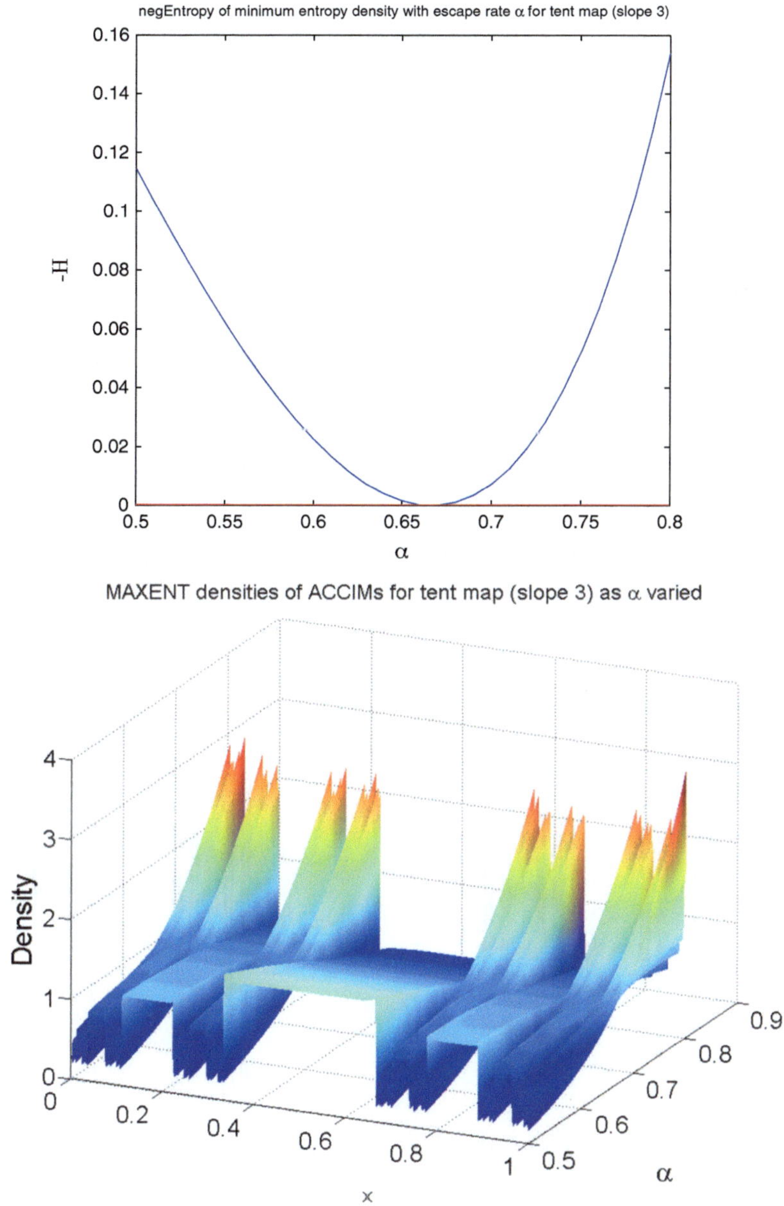

Fig. 5.1 Example 5.3. *Above*: (neg)entropy $-H(f_{n,\alpha})$ of slope 3 tent map ACCIMs, depending on α computed via MAXENT with uniform $n = 1,000$ subinterval partition of [0, 1]. *Below*: densities of the computed ACCIMs as a function of $x \in [0, 1]$ and α

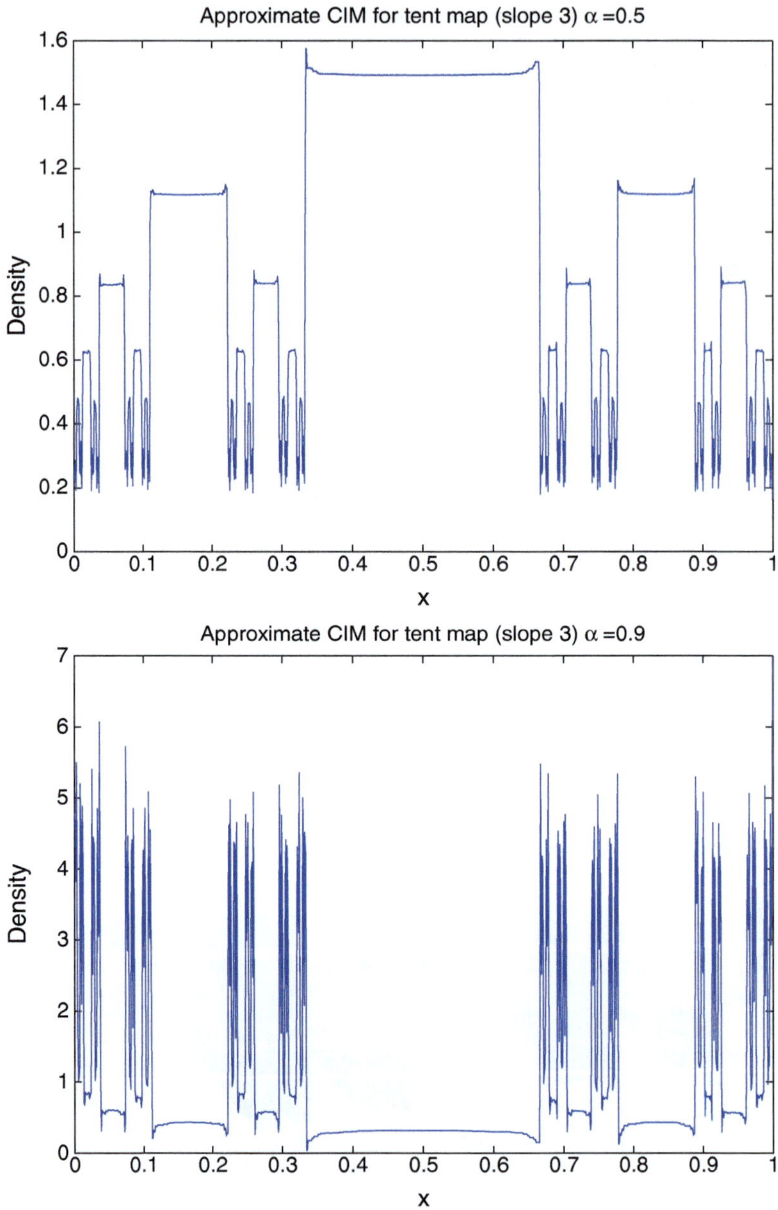

Fig. 5.2 Example 5.3 (compare Fig. 5.1). *Above*: approximate density $f_{1000,0.5}$ of slope 3 tent map; note the concentration of mass on $H_1 = [1/3, 2/3]$ and its preimages. *Below*: approximate density $f_{1000,0.9}$ of slope 3 tent map; note the concentration of mass on the survivor Cantor set A_∞

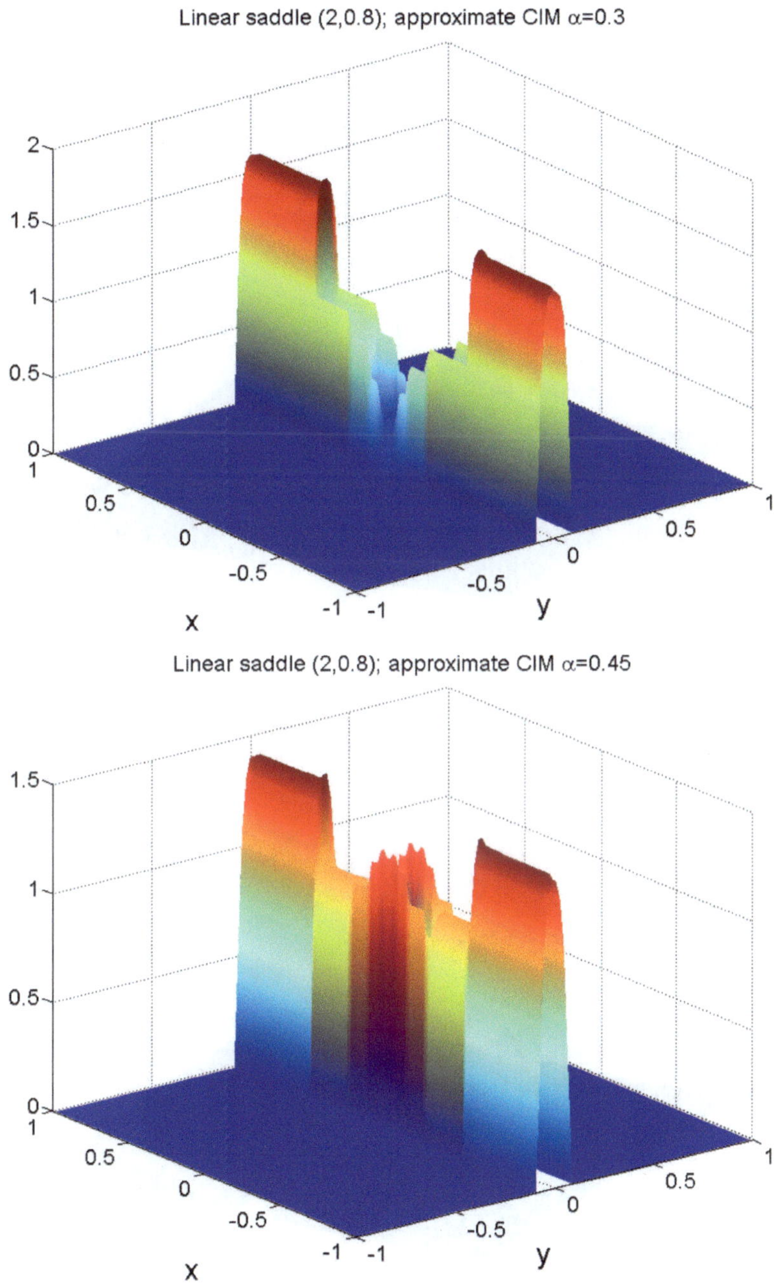

Fig. 5.3 Example 5.4. MAXENT approximations of CIMs for $\alpha = 0.3$ (*above*) and $\alpha = 0.45$ (*below*) for an open system with a simple saddle

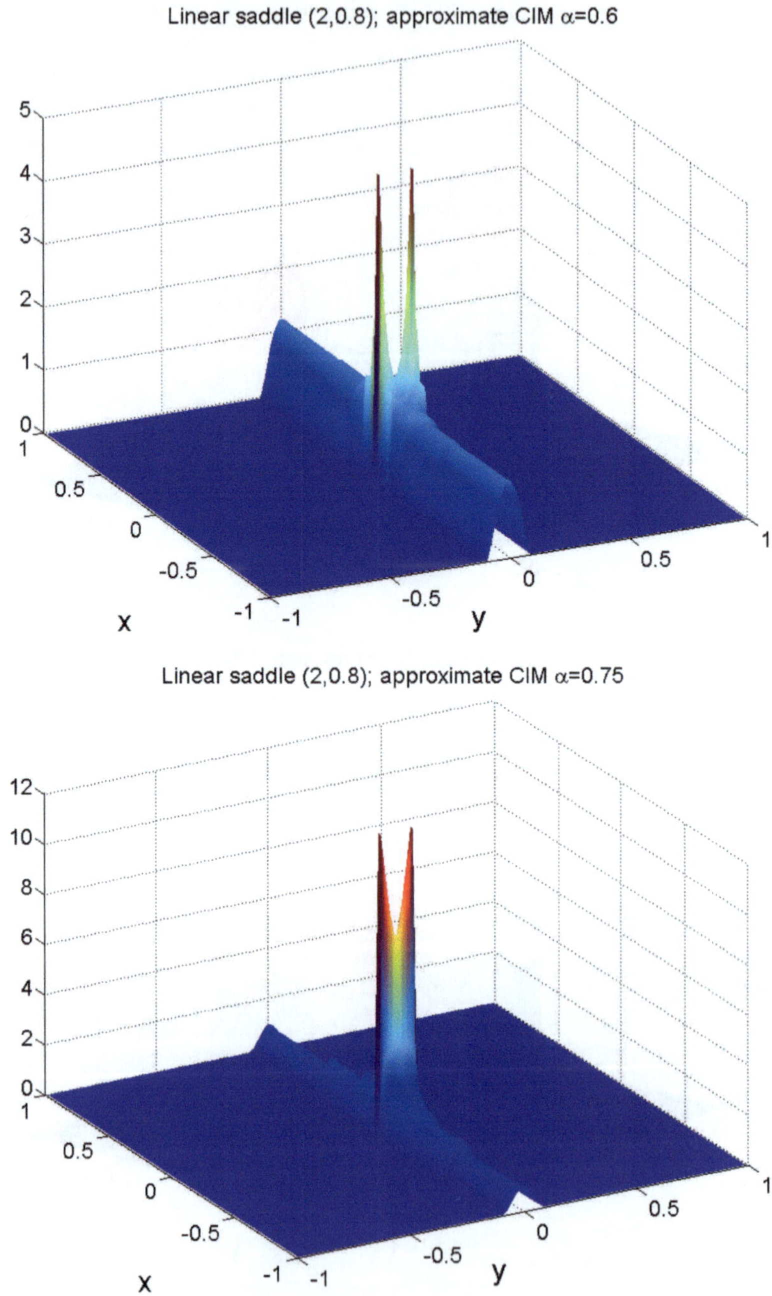

Fig. 5.4 Example 5.4. MAXENT approximations of CIMs for $\alpha = 0.6$ (*above*) and $\alpha = 0.75$ (*below*) for an open system with a simple saddle

are detected by the numerical method. The domain reduction to $\hat{A}$ is nontrivial here, leading to a localisation in support of the MAXENT approximations. Calculations were performed for several α, with 10000 test functions being the characteristic functions of a 100×100 subdivision of A; in this case the set $\hat{A} = [-1, 1] \times [-0.08, 0.08]$. Some CIMs estimates are presented in Figs. 5.3 and 5.4.

5.4 Concluding Remarks

The MAXENT approach to calculating approximate ACCIMs has a sound analytical basis (from optimisation theory) and is easy to implement. With test functions $\{\psi_j\}$ derived from a partition of phase space, the basic dynamical inputs to the computational scheme are the integrals $\int \psi_j \circ T\, \psi_i\, dm$ (which could be estimated from trajectory data). For each choice of test functions, feasibility of the dual optimisation problem depends on reducing the domain of the problem to exclude certain 'backward transient' parts of the phase space. With test functions derived from a partition, the resulting 'reduced domain' covers any recurrent set and local unstable manifolds.

The work reported in this chapter suggests a number of avenues of future enquiry:

- Are entropy-maximising ACCIMs of any particular dynamical relevance?
- Given that the analysis and computation of the variational approach is similar with convex functionals other than $H(\cdot)$, are other choices of objective more appropriate?
- How is the quality of approximation affected by the choice of test functions $\{\psi_k\}$?
- How does the functional $H(f_{n,\alpha})$ depend on α (and n)?
- Can dynamically interesting measures on *unstable manifolds* be recovered from this approach?

References

1. Afraimovich, V.S., Bunimovich, L.A.: Which hole is leaking the most: a topological approach to study open systems. Nonlinearity **23**(3), 643–656 (2010). doi:10.1088/0951-7715/23/3/012
2. Bahsoun, W., Vaienti, S.: Metastability of certain intermittent maps. Nonlinearity **25**(1), 107–124 (2012). doi:10.1088/0951-7715/25/1/107
3. Borwein, J., Lewis, A.: Duality relationships for entropy-like minimization problems. SIAM J. Control. Optim. **26**, 325–338 (1991)
4. Bose, C.J., Murray, R.: Minimum 'energy' approximations of invariant measures for nonsingular transformations. Discrete Contin. Dyn. Syst. **14**(3), 597–615 (2006)
5. Bose, C.J., Murray, R.: Duality and the computation of approximate invariant densities for nonsingular transformations. SIAM J. Optim. **18**(2), 691–709 (electronic) (2007). doi:10.1137/060658163

6. Bruin, H., Demers, M., Melbourne, I.: Existence and convergence properties of physical measures for certain dynamical systems with holes. Ergod. Theor. Dyn. Syst. **30**, 687–728 (2010)
7. Collett, P., Martínez, S., Schmitt, B.: The Yorke–Pianigiani measure and the asymptotic law on the limit cantor set of expanding systems. Nonlinearity **7**, 1437–1443 (1994)
8. Collett, P., Martínez, S., Schmitt, B.: The Pianigiani–Yorke measure for topological markov chains. Israel J. Math. **97**, 61–70 (1997)
9. Demers, M.: Markov extensions and conditionally invariant measures for certain logistic maps with small holes. Ergod. Theor. Dyn. Syst. **25**, 1139–1171 (2005)
10. Demers, M.: Markov extensions for dynamical systems with holes: an application to expanding maps of the interval. Israel J. Math. **146**, 189–221 (2005)
11. Demers, M.: Dispersing billiards with small holes. In: Ergodic Theory, Open Dynamics, and Coherent Structures. Springer, New York (2014)
12. Demers, M., Young, L.S.: Escape rates and conditionally invariant measures. Nonlinearity **19**, 377–397 (2006). doi:10.1088/0951-7715/19/2/008
13. Dolgopyat, D., Wright, P.: The diffusion coefficient for piecewise expanding maps of the interval with metastable states. Stoch. Dyn. **12**(1), 1150,005, 13 (2012). doi:10.1142/S0219493712003547
14. Froyland, G.: Extracting dynamical behavior via Markov models. In: Nonlinear Dynamics and Statistics (Cambridge, 1998), pp. 281–321. Birkhäuser, Boston (2001)
15. Froyland, G., Dellnitz, M.: Detecting and locating near-optimal almost-invariant sets and cycles. SIAM J. Sci. Comput. **24**(6), 1839–1863 (electronic) (2003). doi:10.1137/S106482750238911X
16. Froyland, G., Murray, R., Terhesiu, D.: Efficient computation of topological entropy, pressure, conformal measures, and equilibrium states in one dimension. Phys. Rev. E **76**(3), 036702 (2007). doi:10.1103/PhysRevE.76.036702
17. Froyland, G., Murray, R., Stancevic, O.: Spectral degeneracy and escape dynamics for intermittent maps with a hole. Nonlinearity **24**(9), 2435–2463 (2011). doi:10.1088/0951-7715/24/9/003
18. González-Tokman, C., Hunt, B.R., Wright, P.: Approximating invariant densities of metastable systems. Ergod. Theor. Dyn. Syst. **31**(5), 1345–1361 (2011). doi:10.1017/S0143385710000337
19. Homburg, A.J., Young, T.: Intermittency in families of unimodal maps. Ergod. Theor. Dyn. Syst. **22**(1), 203–225 (2002). doi:10.1017/S0143385702000093
20. Keller, G., Liverani, C.: Rare events, escape rates and quasistationarity: some exact formulae. J. Stat. Phys. **135**(3), 519–534 (2009). doi:10.1007/s10955-009-9747-8
21. Liverani, C., Maume-Deschamps, V.: Lasota-Yorke maps with holes: conditionally invariant probability measures and invariant probability measures on the survivor set. Ann. Inst. H. Poincaré Probab. Stat. **39**, 385–412 (2003)
22. Pianigiani, G., Yorke, J.: Expanding maps on sets which are almost invariant: decay and chaos. Trans. Am. Math. Soc. **252**, 351–366 (1979)
23. Rockafellar, R.T.: Convex Analysis. Princeton University Press, Princeton (1970)
24. Ulam, S.: A Collection of Mathematical Problems. Interscience Publishers, New York (1960)

Chapter 6
Lebesgue Ergodicity of a Dissipative Subtractive Algorithm

Henk Bruin

Abstract We prove Lebesgue ergodicity and exactness of a certain dissipative 2-dimensional subtractive algorithm, completing a partial answer by Fokkink et al. to a question by Schweiger. This implies for Meester's subtractive algorithm in dimension d that there are $d-2$ parameters which completely determine the ergodic decomposition of Lebesgue measure.

6.1 Introduction

Consider a triple $x = x^{(0)} = (x_1, x_2, x_3)$ of positive reals and form a sequence $(x^{(n)})_{n \geq 0}$, by repeatedly subtracting the smallest of the three from the other two. This dynamical system emerged from a percolation problem studied by Meester [5]. Although $(x^{(n)})_{n \geq 0}$ is clearly a decreasing sequence, $x^{\infty} = \lim_{n \to \infty} x^{(n)}$ is different from 0 for Lebesgue-a.e. initial position. Let us write this more formally as iterations of the subtractive map of increasing triples $0 \leq x_1 \leq x_2 \leq x_3$:

$$F(x_1, x_2, x_3) = \mathbf{sort}(x_1, x_2 - x_1, x_3 - x_1),$$

where **sort** stands for putting the coordinates in increasing order. It is obvious that $x_1^{\infty} = x_2^{\infty} = 0$, but also that if $x_3 > x_1 + x_2$, then $\eta := x_3 - (x_1 + x_2)$ is a preserved quantity. This means that once $x_3 > x_1 + x_2$, the third coordinate will always remain the largest, even under the unsorted subtractive algorithm, and in fact $x_3^{\infty} = \eta$. Meester and Nowicki [6] showed that for Lebesgue-a.e. initial vector, there is indeed some $n \geq 0$ such that $\eta = x_3^{(n)} - (x_1^{(n)} + x_2^{(n)}) > 0$.

H. Bruin (✉)
Faculty of Mathematics, University of Vienna, Oskar Morgensternplatz 1,
A-1090 Vienna, Austria
e-mail: henk.bruin@univie.ac.at

W. Bahsoun et al. (eds.), *Ergodic Theory, Open Dynamics, and Coherent Structures*, Springer Proceedings in Mathematics & Statistics 70, DOI 10.1007/978-1-4939-0419-8_6,

© Springer Science+Business Media New York 2014

Therefore F is non-ergodic w.r.t. Lebesgue measure λ: triples with different nonnegative values of η have disjoint orbits and thus belong to 'carriers' of different ergodic components, which can be defined in the usual way even though λ is non-invariant and in fact dissipative. Let us recall these definitions.

Definition 6.1. A transformation $(X, \mathscr{B}, \lambda; T)$ is

- *non-singular* if $\lambda(B) > 0$ implies $\lambda(T(B)) > 0$;
- *ergodic* if $T^{-1}(B) = B$ implies that $\lambda(B) = 0$ or $\lambda(X \setminus B) = 0$;
- *conservative* if for every set $B \in \mathscr{B}$ of positive measure, there is $n \geq 1$ such that $\lambda(T^n(B) \cap B) > 0$;
- *dissipative* if it fails to be conservative, and *totally dissipative*, if there is no invariant subset $X_0 \subset X$ of positive measure on which T is conservative.
- *exact* if $T^{-n} \circ T^n(B) = B$ for all $n \geq 0$ implies that $\lambda(B) = 0$ or $\lambda(X \setminus B) = 0$.

All of these properties can be defined even though λ is not T-invariant.

The result of [6] was generalised by Kraaikamp and Meester [4] to dimension $d \geq 3$. They showed that for the map

$$F_d(x_1, \dots x_d) = \mathbf{sort}(x_1, x_2 - x_1, \dots, x_d - x_1),$$

and Lebesgue-a.e. initial vector x, the quantity $\eta_3 = x_3^{(n)} - (x_1^{(n)} + x_2^{(n)})$ is eventually positive, and so is $\eta_k := x_k^{(n)} - x_{k-1}^{(n)}$ for $k > 3$. Once $\eta_3 > 0$, all η_k are preserved, and, as observed in [3], Lebesgue measure is therefore not ergodic. This answers in the negative a question posed by Schweiger [10]. The natural question, however, is whether the level sets

$$\{x \in \mathbb{R}^d_{\geq 0} : x_k^\infty = \sum_{j=3}^{k} \eta_j \text{ for all } 3 \leq k \leq d\}$$

constitute the ergodic decomposition of Lebesgue measure.

We can rephrase this question by passing from projective space (on which F_d acts) to a fixed simplex $\Delta = \{x = (x_1, \dots, x_d) : 0 \leq x_1 \leq \dots \leq x_d = 1\}$, by scaling the largest coordinate to 1. The map F_d then becomes $f_d : \Delta \to \Delta$, defined as

$$\begin{cases} x' = F_d(x) = \mathbf{sort}(x_1, x_2 - x_1, \dots x_d - x_1), \\ f_d : x \mapsto \frac{1}{x'_d} x'. \end{cases}$$

For $d = 2$, the map reduces to the Farey map

$$x \mapsto \begin{cases} \frac{x}{1-x} & \text{if } x \in [0, \frac{1}{2}]; \\ \frac{1-x}{x} & \text{if } x \in [\frac{1}{2}, 1]. \end{cases} \tag{6.1}$$

In the next simplest case $d = 3$, we know that $\lim_{n\to\infty} f_d^n(x_1, x_2, 1) = (0, 0, 1)$ as soon as $x_1 + x_2 < 1$, so f_d is totally dissipative on the simplex Δ.

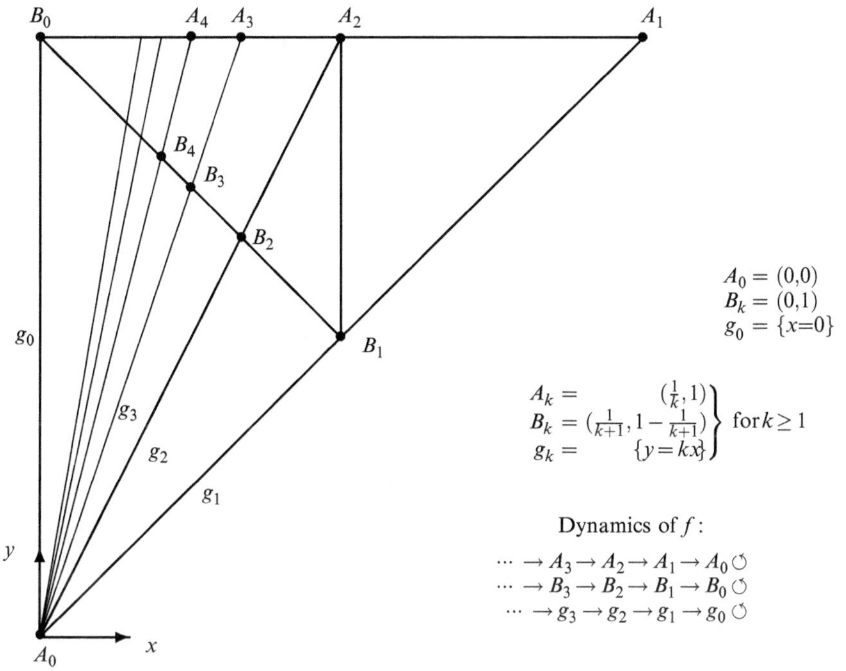

Fig. 6.1 The Markov partition for partition $f : \Delta \to \Delta$ consists of the triangles Δ_L (to the left of the line $g_1 = \{y = 2x\}$), Δ_R (between g_1 and the line $\{x = \frac{1}{2}\}$) and Δ_T (to the right of $\{x = \frac{1}{2}\}$). Each of these triangles is mapped onto Δ by f. Further diagonal lines g_k bound the regions where the first return times to Δ_R are constant (viz. k between g_k and g_{k+1}). The line $\{x + y = 1\}$ is invariant and separates the part where $\eta > 0$ and where η is not yet determined

Nogueira [9] used properties of $GL(2, \mathbb{Z})$ to prove that, although dissipative, the three-dimensional system is Lebesgue ergodic. Further results on the homogeneous (i.e., unscaled) version of these algorithms were obtained by Miernowski and Nogueira [8]. In this paper we use a different method (based on a transient random walk argument with a Lebesgue typical speed of 'convergence to 0', combined with distortion estimates) to reprove ergodicity. Our method also yields Lebesgue exactness and is, we hope, adaptable to similar (higher-dimensional) systems as well; see also Remark 6.1.

Theorem 6.1. *Partition the triangle* $\Delta = \{(x, y) : 0 \leq x \leq y \leq 1\}$ *into* $\Delta_L = \{(x, y) : 0 \leq 2x \leq y \leq 1\}$, $\Delta_R = \{(x, y) : 0 < x \leq y < 2x \leq 1\}$ *and* $\Delta_T = \{(x, y) : \frac{1}{2} < x \leq y \leq 1\}$.

Then with respect to the map $f : \Delta \to \Delta$ *defined as*

$$f(x, y) = \begin{cases} (\frac{y-x}{x}, \frac{1-x}{x}) & \text{if } (x, y) \in \Delta_T, \\ (\frac{y-x}{1-x}, \frac{x}{1-x}) & \text{if } (x, y) \in \Delta_R, \\ (\frac{x}{1-x}, \frac{y-x}{1-x}) & \text{if } (x, y) \in \Delta_L, \end{cases}$$

see Fig. 6.1, Lebesgue measure is totally dissipative, ergodic and exact.

It follows from [4] that for $d \geq 3$ and Lebesgue-a.e. initial vector x, there is $n \in \mathbb{N}$ such that for $x_3^{(n)} > x_2^{(n)} + x_1^{(n)}$, so this case reduces to Theorem 6.1 as well. In fact, we have the corollary:

Corollary 6.1. *For each $\eta_3 > 0$ and $\eta_4, \dots, \eta_d \geq 0$, the map F_d restricted to the invariant set $\{x \in \mathbb{R}^d_{\geq 0} : x_k^\infty = \sum_{j=3}^k \eta_j$ for $3 \leq k \leq d\}$ is ergodic and exact w.r.t. Lebesgue measure.*

Proof. Since $\eta_3 > 0$, we can divide the space $\{x : x_3^\infty = \eta_3\}$ into a countable union $\cup_{\tau \geq 0} X_\tau$ where $\tau = \min\{n \geq 0 : F_d^n(x)_3 > F_d^n(x)_1 + F_d^n(x)_2\}$. That is, after τ iterations, the order of the coordinates $F_d^\tau(x)_k$ for $3 \leq k \leq d$ will not change anymore under further iteration. (In fact $F_d^\tau(x)_k = \sum_{j=3}^k \eta_j + F_d^\tau(x)_1 + F_d^\tau(x)_2$.) So from this iterate onwards, we can scale so that $F_d^\tau(x)_3 = 1$ and restrict our attention to the first two coordinates. Theorem 6.1 applies to them. □

Remark 6.1. Meester and Nowicki's result was generalised by Fokkink et al. [3] to a two-parameter setting, called Schweiger's fully subtractive algorithm; see [10, Chap. 9]:

$$F_{ad}(x_1, \dots x_d) = \mathbf{sort}(x_1, \dots, x_a, x_{a+1} - x_a, \dots, x_d - x_a).$$

Analogous quantities η_k for $k \geq a + 2$ are still preserved as soon as $\eta_{a+2} \geq 0$, and [3] show that this happens almost surely. The present paper shows Lebesgue ergodicity and exactness of the level sets of $(\eta_2, \dots, \eta_d)$ for $F_{1,d}$ and all $d \geq 3$. It is hoped that the techniques will be useful to understand F_{ad} for general $a \in \{1, 2, \dots, d-2\}$.

6.2 The Proof of Theorem 7.3

6.2.1 Finding Convenient Coordinates

To start the proof, it helps to recall from [3] the Markov partition of Δ that f possesses; see Fig. 6.1. The Markov partition $\Delta = \Delta_L \cup \Delta_R \cup \Delta_T$ consists of three full branches. In fact, f extends to a diffeomorphism $f : \overline{\Delta_i} \to \overline{\Delta}$ for $i = L, R, T$. The region Y under the line $x + y = 1$ is invariant; it is here that $\eta = 1 - x - y > 0$, and $f^n(x, y) \to (0, 0)$ for every $(x, y) \in Y$. Clearly $f(\Delta_T) \supset Y$, and an additional distortion argument ensures that Lebesgue-a.e. (x, y) eventually falls into Y. Therefore f is totally dissipative.

The question is whether the convergence to $(0, 0)$ is so chaotic that $\lambda|_Y$ is in fact ergodic or even exact. Let us restrict our Markov partition to

$$\{Y_L = \Delta_L \cap Y,\ Y_R = \Delta_R \cap Y\},$$

and study the first entry map $G : Y \to Y_R$ in a new set of coordinates. First, note that the lines $g_k = \{(x,y) \in Y : y = kx\}$, $k \geq 1$, and $g_0 = \{(x,y) \in Y : x = 0\}$ satisfy $f(g_k) = g_{k-1}$ for $k \geq 1$ and g_0 consists of neutral fixed points. Hence the return time to Y_R on the region between g_{k+1} and g_{k+2} is exactly k for $k \geq 1$. For fixed $t \geq 0$, the lines $\ell(p,t) = \{(x,y) \in Y : y = p - tx\}$, $0 < p \leq 1$, foliate Y and

$$f(\ell(p,t) \cap Y_L) = \ell(p, t+1-p), \quad f(\ell(p,t) \cap Y_R) = \ell\left(\frac{p}{t+1-p}, \frac{1}{t+1-p}\right).$$

Therefore, if $A_n(p,t) \subset \ell(p,t) \cap Y_R$ is a maximal arc on which the first return time is n, then

$$G_n(p,t) := G(A_n(p,t)) = \ell\left(\frac{p}{t+1-p}, \frac{n + (n-1)t - 2(n-1)p}{t+1-p}\right) \cap Y_R.$$

Remark 6.2. The point $(0,0)$ is attracting under G, but not quite under f itself. Namely, on Y_L,

$$Df|_{Y_L}(0,0) = \begin{pmatrix} 1 & 0 \\ -1 & 1 \end{pmatrix},$$

which is a nilpotent shear, whereas on Y_R,

$$Df|_{Y_R}(0,0) = \begin{pmatrix} -1 & 1 \\ 1 & 0 \end{pmatrix},$$

which is hyperbolic with stable eigenvalue $\lambda_s = \frac{1}{2}(\sqrt{5} - 1)$ on stable eigenspace $E_s = \text{span}(\lambda_s, 1)^T$ (where ^{T} stands for the transpose) and unstable eigenvalue $\lambda_u = -\frac{1}{2}(\sqrt{5} + 1) < -1$ on unstable eigenspace $E_u = \text{span}(\lambda_u, 1)^T$. Therefore, if

$$(p_k, t_k) = G_{n_1 \dots n_k} := G_{n_k} \circ G_{n_{k-1}} \circ \cdots \circ G_{n_1}(p,t)$$

for successive return times $(n_k)_{k \in \mathbb{N}}$, then $t_k \to \frac{1}{2}(\sqrt{5} + 1)$ as $k \to \infty$ and $n_j = 1$ for all large j, whereas t_k immediately becomes large if n_k is large.

Remark 6.3. For each (p,t), the length of $A_n(p,t)$ is $1/n(n+1)$ times the length of $\ell(p,t) \cap Y_R$. Let

$$A_{n_1 \dots n_k}(p,t) = \{x \in \ell(p,t) \cap Y_R : \text{the first } k \text{ return times to } Y_R \text{ are } n_1, \dots, n_k\}.$$

Its length is approximately $\prod_{i=1}^k n_i^{-2}$. Each map $G^k : A_{n_1 \dots n_k}(p,t) \to Y_R$ acts as the Gauss map with corresponding uniform distortion control; see Lemma 6.2. Therefore, the conditional probability $\mathbb{P}(n_{k+1} = n \mid n_1 \dots n_k\}) \sim n^{-2}$, uniformly in k and the history $n_1, \dots, n_k$. The process $(S_k)_{k \in \mathbb{N}}$ given by $S_k(x) = n_1 + \cdots + n_k$

if $x \in A_{n_1 \dots n_k}$ (which is a cone over $A_{n_1 \dots n_k}(1,1)$) is a deterministic version of the one-sided discrete Cauchy walk. Taking the difference of two sample paths of such a walk, we obtain a symmetric two-sided Cauchy walk, i.e. a random walk where the steps are distributed according to $\mathbb{P}(X_k = n) = \mathbb{P}(X_k = -n) \sim cn^{-2}$. This walk is recurrent, as follows from a more general theory on stable laws (see [2, Theorem 2.9][1]), so for λ-a.e. pair $(z, z') \in Y_R^2$, there are infinitely many k, such that their respective sums $S_k = S'_k$, i.e. $f^k(z)$ and $f^k(z')$ both belong to Y_R. For our proof, however, it suffices to have the somewhat weaker result proved in Proposition 6.1.

Let us write $p = \frac{p}{\alpha+\beta t+\gamma p}$ and $t = \frac{\hat{\alpha}+\hat{\beta}t+\hat{\gamma}p}{\alpha+\beta t+\gamma p}$, for integers $\alpha, \beta, \gamma, \hat{\alpha}, \hat{\beta}.\hat{\gamma}$, so the initial values are $\alpha = \hat{\beta} = 1$ and $\hat{\alpha} = \beta = \gamma = \hat{\gamma} = 0$. Direct computation gives

$$G_n\left(\ell\left(\frac{p}{\alpha+\beta t+\gamma p}, \frac{\hat{\alpha}+\hat{\beta}t+\hat{\gamma}p}{\alpha+\beta t+\gamma p}\right)\right) =$$

$$Y_R \cap \ell\left(\frac{p}{\alpha+\hat{\alpha}+(\beta+\hat{\beta})t+(\gamma+\hat{\gamma}-1)p},\right.$$

$$\left.\frac{n\alpha+(n-1)\hat{\alpha}+(n\beta+(n-1)\hat{\beta})p+(n\gamma+(n-1)\hat{\gamma}-2(n-1))p}{\alpha+\hat{\alpha}+(\beta+\hat{\beta})t+(\gamma+\hat{\gamma}-1)p}\right).$$

This means that the iteration of G, for initial values $p \in (0,1]$ and $t \geq p$, we find that we can represent the iterations

$$(p_k, t_k) = G_{n_1 \dots n_k}(p,t) = \left(\frac{p}{\alpha_k+\beta_k t+\gamma_k p}, \frac{\hat{\alpha}_k+\hat{\beta}_k t+\hat{\gamma}_k p}{\alpha_k+\beta_k t+\gamma_k p}\right) \tag{6.2}$$

by affine transformations on the integer vectors $(\alpha, \hat{\alpha}, \beta, \hat{\beta}, \gamma, \hat{\gamma})^T$:

$$\begin{pmatrix}\alpha\\ \hat{\alpha}\\ \beta\\ \hat{\beta}\\ \gamma\\ \hat{\gamma}\end{pmatrix} \mapsto \begin{pmatrix}1 & 1 & 0 & 0 & 0 & 0\\ n & n-1 & 0 & 0 & 0 & 0\\ 0 & 0 & 1 & 1 & 0 & 0\\ 0 & 0 & n & n-1 & 0 & 0\\ 0 & 0 & 0 & 0 & 1 & 1\\ 0 & 0 & 0 & 0 & n & n-1\end{pmatrix} \cdot \begin{pmatrix}\alpha\\ \hat{\alpha}\\ \beta\\ \hat{\beta}\\ \gamma\\ \hat{\gamma}\end{pmatrix} - \begin{pmatrix}0\\ 0\\ 0\\ 0\\ 1\\ 2(n-1)\end{pmatrix}$$

[1]In fact, the Cauchy distribution models the position on the horizontal axis where a standard random walk on $\mathbb{Z}^2$, starting from $(0,0)$ returns to the horizontal axis. Since the standard random walk on $\mathbb{Z}^2$ is recurrent, the Cauchy walk is recurrent as well.

with initial value $(1,0,0,1,0,0)^T$ mapping to $(1,n,1,n-1,-1,-2(n-1))^T$, etc. It is easy to check by induction that

$$\begin{cases} \alpha_k + \beta_k + \gamma_k = \hat{\alpha}_k + \hat{\beta}_k + \hat{\gamma}_k = 1 & \text{for all } k \geq 0, \\ \beta_k \leq \alpha_k \leq 2\beta_k, & \text{for all } k \geq 0, \\ \hat{\beta}_k \leq \hat{\alpha}_k \leq 2\hat{\beta}_k, & \text{except that } \hat{\beta}_1 = 0 \\ & \quad \text{when } n_1 = 1, \\ \alpha_k \leq \hat{\alpha}_k \leq 2\alpha_k & \text{when } n_k = 2. \end{cases} \tag{6.3}$$

Therefore, as far as asymptotics are concerned, it suffices to keep track of α_k and $\hat{\alpha}_k$ (or just of α_k whenever $n_k = 2$), cf. Proposition 6.1, so it makes sense to focus just on the recursive relation

$$\begin{cases} \alpha_{k+1} = \alpha_k + \hat{\alpha}_k, \\ \hat{\alpha}_{k+1} = n_{k+1}\alpha_k + (n_{k+1} - 1)\hat{\alpha}_k, \end{cases} \qquad \alpha_0 = 1,\ \hat{\alpha}_0 = 0. \tag{6.4}$$

In fact, there is $\Lambda = \Lambda(p,t)$, but independent of k, such that

$$1 \leq \frac{\alpha_k + \beta_k t + \gamma_k p}{\alpha_k}, \frac{\hat{\alpha}_k + \hat{\beta}_k t + \hat{\gamma}_k p}{\hat{\alpha}_k} \leq \Lambda, \tag{6.5}$$

whenever $t \geq p$ and $n_k = 2$.

6.2.2 Distortion Results

Given intervals $J' \subset J$, we say that J is a δ-*scaled* neighbourhood of J' if both component of $J \setminus J'$ have length $\geq \delta|J'|$. The following Koebe distortion property is well-known; see [7, Sect. IV.1]: If $g : I \to J$ is a diffeomorphism with *Schwarzian derivative* $Sg := g'''/g' - 3/2(g''/g')^2 \leq 0$, then for every $I' \subset I$ such that J is a δ-scaled neighbourhood of $J' := g(I')$, the distortion

$$\sup_{x,y \in I'} \left| \frac{g'(x)}{g'(y)} \right| \leq K(\delta) := \left(\frac{1+\delta}{\delta} \right)^2. \tag{6.6}$$

Möbius transformations g have zero Schwarzian derivative, so (6.6) holds for g and g^{-1} alike.

Lemma 6.1. *The foliation of Y into radial lines*

$$h_\theta = \{(r\cos\theta, r\sin\theta) : 0 \leq r \leq (\sin\theta + \cos\theta)^{-1}\}$$

with $\theta \in [\pi/4, \pi/2]$ is invariant. Moreover, the distortion of $G^k : h_\theta \to h_{\theta_k}$ is bounded in the sense of (6.6) *uniformly in $\theta \in [0, \pi/2]$ and $k \in \mathbb{N}$.*

Proof. Since f preserves lines and $(0,0)$ is fixed, the invariance of the foliation is immediate.

Let t_k be as in (6.2) and θ_k the angle of the image of h_θ under $G_{n_1...n_k}$. The lines $\ell(1,t_k)$ and h_{θ_k} intersect at a point $(R_k \cos\theta_k, R_k \sin\theta_k)$ for $R_k = (\cos\theta_k + t_k \sin\theta_k)^{-1}$. Using (6.2) again, we see that $G_{n_1...n_k}$ acts on the parameter r as a Möbius transformation

$$M_k : r \mapsto R_k \frac{r}{1+\beta_k(1-r)},$$

which has zero Schwarzian derivative, and so has its inverse. Therefore, within an interval $J \Subset [0, R_0]$ such that both components of $[0, R_0] \setminus J$ have length $\delta|J|$, the distortion $\sup_{r_0,r_1 \in J} |M_k'(r_0)|/|M_k'(r_1)|$ is bounded by $K(\delta)$ uniformly in k and $n_1, \dots, n_k$. □

The following lemma is straightforward, using $d = 1$ in (6.6).

Lemma 6.2. *The map f preserves the line $\ell(1,1) = \{(x,y) : x+y=1\}$ and acts on it like the Farey map* (6.1). *Hence the return map G acts like the Gauss map, and the distortion of every branch $G_{n_1...n_k} : \cup_{n \geq n_k} A_{n_1...n_k n} \to \ell(1,1)$ is uniformly bounded by $K = 4$.*

6.2.3 Growth of α_k and $\hat{\alpha}_k$ at Different Points

Let $\alpha_k(x)$ and $\hat{\alpha}_k(x)$ be as in (6.4). The first component of the expression (6.2), together with (6.5), shows that the $\alpha_k(x)$ roughly dictate the distance between $F^k(x)$ and the origin. Hence the following proposition should be interpreted as typical pairs of points infinitely often visit regions of similar distance to the origin.

Proposition 6.1. *There is $L \geq 10$ such that for Lebesgue-a.e. $(x,y) \in Y_R^2$,*

$$\frac{1}{L} \leq \frac{\alpha_k(x)}{\alpha_l(y)}, \frac{\hat{\alpha}_k(x)}{\hat{\alpha}_l(y)} \leq L \qquad \text{for infinitely many } k, l \in \mathbb{N}. \tag{6.7}$$

Proof. The heuristics behind proving (6.7) is that the numbers $\log \alpha_k$ are dominated by random variables

$$X_k = \sum_{j=1}^{k} \lceil 3 \log n_j \rceil.$$

This follows immediately from (6.4). The probabilities $\mathbb{P}(\lceil 3 \log n_k \rceil = t) = O(e^{-t/3})$ for all k and t, so X_k is the sum of k random variables of finite expectation μ. Standard probability theory (see, e.g. [2, Theorem 4.1]) gives that $\frac{1}{r}\#\{k : X_k \in [0,r]\} \to 1/\mu > 0$ as $r \to \infty$. Therefore, almost every sample path of $\{\Gamma_k\}_{k\in\mathbb{N}}$ is a sequence with positive density, and since $\log\alpha_k \leq X_k$ also for λ-a.e. x, the sequence $(\log\alpha_k)_{k\in\mathbb{N}}$ has positive density. It follows that there is L_0 such that for $\lambda\times\lambda$-a.e. pair (x,y), there are infinitely many integers k,l such that $|\log\alpha_k(x) - \log\alpha_l(y)| \leq L_0$. Taking the exponential function, we obtain the required result for α_k in (6.7). Since $\hat{\alpha}_k = n_k\alpha_{k-1} + (n_k - 1)\hat{\alpha}_{k-1}$ and the event $\{n_k = 2\}$ is basically independent of the previous choices of n_j, the result for $\hat{\alpha}_k$ in (6.7) follows as well. □

6.2.4 The Main Proof

The total dissipativity of f already follows from [3]; it is a direct consequence of $f^n(x,y)\to(0,0)$ Lebesgue-a.e. We will now finish the proof of Theorem 7.3.

Proof. Assume that $A, A' \subset Y_R$ are sets of positive measure such that $f^{-1}(A) = A$ and $f^{-1}(A') = A'$. To prove ergodicity, we will find some $i,j\in\mathbb{N}$ such that $f^i(A)\cap f^j(A')\neq\emptyset$, so A and A' cannot be disjoint.

Use coordinates $u\in[0,1]$, $v\in[0,p]$ to indicate points below the line $\ell(p,1)$: $(x,y) = (uv, u(p-v))$. First take $a = (v_A, p_A - v_A)$ a density point of A, where it is not restrictive to assume that $p_A\in(0,1)$. By Fubini's theorem, we can find $\varepsilon\in(0,1-p_A)$ such that, letting $H_0(A)$ be the strip between parallel lines $\ell(p_A,1)$ and $\ell(p_A(1-\varepsilon),1)$ (see Fig. 6.2, left), there is a set $V_A\in[0,p_A]$ of positive measure such that $\{u\in[1-\varepsilon,1] : (uv, u(p_A - v))\notin A\}$ has measure $\leq\varepsilon/(10KL)$ for every $v\in V_A$ and K as in (6.6) and L as in Proposition 6.1.

Since a 1-scaled neighbourhood of $[p_A(1-\varepsilon), p_A]$ is still contained in $[0,1]$, we can choose $K = 4$ here as the common distortion bound in Lemmas 6.1 and 6.2. We can also assume that v_A is a density point of V_A.

We do the same for A', finding a point $p_{A'}\in(0,1)$, a set $V_{A'}\subset[0,p_{A'}]$ of positive measure and a density point $a' = (v_{A'}, p_{A'} - v_{A'})$ of $V_{A'}$.

By Proposition 6.1, it is not restrictive to assume that $a = (v_A, p_A - v_A)$ and $a' = (v_{A'}, p_{A'} - v_{A'})$ satisfy

$$\frac{1}{L} \leq \frac{\alpha_k(a)}{\alpha_l(a')}, \frac{\hat{\alpha}_k(a)}{\hat{\alpha}_l(a')} \leq L \quad \text{and} \quad n_k = n'_l = 2$$

for infinitely many $k,l\in\mathbb{N}$. Let $Z_{n_1\dots n_k}\ni a$ denote the k-cylinder set containing a, intersected with $H_0(A)$. Then $G^k(Z_{n_1\dots n_k}) = H_k(A)\cap Y_R$, where $H_k(A)$ is the strip between the lines $G^k(\ell(p_A,1)$ and $G^k(\ell(p_A(1-\varepsilon),1))$. Due to the small difference between initial values p_A and $p_A(1-\varepsilon)$, formula (6.2) gives that these lines are roughly parallel.

Applying (6.4) twice we get

$$\begin{cases} \alpha_{k+2} = (n_{k+1}+1)\alpha_k + n_{k+1}\hat{\alpha}_k, \\ \hat{\alpha}_{k+2} = (n_{k+2}n_{k+1} + n_{k+2} - n_{k+1})\alpha_k + (n_{k+2}n_{k+1} - n_{k+1} + 1)\hat{\alpha}_k. \end{cases} \tag{6.8}$$

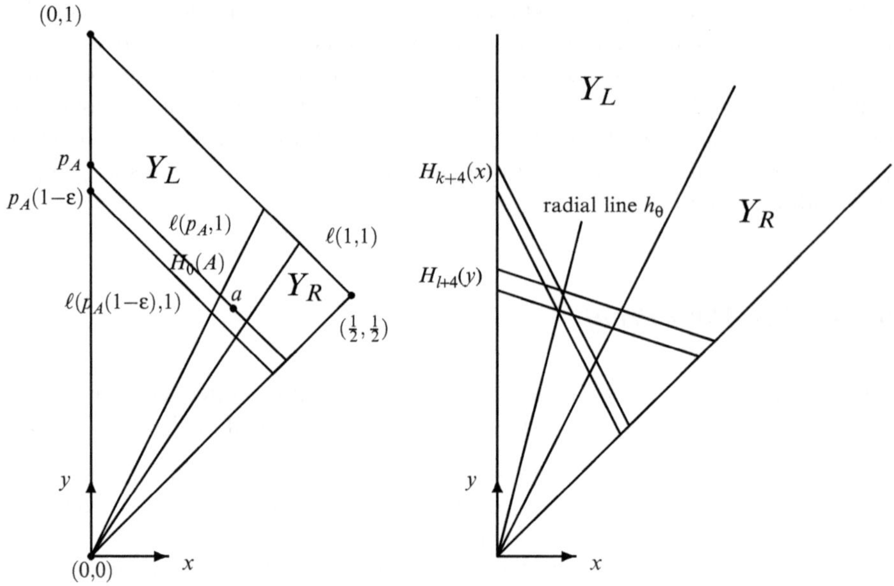

Fig. 6.2 *Left*: The lines $\ell(p_A, 1)$ and $\ell(p_A(1-\varepsilon), 1)$ enclose $H_0(A)$ and the area of large density near a. *Right*: The strips $H_{k+4}(x)$ and $H_{l+4}(y)$ must intersect

For $x \in Z_{n_1 \dots n_k}$, the variables $\alpha_k(x), \hat{\alpha}_k(x), \beta_k(x), \hat{\beta}_k(x), \gamma(x)$ and $\hat{\gamma}_k(x)$ are all well-defined and constant. By choosing $x \in Z_{n_1 \dots n_k}(a)$ so that $n_{k+2}(x) = n_{k+1}(x) = 1$ (which corresponds to choosing a $k+2$-subcylinder $Z_{n_1 \dots n_k 11}(x)$), formula (6.8) simplifies to

$$\begin{cases} \alpha_{k+2} = 2\alpha_k + \hat{\alpha}_k, \\ \hat{\alpha}_{k+2} = \alpha_k + \hat{\alpha}_k, \end{cases}$$

and we have $\hat{\alpha}_k(x) \leq \alpha_{k+2}(x) \leq 2\alpha_{k+2}(x)$ for each x in this subcylinder. In view of (6.2) and (6.5), this means that the slope of the strip $H_{k+2}(a)$ is between Λ and $1/\Lambda$. More precisely,

$$\frac{1}{\Lambda} \leq t_{k+2}(x) \leq \Lambda \qquad \text{for each } x \in Z_{n_1 \dots n_k 11}.$$

Similarly for cylinder $Z_{n'_1 \dots n'_l} \ni a'$, choosing also $n'_{l+1} = n'_{l+2} = 1$ and taking a similar $l+2$-subcylinder $Z_{n'_1 \dots n'_l}$, we find $\Lambda' = \Lambda'(p_{A'}, \varepsilon)$ such that $\frac{1}{\Lambda'} \leq t_{k+2}(y) \leq \Lambda'$ for each $y \in Z_{n_1 \dots n_l 11}$.

Furthermore,

$$\frac{1}{L} \leq \frac{\alpha_{k+2}(x)}{\alpha_{l+2}(y)}, \frac{\hat{\alpha}_{k+2}(x)}{\hat{\alpha}_{l+2}(y)} \leq L,$$

which implies that

$$\frac{1}{\Lambda L} \le \frac{p_{k+2}(x)}{p_{l+2}(y)} \le \Lambda L \qquad \text{for all } x \in Z_{n_1\dots n_k 11} \text{ and } y \in Z_{n'_1\dots n'_k 11}.$$

In other words, $H_{k+2}(x)$ and $H_{l+2}(y)$ are two strips of roughly the same slope and ordinates $p_{k+2}(x)$ and $p_{l+2}(y)$ differing by no more than a uniform factor ΛL.

The next step is to choose a $k+4$-subcylinder of $Z_{n_1\dots n_k 11}$ and a $l+4$-subcylinder of $Z_{n'_1\dots n'_l 11}$ so that their images $H_{k+2}(x)$ and $H_{l+2}(y)$ must intersect. We use (6.3) and (6.8) for $k+4$ instead of $k+2$ to find

$$\begin{aligned} p_{k+4} &= \frac{p_{k+4}}{p_{k+2}} p_{k+2} = \frac{\alpha_{k+2} + \beta_{k+2}t + \gamma_{k+2}p}{\alpha_{k+4} + \beta_{k+4}t + \gamma_{k+4}p} p_{k+2} \\ &= \frac{\alpha_{k+2} + \beta_{k+2}t + \gamma_{k+2}p}{(n_{k+3}+1)(\alpha_{k+2} + \beta_{k+2}t + \gamma_{k+4}p) + n_{k+3}(\hat\alpha_{k+2} + \hat\beta_{k+2} + \hat\gamma_{k+2}p)} p_{k+2} \\ &\sim \frac{p_{k+2}}{n_{k+3}} \end{aligned}$$

and

$$\begin{aligned} t_{k+4} &= \frac{t_{k+4}}{t_{k+2}} t_{k+2} = \frac{\hat\alpha_{k+4} + \hat\beta_{k+4}t + \hat\gamma_{k+4}p}{\alpha_{k+4} + \beta_{k+4}t + \gamma_{k+4}p} \frac{\alpha_{k+2} + \beta_{k+2}t + \gamma_{k+2}p}{\hat\alpha_{k+2} + \hat\beta_{k+2}t + \hat\gamma_{k+2}p} t_{k+2} \\ &= \frac{\begin{array}{r}(n_{k+4}n_{k+3} + n_{k+4} - n_{k+3})(\alpha_{k+2} + \beta_{k+2}t + \gamma_{k+2}p) + \\ (n_{k+4}n_{k+3} - n_{k+3} + 1)(\hat\alpha_{k+2} + \hat\beta_{k+2}t + \hat\gamma_{k+2}p\end{array}}{(n_{k+3}+1)(\alpha_{k+2} + \beta_{k+2}t + \gamma_{k+4}p) + n_{k+3}(\hat\alpha_{k+2} + \hat\beta_{k+2}t + \hat\gamma_{k+2}p)} \\ &\quad \cdot \frac{\alpha_{k+2} + \beta_{k+2}t + \gamma_{k+2}p}{\hat\alpha_{k+2} + \hat\beta_{k+2}t + \hat\gamma_{k+2}p} \cdot t_{k+2} \\ &\sim \frac{n_{k+4}}{n_{k+3}} \end{aligned}$$

By interchanging the role of A and A', we can assume that

$$p_{l+2}(y) \le p_{k+2}(x) \le \Lambda L p_{l+2}(y)$$

for and $x \in Z_{n_1\dots n_{k+4}}$, $y \in Z_{n'_1\dots n'_{l+4}}$. Next choose $10 < M < 2\Lambda L$ and

$$n'_{l+3} = n_{k+4} = 2, \quad n'_{l+4} = n_{k+3} = 8M,$$

so that $\sum_{j=1}^4 n_{k+j} = \sum_{j=1}^4 n'_{l+j} = 4 + 8M$. Furthermore, for x in the corresponding $k+4$-subcylinder of $Z_{n_1\dots n_k}$ and y in the corresponding $l+4$-subcylinder of $Z_{n'_1\dots n'_l}$, we have

$$t_{k+4}(x) \sim \frac{1}{4M}, \quad t_{l+4}(y) \sim 4M.$$

Let $H_{k+4}(x)$ be entire strip between $\ell(p_{k+4}(x), t_{k+4}(x))$ and $\ell(p_{k+4}(x(1-\varepsilon)), t_{k+4}(x(1-\varepsilon)))$, and similarly for $H_{l+4}(y)$. By the above estimates on p_{k+4} and t_{k+4}, we see that $H_{k+4}(x)$ and $H_{l+4}(y)$ intersect, see Fig. 6.2, right.

The foliation of Y into radial lines h_θ is invariant, see Lemma 6.1. There is an interval Θ, depending only on ε and M, such that if $\theta \in \Theta$, then the radial line h_θ intersects $H_{k+4}(x) \cap H_{l+4}(y)$. More precisely, the length

$$|h_\theta \cap H_{k+4}(x) \cap H_{l+4}(y)| \geq \frac{p_{k+4}(x)\varepsilon}{4M} \geq \frac{1}{8L} \min\{|h_\theta \cap H_{l+4}(y)|, |h_\theta \cap H_{k+4}(x)|\}.$$

Write $h(v)$ for the radial line intersecting the point $(v, p_A - v)$, and similarly for $h(w)$. If these lines are chosen such that both $G^{k+4}(h(v))$ and $G^{l+4}(h(w))$ are subset of h_θ, and $v \in V_A$, $w \in V_{A'}$, then we derive from the definition of V_A and $V_{A'}$, using the distortion bound K in Lemma 6.1, that

$$G^{k+4}(h(v) \cap A) \cap G^{l+4}(h(w) \cap A') \neq \emptyset.$$

Since v_A and $v_{A'}$ are density points of V_A and $V_{A'}$, respectively, we can assume that k and l are so large that the relative measure of $V_{A'}$ in $\cup_{n' \geq M} Z_{n'_1\dots n'_{l+2}n'} \cap \ell(p_{A'}, 1)$ is at least $1 - |\Theta|/2K$ and, similarly, the relative measure of V_A in $\cup_{n\geq 1} Z_{n_1\dots n_{k+2}n} \cap \ell(p_A, 1)$ is at least $1 - |\Theta|/2K$.

Recall that $K = 4$ is also the uniform distortion bound for iterates of the Gauss map in Lemma 6.2 and that $G|_{\ell(1,1)}$ acts as the Gauss map. Thus expressed in terms of polar angle $\theta \in [\pi/4, \pi/2]$, the distortion bound is similar.

From this we can conclude that for each θ in a subset of Θ of positive measure, h_θ indeed intersects both $G^{k+4}(H_0(A) \cap h(v))$ for some $v \in V_A$ and $G^{l+4}(H_0(A') \cap h(w))$ for some $w \in V_{A'}$. Therefore $h_\theta \cap G^k(A) \cap G^l(A') \neq \emptyset$, proving that $f^i(A) \cap f^j(A') \neq \emptyset$ for some $i, j \geq 0$. This concludes the ergodicity proof.

Now to prove exactness, we invoke [1, Proposition 2.1], which states that a non-singular ergodic transformation $(X, \mathcal{B}, \lambda; T)$ is exact if and only if for every set $A \in \mathcal{B}$ of positive measure, there is $n \in \mathbb{N}$ such that $\lambda(T^{n+1}(A) \cap T^n(A)) > 0$. Choosing $a = (v_A, p_A - v_A)$ for density point $v_A \in V_A$ and $\varepsilon \in (0, 1 - p_A)$ as before, we can assume that $(n_i(a))_{i\in\mathbb{N}}$ contains infinitely many k such that $n_k(a) = n_{k+1}(a) = 1$. Let us consider the $k+2$-subcylinder $Z_{n_1\dots n_{k+1}1}$ as the set A' with $a' = (v_{A'}, p_{A'} - v_{A'})$ for $p_{A'} = p_B$ and $v_{A'}$ a density point of $V_{A'} = V_A$. Also set $l = k+1$. Then the above methods show that $G^{k+4}(A') \cap G^{l+4}(A)$ intersect, and since $A' \subset A$, we have verified the above condition for exactness with $n = k+4$, $n+1 = l+4$. □

Acknowledgements The hospitality of the Mittag-Leffler Institute in Stockholm (2010 Spring programme on Dynamics and PDEs) is gratefully acknowledged. This paper grew out of a project on subtractive algorithm for which I would like to thank Delft University of Technology for its hospitality over the summer of 2009.

References

1. Bruin, H., Hawkins, J.: Exactness and maximal automorphic factors of unimodal interval maps. Ergod. Th. Dyn. Sys. **21**, 1009–1034 (2001)
2. Durrett, T.: Probability. Theory and Examples. The Wadsworth & Brooks/Cole Statistics/Probability Series. Wadsworth & Brooks/Cole Advanced Books & Software, Pacific Grove (1991)
3. Fokkink, R., Kraaikamp, C., Nakada, H.: On Schweiger's problems on fully subtractive algorithms. Israel J. Math. **186**, 285–296 (2011)
4. Kraaikamp, C., Meester, R.: Ergodic properties of a dynamical system arising from percolation theory. Ergod. Th. Dyn. Sys. **15**(4), 653–661 (1995)
5. Meester, R.: An algorithm for calculating critical probabilities and percolation functions in percolation models defined by rotations. Ergod. Th. Dyn. Sys. **9**, 495–509 (1989)
6. Meester, R., Nowicki, T.: Infinite clusters and critical values in two-dimensional circle percolation. Israel J. Math. **68**(1), 63–81 (1989)
7. de Melo, W., van Strien, S.: One-Dimensional Dynamics. Springer, Berlin (1993)
8. Miernowski, T., Nogueira, A.: Absorbing sets of homogeneous subtractive algorithms, Monatsh. Math. **167**, 547–567 (2012)
9. Nogueira, A.: The three-dimensional Poincaré continued fraction algorithm. Israel J. Math. **90**, 373–401 (1995)
10. Schweiger, F.: Multidimensional Continued Fractions. Oxford University Press, Oxford (2000)

Chapter 7
Improved Estimates of Survival Probabilities via Isospectral Transformations

L.A. Bunimovich and B.Z. Webb

Abstract We consider open systems generated from one-dimensional maps that admit a finite Markov partition and use the recently developed theory of isospectral graph transformations to estimate a system's survival probabilities. We show that these estimates are better than those obtained through a more direct approach.

7.1 Introduction

Recently, a nontrivial relation between the dynamics of networks and the dynamics of open systems has been found. This discovery has lead to advances in the analysis of network dynamics and has also introduced a new research direction concerned with the finite time properties of open systems with finite sized holes.

The main idea behind this approach is that network dynamics can be broken down into three parts: (1) the network's graph structure, often referred to as its *topology*, (2) the local or intrinsic dynamics of the network elements, and (3) the network interactions between these elements. To each of (1)–(3) there is an associated dynamical system, which together can be used to characterize the dynamics of the network [2, 3, 5, 8]. The same approach can be used to study the topological properties of open dynamical systems [1,4,9].

One of the fundamental concerns in the study of networks is understanding the relation between a network's structure and its dynamics. However, the networks we

L.A. Bunimovich (✉)
ABC Math Program and School of Mathematics, Georgia Institute of Technology, 686 Cherry Street, Atlanta, GA 30332, USA
e-mail: bunimovih@math.gatech.edu

B.Z. Webb
Laboratory of Statistical Physics, Rockefeller University, 1230 York Avenue, New York, NY 10065, USA
e-mail: bwebb@rockefeller.edu

W. Bahsoun et al. (eds.), *Ergodic Theory, Open Dynamics, and Coherent Structures*, Springer Proceedings in Mathematics & Statistics 70, DOI 10.1007/978-1-4939-0419-8_7,

© Springer Science+Business Media New York 2014

often encounter in either nature or engineering are typically very large, i.e., have a large number of elements. It is therefore tempting to want to reduce such networks by excluding some subset of these elements while preserving some important characteristic(s) of the original network.

Since real networks are typically dynamic, one might first consider the spectrum of the network's weighted adjacency matrix to be one such a characteristic worth preserving. That is, one could hope to find a way of reducing a network while maintaining its spectrum. However, it seems immediately clear that such an "isospectral reduction" is impossible based on the Fundamental Theorem of Algebra since the matrix $A \in \mathbb{C}^{n \times n}$ has more eigenvalues than $B \in \mathbb{C}^{m \times m}$ if $m < n$.

In fact, it is possible to reduce the size of a network (or matrix) while preserving its spectrum. This theory of isospectral reductions can be found in [6] and is part of a larger theory of isospectral transformations introduced in the paper. Such transformations were used in [7] to improve each of the classical eigenvalue estimates associated with Gershgorin, et al. [10].

In the present paper we make use of this interplay between dynamical networks and open dynamical systems. We obtain estimates of survival probabilities corresponding to Lebesgue measure for a class of one-dimensional maps which admit a finite Markov partition. Most importantly, we have shown that any isospectral transformation corresponding to an open system leads to sharper estimates of the system's survival probabilities.

7.2 Open and Closed Dynamical Systems

Let $f : I \to I$ where $I = [0, 1]$. For $0 = q_0 < q_1 < \cdots < q_{m-1} < q_m = 1$, we let $\xi_i = (q_{i-1}, q_i]$ for $1 \leq i \leq m$ and assume that the following hold. First, the function $f|_{\xi_i}$ is differentiable for each $1 \leq i \leq m$. Second, the sets $\xi_i = (q_{i-1}, q_i]$ form a *Markov partition* $\xi = \{\xi_i\}_{i=1}^m$ of f. That is, for each $1 \leq i \leq m$ the closure $cl(f(\xi_i))$ is the interval $[q_j, q_{j+k}]$ for some $k \geq 1$ and j that depends on i.

We consider the situation where orbits of $f : I \to I$ escape through an element of the Markov partition $\xi = \{\xi_i\}_{i=1}^m$ or, more generally, some union H of these partition elements. Equivalently, we can modify the function f so that orbits cannot leave the set H once they have entered it. Here, orbits that enter H are considered to have escaped from the system. This later approach of modifying $f|_H$ turns out to be more convenient for our discussion and will be the direction we take. In what follows we let $M = \{1, \ldots m\}$.

Definition 7.1. Let $H = \bigcup_{i \in \mathscr{I}} \xi_i$ where $\mathscr{I} \subset M$. We introduce the new map $f_H : I \to I$ defined by

$$f_H(x) = \begin{cases} f(x) & \text{if } x \notin H \\ x & \text{otherwise} \end{cases}.$$

We call the set H a *hole* and the function $f_H : I \to I$ the *open dynamical system* generated by the (closed) dynamical system $f : I \to I$ over H.

The partition ξ will remain a Markov partition of the open dynamical system $f_H : I \to I$, but the dynamics of the original system $f : I \to I$ will have been modified such that each point in H is now a fixed point.

For $n \geq 0$ let

$$\begin{aligned} X^n(f_H) &= \{x \in I : f^n(x) \in H,\ f^k(x) \notin H,\ 0 \leq k < n\} \\ &= \{x \in I : f_H^n(x) \in H,\ f_H^k(x) \notin H,\ 0 \leq k < n\} \text{ and} \\ Y^n(f_H) &= \{x \in I : f^k(x) \in H,\ \text{for some } k,\ 0 \leq k \leq n\} \\ &= \{x \in I : f_H^k(x) \in H,\ \text{for some } k,\ 0 \leq k \leq n\}. \end{aligned}$$

The set $X^n(f_H)$ consists of those points that escape through the hole H at time n, while $Y^n(f_H)$ are those points that escape through H before time $n+1$.

For the moment suppose μ is a probability measure on I, i.e. $\mu(I) = 1$. Then $\mu(X^n(f_H))$ can be treated as the probability that an orbit of f enters H for the first time at time n and $\mu(Y^n(f_H))$ the probability that an orbit of f enters H before time $n+1$. In this regard,

$$P^n(f_H) = 1 - \mu(Y^n(f_H))$$

represents the probability that a typical point of I does not fall into the hole H by time n. For this reason, the quantity $P^n(f_H)$ is called the *survival probability* at time n of the dynamical system f_H for the measure μ.

One of the fundamental problems in the theory of open systems is determining or finding ways of approximating $\mu(X^n(f_H))$ and $\mu(Y^n(f_H))$ for finite $n \geq 0$. In the following section we give exact formulae for these quantities in the case where f_H is a piecewise linear function with nonzero slope and μ is Lebesgue measure. In Sect. 7.3 we remove the assumption that f_H is piecewise linear and present a method for estimating $\mu(X^n(f_H))$ and $\mu(Y^n(f_H))$ for functions that are piecewise nonlinear.

7.3 Piecewise Linear Functions

As a first step, we consider those open systems $f_H : I \to I$ that are linear when restricted to the elements of the partition ξ. More formally, suppose $H = \bigcup_{i \in \mathscr{I}} \xi_i$ for some $\mathscr{I} \subset M$. Let $\mathscr{L}$ be the set of all open systems $f_H : I \to I$ such that

$$|f_H'(x)| = c_i > 0 \ \text{ for } \ x \in \xi_i \ \text{ and } \ i \notin \mathscr{I}$$

where each $c_i \in \mathbb{R}$. The set $\mathscr{L}$ consists of all open systems that have a nonzero constant slope when restricted to any $\xi_i \not\subseteq H$.

To each $f_H \in \mathscr{L}$ there is an associated matrix that can be used to compute $\mu(X^n(f_H))$ and $\mu(Y^n(f_H))$. To define this matrix let

$$\xi_{ij} = \xi_i \cap f^{-1}(\xi_j) \text{ for } 1 \le i, j \le m. \tag{7.1}$$

Definition 7.2. Let $f_H \in \mathscr{L}$ where $H = \bigcup_{i \in \mathscr{I}} \xi_i$ for some $\mathscr{I} \subset M$. The matrix $A_H \in \mathbb{R}^{m \times m}$ given by

$$(A_H)_{ij} = \begin{cases} |f'(x)|^{-1} & \text{for } x \in \xi_{ij} \neq \emptyset,\ i \notin \mathscr{I}, \\ 0 & \text{otherwise} \end{cases} \quad 1 \le i, j \le m$$

is called the *weighted transition matrix* of f_H.

Associated with the open system $f_H : I \to I$, there is also a directed graph $\Gamma = (V, E_H)$ with *vertices* V and *edges* E_H. For $V = \{v_1, \ldots, v_m\}$, we let e_{ij} denote the edge from vertex v_i to v_j.

Definition 7.3. Let $f_H : I \to I$ where $H = \bigcup_{i \in \mathscr{I}} \xi_i$ for some $\mathscr{I} \subset M$. We define $\Gamma_H = (V, E_H)$ to be the graph with

(a) vertices $V = \{v_1, \ldots, v_m\}$.
(b) edges $E_H = \{e_{ij} : cl(\xi_j) \subseteq cl(f(\xi_i)),\ i \notin \mathscr{I}\}$.

The graph Γ_H is called the *transition graph* of f_H.

The vertex set V of Γ_H represents the elements of the Markov partition $\xi = \{\xi_i\}_{i=1}^m$ and the edge set E_H, the possible transitions between the elements of ξ. Hence, $e_{ij} \in E_H$ only if there is an $x \in \xi_i \not\subseteq H$ such that $f_H(x) \in \xi_j$, i.e. it is possible to make a transition from ξ_i to ξ_j. We note that as $H = \emptyset$ is a possible hole, the original (closed) system $f : I \to I$ has a well-defined transition graph which we denote by Γ.

Note that the matrix A_H and the graph Γ_H do not carry the same information as Γ_H only designates how orbits can make a transition between elements of ξ whereas A_H additionally gives each of these transitions a weight. However, the graph Γ_H gives us a way of visualizing how orbits escape from the system, which will be useful in the following sections. An open system and its transition graph are demonstrated in the following example.

Example 7.1. Let the function $f : I \to I$ be the tent map

$$f(x) = \begin{cases} 2x & 0 \le x \le 1/2, \\ 2 - 2x & 1/2 < x \le 1 \end{cases}$$

with Markov partition $\xi = \{(0, 1/4], (1/4, 1/2], (1/2, 3/4], (3/4, 1]\}$. Here we consider the hole $H = (0, 1/4]$. The open system $f_H : I \to I$ is shown in Fig. 7.1 (left) with the graph Γ_H (right).

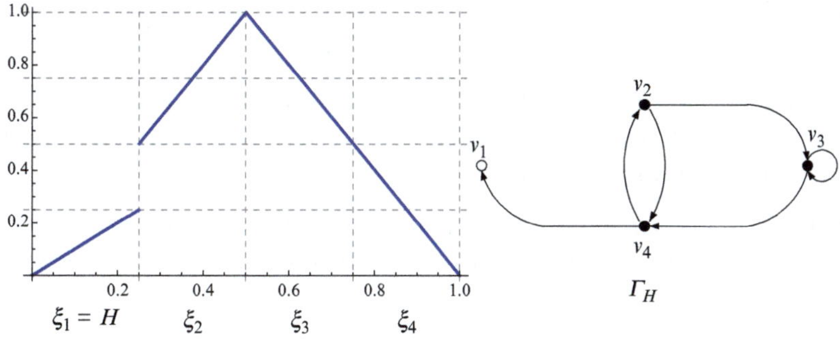

Fig. 7.1 The transition graph Γ_H (*right*) of the open system $f_H : I \to I$ (*left*) in Example 7.1

As $H = \xi_1$ we emphasize this in Γ_H by drawing the vertex v_1 as an open circle, i.e., as a hole. We note that the only difference between the transition graph Γ of $f : I \to I$ and Γ_H is that there are no edges originating from v_1 in Γ_H. In this sense a hole H is an absorbing state since nothing leaves H once it enters.

Let $\mathbf{1} = [1, \ldots, 1]$ be the $1 \times m$ vector of ones and $\mathbf{e}_H$ the $m \times 1$ vector given by

$$(\mathbf{e}_H)_i = \begin{cases} \mu(\xi_i) & \text{if } i \in \mathscr{I} \\ 0 & \text{otherwise} \end{cases}.$$

Theorem 7.1. *If $f_H \in \mathscr{L}$ and $n \geq 0$ then*

$$\mu(X^n(f_H)) = \mathbf{1}A_H^n\mathbf{e}_H; \text{ and} \tag{7.2}$$

$$\mu(Y^n(f_H)) = \mathbf{1}\left(\sum_{i=0}^{n} A_H^i\right)\mathbf{e}_H. \tag{7.3}$$

Instead of giving a proof of Theorem 7.1 we note that in Sect. 7.5 the main result, Theorem 7.3, implies Theorem 7.2 in Sect. 7.4, which in turn implies Theorem 7.1. We therefore omit the proofs of Theorems 7.1 and 7.2.

For the matrix $B \in \mathbb{R}^{m \times m}$ let $\sigma(B)$ and $\rho(B)$ denote the set of all *eigenvalues* and *spectral radius* of B, respectively. We then have the following corollaries to Theorem 7.1.

Corollary 7.1. *If $f_H \in \mathscr{L}$ and $\rho(A_H) < 1$, then*

$$\mu(Y^n(f_H)) = \mathbf{1}(I - A_H)^{-1}(I - A_H^{n+1})\mathbf{e}_H \text{ and}$$
$$\lim_{n\to\infty} \mu(Y^n(f_H)) = \mathbf{1}(I - A_H)^{-1}\mathbf{e}_H,$$

where I is the identity matrix.

Corollary 7.2. *Suppose $f_H \in \mathscr{L}$. If $0 < \rho(A_H) < 1$, then $\lim_{n\to\infty} P^n(f_H) = 0$. If $\rho(A_H) = 0$, then $\mu(Y^n(f_H)) = 1$ for some $n < \infty$.*

A matrix $B \in \mathbb{R}^{m\times m}$ is called *defective* if it does not have an eigenbasis, i.e., if there are not enough linearly independent eigenvectors of B to form a basis of $\mathbb{R}^m$. A matrix with an eigenbasis is called *nondefective*.

Corollary 7.3. *Let $f_H \in \mathscr{L}$ and suppose the matrix A_H is nondefective with eigenpairs $\{(\lambda_1, \mathbf{v}_1), \dots, (\lambda_k, \mathbf{v}_k)\}$ with no eigenvalue equal to 1. Then $\mathbf{e}_H = \sum_{i=1}^k c_i \mathbf{v}_i$ for some $c_1, \dots, c_k \in \mathbb{C}$ and*

$$\mu(X^n(f_H)) = \sum_{i=1}^{k} c_i s_i \lambda_i^n \tag{7.4}$$

$$\mu(Y^n(f_H)) = \sum_{i=1}^{k} c_i s_i \left(\frac{1-\lambda_i^{n+1}}{1-\lambda_i} \right) \tag{7.5}$$

where $s_i = \mathbf{1}\mathbf{v}_i$.

Example 7.2. Let the function $f : I \to I$ be the tent map considered in Example 7.1 and let H=(0,1/4]. As $f_H \in \mathscr{L}$, one can calculate that f_H has the weighted transition matrix

$$A_H = \begin{bmatrix} 0 & 0 & 0 & 0 \\ 0 & 0 & 1/2 & 1/2 \\ 0 & 0 & 1/2 & 1/2 \\ 1/2 & 1/2 & 0 & 0 \end{bmatrix}.$$

The matrix A_H is nondefective as its eigenvalues $\sigma(A_H) = \{\frac{1+\sqrt{5}}{4}, \frac{1-\sqrt{5}}{4}, 0, 0\}$ correspond, respectively, to the linearly independent eigenvectors

$$\mathbf{v}_1 = \begin{bmatrix} 0 \\ \frac{1+\sqrt{5}}{2} \\ \frac{1+\sqrt{5}}{2} \\ 1 \end{bmatrix}, \mathbf{v}_2 = \begin{bmatrix} 0 \\ \frac{1-\sqrt{5}}{2} \\ \frac{1-\sqrt{5}}{2} \\ 1 \end{bmatrix}, \mathbf{v}_3 = \begin{bmatrix} 0 \\ 0 \\ -1 \\ 1 \end{bmatrix}, \mathbf{v}_4 = \begin{bmatrix} -1 \\ 1 \\ 0 \\ 0 \end{bmatrix}.$$

Since the vector $\mathbf{e}_H = [1/4, 0, 0, 0]^T$ can be written as

$$\mathbf{e}_H = \frac{5-\sqrt{5}}{20+20\sqrt{5}}\mathbf{v}_1 - \frac{3+\sqrt{5}}{8\sqrt{5}}\mathbf{v}_2 + \frac{1}{4}\mathbf{v}_3 - \frac{1}{4}\mathbf{v}_4,$$

Eqs. (7.4) and (7.5) in Corollary 7.3 imply

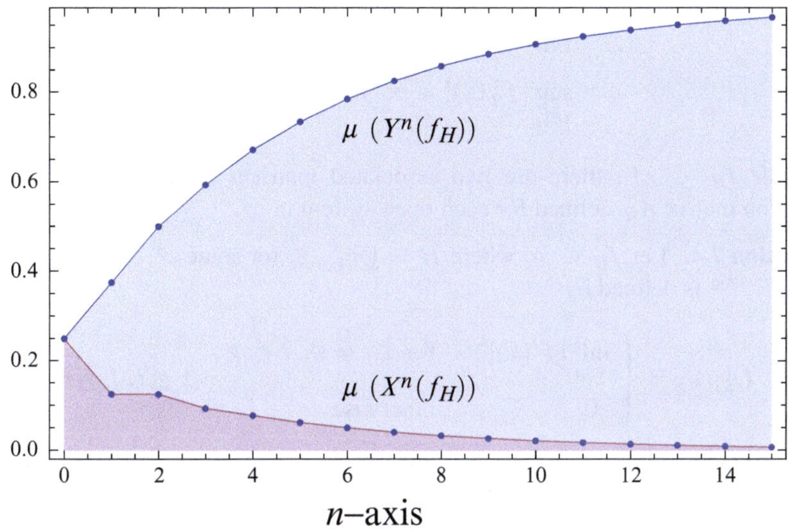

Fig. 7.2 Plots of $\mu(X^n(f_H))$ and $\mu(Y^n(f_H))$ for $f_H : I \to I$ in Example 7.2

$$\mu(X^n(f_H)) = \frac{1}{40}(5+\sqrt{5})\lambda_1^n + \frac{1}{40}(5-\sqrt{5})\lambda_2^n \text{ and}$$

$$\begin{aligned}\mu(Y^n(f_H)) &= \frac{1}{40}(5+\sqrt{5})\frac{1-\lambda_1^{n+1}}{1-\lambda_1} + \frac{1}{40}(5-\sqrt{5})\frac{1-\lambda_2^{n+1}}{1-\lambda_2} \\ &= 1 - \left(\frac{1}{2}+\frac{1}{\sqrt{5}}\right)\lambda_1^{n+1} - \left(\frac{1}{2}-\frac{1}{\sqrt{5}}\right)\lambda_2^{n+1}.\end{aligned}$$

Note that as $\rho(A_H) < 1$ then $\lim_{n\to\infty} P^n(f_H) = 0$. Hence, the probability of surviving indefinitely in this system for a typical $x \in I$ is in fact zero. This can be seen in Fig. 7.2 where both $\mu(X^n(f_H))$ and $\mu(Y^n(f_H))$ are plotted.

7.4 Nonlinear Estimates

We now consider the open systems $f_H : I \to I$ where f is allowed to be a nonlinear but differentiable function when restricted to the elements of ξ. The formulae we derive in this section allow us to give upper and lower bounds on $\mu(X^n(f_H))$ and $\mu(Y^n(f_H))$ for any finite time $n \geq 0$.

Suppose $H = \bigcup_{i\in\mathscr{I}} \xi_i$ for some $\mathscr{I} \subset M$. Let $\mathscr{N}$ be the open systems $f_H : I \to I$ such that

$$\inf_{x\in\xi_i}|f_H'(x)|>0 \text{ for } i\notin\mathscr{I} \text{ and}$$

$$\sup_{x\in\xi_i}|f_H'(x)|<\infty \text{ for } i\notin\mathscr{I}.$$

To each $f_H\in\mathscr{N}$, there are two associated matrices similar to the weighted transition matrix A_H defined for each open system in $\mathscr{L}$.

Definition 7.4. Let $f_H\in\mathscr{N}$ where $H=\bigcup_{i\in\mathscr{I}}\xi_i$ for some $\mathscr{I}\subset M$. The matrix $\underline{A}_H\in\mathbb{R}^{m\times m}$ is defined by

$$(\underline{A}_H)_{ij}=\begin{cases}\inf\limits_{x\in\xi_{ij}}|f'(x)|^{-1} & \text{for } \xi_{ij}\neq\emptyset,\ i\notin\mathscr{I},\\ 0 & \text{otherwise}\end{cases}\qquad 1\le i,j\le m.$$

Similarly, define the matrix $\overline{A}_H\in\mathbb{R}^{m\times m}$ by

$$(\overline{A}_H)_{ij}=\begin{cases}\sup\limits_{x\in\xi_{ij}}|f'(x)|^{-1} & \text{for } \xi_{ij}\neq\emptyset,\ i\notin\mathscr{I}\\ 0 & \text{otherwise}\end{cases}\qquad 1\le i,j\le m.$$

For $f_H\in\mathscr{N}$ and $n\ge 0$ let

$$\underline{X}^n(f_H)=\mathbf{1}\underline{A}_H^n\mathbf{e}_H \text{ and } \overline{X}^n(f_H)=\mathbf{1}\overline{A}_H^n\mathbf{e}_H;$$

$$\underline{Y}^n(f_H)=\mathbf{1}\left(\sum_{i=0}^m\underline{A}_H^i\right)\mathbf{e}_H \text{ and } \overline{Y}^n(f_H)=\mathbf{1}\left(\sum_{i=0}^m\overline{A}_H^i\right)\mathbf{e}_H.$$

Theorem 7.2. *If $f_H\in\mathscr{N}$ and $n\ge 0$ then $\underline{X}^n(f_H)\le\mu(X^n(f_H))\le\overline{X}^n(f_H)$ and $\underline{Y}^n(f_H)\le\mu(Y^n(f_H))\le\overline{Y}^n(f_H)$.*

Theorem 7.2 allows us to bound the amount of phase space that escapes through H at time n and before time $n+1$. If the matrices $\underline{A}_H$ and $\overline{A}_H$ are nondefective, then we have the following result similar to Corollary 7.3.

Corollary 7.4. *Let $f_H\in\mathscr{N}$ and suppose both $\underline{A}_H$ and $\overline{A}_H$ are nondefective with eigenpairs $\{(\underline{\lambda}_1,\underline{\mathbf{v}}_1),\dots,(\underline{\lambda}_k,\underline{\mathbf{v}}_k)\}$ and $\{(\overline{\lambda}_1,\overline{\mathbf{v}}_1),\dots,(\overline{\lambda}_k,\overline{\mathbf{v}}_k)\}$ with no eigenvalue equal to 1. Then for each $n>0$*

$$\sum_{i=1}^k\underline{c}_i\underline{s}_i\underline{\lambda}_i^n\le\mu(X^n(f_H))\le\sum_{i=1}^k\overline{c}_i\overline{s}_i\overline{\lambda}_i^n \text{ and} \tag{7.6}$$

$$\sum_{i=1}^k\underline{c}_i\underline{s}_i\left(\frac{1-\underline{\lambda}_i^{n+1}}{1-\underline{\lambda}_i}\right)\le\mu(Y^n(f_H))\le\sum_{i=1}^k\overline{c}_i\overline{s}_i\left(\frac{1-\overline{\lambda}_i^{n+1}}{1-\overline{\lambda}_i}\right) \tag{7.7}$$

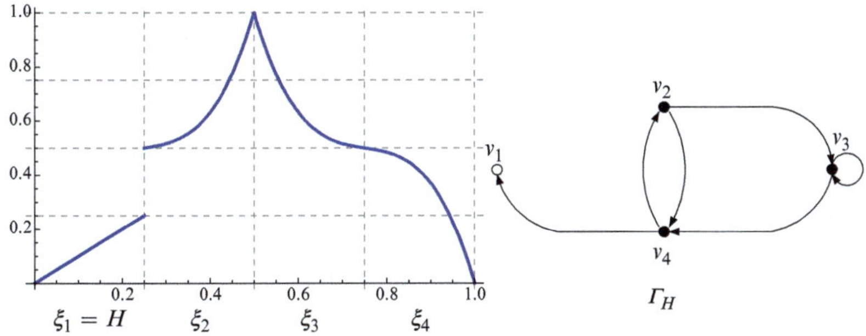

Fig. 7.3 The transition graph Γ_H *(right)* of the open system $g_H : I \to I$ *(left)* in Example 7.3

where $\underline{s}_i = \mathbf{1}\underline{\mathbf{v}}_i$, $\overline{s}_i = \mathbf{1}\overline{\mathbf{v}}_i$, $\mathbf{e}_H = \sum_{i=1}^{n} \underline{c}_i \underline{\mathbf{v}}_i$, *and* $\mathbf{e}_H = \sum_{i=1}^{n} \overline{c}_i \overline{\mathbf{v}}_i$.

The upper and lower bounds given in (7.6) are $\underline{X}^n(f_H)$ and $\overline{X}^n(f_H)$, respectively. The upper and lower bounds given in (7.7) are $\underline{Y}^n(f_H)$ and $\overline{Y}^n(f_H)$, respectively.

Remark 7.1. If $f \in \mathscr{L}$, then the dynamics of f_H over ξ reduces to Markov chain, which allows for the exact formulas for $\mu(X^n(f_H))$ and $\mu(Y^n(f_H))$ in Theorem 7.1. However, if $f \in \mathscr{N}$, then this does not in general mean that the dynamics of f_H over ξ has this property. The result is that Theorem 7.2 contains bounds instead of equations.

Example 7.3. Consider the function $g : I \to I$ given by

$$g(x) = \begin{cases} \frac{11}{2}x - 21x^2 + 28x^3 & 0 \leq x \leq 1/2 \\ \frac{11}{2}(1-x) - 21(1-x)^2 + 28(1-x)^3 & 1/2 < x \leq 1 \end{cases}$$

with Markov partition $\xi = \{(0,1/4],(1/4,1/2],(1/2,3/4],(3/4,1]\}$ and $H = (0,1/4]$. The function $g : I \to I$ can be considered to be a nonlinear version of the tent map $f : I \to I$ in Example 7.1. In fact both systems have the same transition graph (see Fig. 7.3).

For $g_H : I \to I$ one can compute $\xi_{23} = (1/4, .44]$, $\xi_{24} = (.44, 1/2]$, $\xi_{34} = (1/2, .55]$, $\xi_{33} = (.55, .3/4]$, $\xi_{42} = (3/4, .94]$, and $\xi_{41} = (.94, 1]$. From this we find

$$\underline{A}_H = \begin{bmatrix} 0 & 0 & 0 & 0 \\ 0 & 0 & 0.29 & 2/11 \\ 0 & 0 & 0.29 & 2/11 \\ 2/11 & 0.29 & 0 & 0 \end{bmatrix} \text{ and } \overline{A}_H = \begin{bmatrix} 0 & 0 & 0 & 0 \\ 0 & 0 & 4 & 0.29 \\ 0 & 0 & 4 & 0.29 \\ 0.29 & 4 & 0 & 0 \end{bmatrix}.$$

As $\underline{A}_H$ and $\overline{A}_H$ are nondefective, then using Corollary 7.4 one can compute that

$$0.097\underline{\lambda}_1^n - 0.034\underline{\lambda}_2^n \leq \mu(X^n(g_H)) \leq 0.003\overline{\lambda}_1^n - 0.215\overline{\lambda}_2^n \tag{7.8}$$

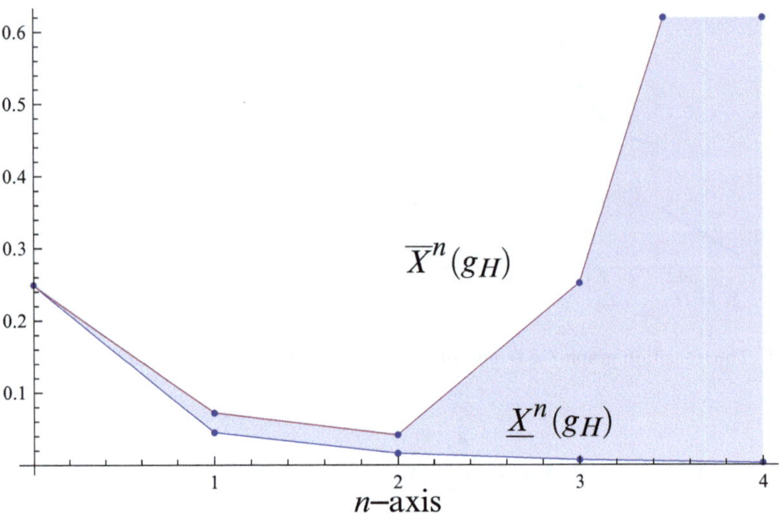

Fig. 7.4 The upper bounds $\overline{X}^n(g_H)$ and lower bounds $\underline{X}^n(g_H)$ for $\mu(X^n(g_H))$ are shown for the open system $g_H : I \to I$ in Example 7.3

where $\underline{\lambda}_1 = 0.41$, $\underline{\lambda}_2 = -0.12$, $\overline{\lambda}_1 = 4.27$, and $\overline{\lambda}_2 = -0.27$. Plotting the inequalities in (7.8) yields the picture in Fig. 7.4. Here the shaded area indicates the region in which $\mu(X^n(g_H))$ must lie.

7.5 Improved Escape Estimates

In this section we define a delayed first return map of an open system $f_H \in \mathscr{N}$, which we will use to improve the escape estimates given in Theorem 7.2. A key step in this procedure is to choose a particular vertex set of Γ_H over which this map will be defined. This requires that we know the cycle structure of Γ_H.

A *path* $\mathscr{P}$ in the graph $\Gamma_H = (V, E_H)$ is an ordered sequence of distinct vertices $v_1, \dots, v_k \in V$ such that $e_{i,i+1} \in E$ for $1 \leq i \leq k-1$. If the vertices v_1 and v_k are the same, then $\mathscr{P}$ is a *cycle*. If $S \subseteq V$ where V is the vertex set of the graph, we will write $\bar{S} = V \setminus S$.

Definition 7.5. Let $H = \bigcup_{i \in \mathscr{I}} \xi_i$ for some $\mathscr{I} \subset M$ and let $\Gamma_H = (V, E_H)$. The set $S \subseteq V = \{v_1, \dots, v_m\}$ is an *open structural set* of Γ_H if $v_i \in S$ for $i \in \mathscr{I}$ and $\Gamma_H|_{\bar{S}}$ has no cycles.

Structural sets were first defined in [6] where they were used to gain improved estimates of a dynamical network's stability. Later in [7] they were used to improve the eigenvalue estimates of Gershgorin et. al. Here, our goal is to extend their use to improve our estimates of $\mu(X^n(f_H))$ and $\mu(Y^n(f_H))$.

For the open system $f_H : I \to I$ we let $st(\Gamma_H)$ denote the set of all open structural sets of Γ_H. If $S \in st(\Gamma_H)$, we let $\mathscr{I}_S = \{i \in M : v_i \in S\}$ be the *index set* of S and $\xi_S = \bigcup_{i \in \mathscr{I}_S} \xi_i$.

Definition 7.6. Let $S \in st(\Gamma_H)$. For $x \in I$ we let $\gamma(x) = i_0 i_1 \ldots i_t$ where $i_j = k$ if $f_H^j(x) \in \xi_k$ and t is the smallest $k > 0$ such that $f_H^k(x) \in \xi_S$. The set

$$\Omega_S = \{\gamma : \gamma = \gamma(x) \text{ for some } x \in I \setminus H\}$$

are the *admissible sequences* of f_H with respect to S.

For $x \in I$ we say $\gamma(x) = i_0 i_1 \ldots i_t$ has length $|\gamma(x)| = t$. The reason $|\gamma(x)| < \infty$ is that the graph $\Gamma_H|_{\bar{S}}$ has no cycles. Hence, after a finite number of steps $f_H^t(x)$ must enter ξ_S.

Definition 7.7. Let $S \in st(\Gamma_H)$. For $x_0 \in I$ and $k \geq 0$, we inductively define $x_{k+1} = Rf_S(x_k, \ldots, x_0)$ where

$$x_{k+1} = \begin{cases} f_H^{|\gamma(x_k)|}(x_k) & \text{if } x_{k-i} = x_k \text{ for each } 0 \leq i \leq |\gamma(x_k)| - 1, \\ x_k & \text{otherwise.} \end{cases}$$

The function $Rf_S : I^{k+1} \to I$ is called the *delayed first return map* of f_H with respect to S. The sequence $x_0, x_1, x_2, \ldots$ is the *orbit* of x_0 under Rf_S.

If $T = \max_{x \in I} |\gamma(x)|$, then strictly speaking $x_{k+1} = Rf_S(x_k, \ldots, x_\tau)$ for some $\tau < T$. The map Rf_S acts almost like a first return map of f_H to the set ξ_S. The difference is that a return to ξ_S does not happen instantaneously (as it would happen in the case of a first return map) but is delayed so that the trajectory of a point under f_H and Rf_S coincides after a return to ξ_S.

Example 7.4. Consider the open system $g_H : I \to I$ given in Example 7.3. Observe that the vertex set $S = \{v_1, v_3, v_4\}$ is an open structural set of Γ_H since v_1 is the vertex that corresponds to H and the graph $\Gamma_H|_{\bar{S}} = \{v_2\}$ has no cycles (see Fig. 7.3).

The delayed first return map Rg_S can be written as

$$Rg_S(x_k) = \begin{cases} g_H(x_k) & \text{if } x_k \notin \xi_{42}, \\ g_H^2(x_k) & \text{if } x_k, x_{k-1} \in \xi_{42}, \\ x_k & \text{otherwise.} \end{cases}$$

The map Rg_S is shown in Fig. 7.5 as a one-dimensional map, which is delayed on the set ξ_{42}. Specifically, $Rg_S = g_H(x)$ for $x \notin \xi_{42}$ and $Rg_S = g_H^2(x)$ for $x \in \xi_{42}$ where the trajectories of Rg_H stay in ξ_{42} for two time-steps before leaving.

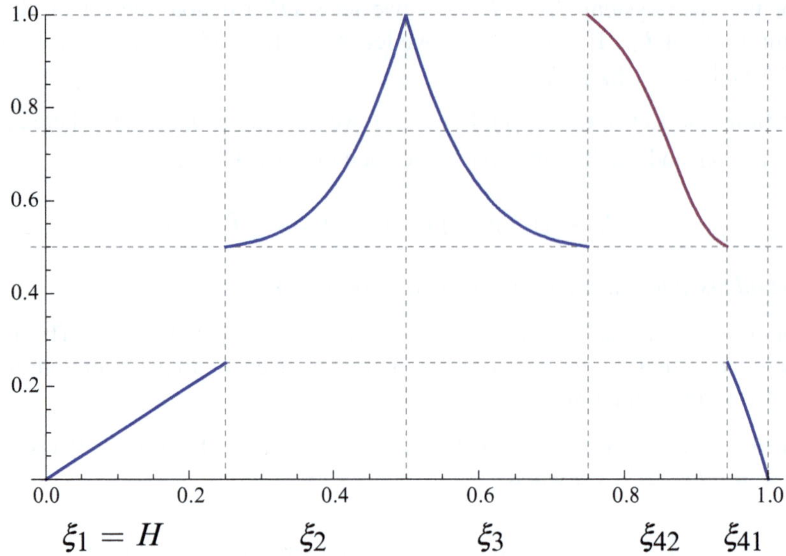

Fig. 7.5 The delayed first return map $Rg_S : I \to I$ in Example 7.4 where Rg_S is delayed on the set ξ_{42}

For $n \geq 0$ we let $Rf_S^n(x_0) = x_n$ and define

$$X^n(Rf_S) = \{x \in I : Rf_S^n(x) \in H,\ Rf_S^k(x) \notin H,\ 0 \leq k < n\} \text{ and}$$

$$Y^n(Rf_S) = \{x \in I : Rf_S^k(x) \in H,\ \text{for some } k,\ 0 \leq k \leq n\}.$$

Lemma 7.1. *If $S \in st_\xi(\Gamma_{f_H})$ and $n \geq 0$ then $X^n(f_H) = X^n(Rf_S)$.*

Proof. For $x_0 \in I$ let $\tilde{\gamma}(x_0) = i_0 i_1 \dots$ where $i_j = k$ if $f_H^j(x_0) \in \xi_k$. Choosing $S \in st(\Gamma_H)$ let $\tilde{\gamma}_S(x_0) = \ell_0 \ell_1 \dots$ where $\ell_j = k$ if $Rf_S^j(x_0) \in \xi_k$. Let $t > 0$ be the smallest number such that $i_t \in \mathscr{I}_S$. Then $\gamma(x_0) = i_0 i_1 \dots i_t$ and Definition 7.7 implies $Rf_S^t(x_0) = f_H^t(x_0)$. Therefore, $i_t = \ell_t$ where $t \in \mathscr{I}_{\mathscr{S}}$.

Continuing in this manner it follows that $i_j = \ell_j$ for each $j \in \mathscr{I}_S$. Since $H \subseteq \xi_S$ the point x_0, if it escapes, will escape for both f_H and Rf_S at exactly the same time. This implies the result. □

The major idea in this section is that one can use Rf_S to study the escape of f_H through H. However, the weighted transition matrix of $Rf_S : I^k \to I$ cannot be defined in the same way that we have defined either $\underline{A}_H$ or $\overline{A}_H$. To define a transition matrix for Rf_S we require the following.

For $S \in st(\Gamma_H)$ let

$$M_S = M \cup \{\gamma; i : \gamma \in \Omega_S, 0 < i < |\gamma|\}$$

If $\gamma = i_0 \dots, i_t$ we identify the index $\gamma;0$ with i_0 and the index $\gamma;t$ with i_t. We also let $\xi_\gamma = \{x \in I : \gamma(x) = \gamma\}$ for each admissible sequence $\gamma \in \Omega_S$, which simply extends our notation given by (7.1) in Sect. 7.3.

Definition 7.8. For $S \in st(\Gamma_H)$ let $\underline{A}_S$ be the matrix with rows and columns indexed by elements of M_S where

$$(\underline{A}_S)_{ij} = \begin{cases} \inf\limits_{x\in\xi_\gamma} |(f^{|\gamma|}(x))'|^{-1} & \text{if } i = \gamma;|\gamma|-1,\ j = \gamma;|\gamma|, \text{ for some } \gamma \in \Omega_S \\ 1 & \text{if } i = \gamma;k-1,\ j = \gamma;k,\ k \neq |\gamma|, \text{ for some } \gamma \in \Omega_S \\ 0 & \text{otherwise.} \end{cases} \tag{7.9}$$

We call $\underline{A}_S$ the *lower transition matrix* of Rf_S. The matrix $\overline{A}_S$ defined by replacing the infimum in (7.9) by a supremum is the *upper transition matrix* of Rf_S.

Let $\mathbf{1}_S$ be the $1 \times |M_S|$ vector given by

$$(\mathbf{1}_S)_i = \begin{cases} 1 & \text{if } i \in M, \\ 0 & \text{otherwise.} \end{cases}$$

Let $\mathbf{e}_S$ be the $|M_S| \times 1$ vector given by

$$(\mathbf{e}_S)_i = \begin{cases} \mu(\xi_i) & \text{if } i \in \mathscr{I}, \\ 0 & \text{otherwise.} \end{cases}$$

Lastly, for $n \geq 0$ let

$$\underline{X}^n(Rf_S) = \mathbf{1}_S \underline{A}_S^n \mathbf{e}_S \text{ and } \overline{X}^n(Rf_S) = \mathbf{1}_S \overline{A}_S^n \mathbf{e}_S;$$

$$\underline{Y}^n(Rf_S) = \mathbf{1}_S \left(\sum_{i=0}^{n} \underline{A}_S^i\right) \mathbf{e}_S \text{ and } \overline{Y}^n(Rf_S) = \mathbf{1}_S \left(\sum_{i=0}^{n} \overline{A}_S^i\right) \mathbf{e}_S.$$

Using these quantities we give the following improved escape estimates.

Theorem 7.3. *Let $f_H \in \mathscr{N}$ and suppose $S \in st(\Gamma_H)$. If $n \geq 0$, then*

$$\underline{X}^n(f_H) \leq \underline{X}^n(Rf_S) \leq \mu(X^n(f_H)) \leq \overline{X}^n(Rf_S) \leq \overline{X}^n(f_H) \text{ and}$$

$$\underline{Y}^n(f_H) \leq \underline{Y}^n(Rf_S) \leq \mu(Y^n(f_H)) \leq \overline{Y}^n(Rf_S) \leq \overline{Y}^n(f_H).$$

Theorem 7.3 together with Lemma 7.1 implies that the escape of f_H through H is better approximated by considering any of its delayed first return maps Rf_S than f_H itself. We now give a proof of Theorem 7.3.

Proof. For $S \in st(\Gamma_H)$ suppose $i \in M \setminus \mathscr{I}$ and $j \in \mathscr{I}$. Then

$$(\underline{A}_S)_{ij}(\mathbf{e}_S)_j = \begin{cases} \inf_{x\in\xi_{ij}} |f'(x)|^{-1}\mu(\xi_j) & \text{if } \xi_{ij} \neq \emptyset, \\ 0 & \text{otherwise} \end{cases} \leq \mu\{x \in \xi_i : f_H(x) \in \xi_j\}.$$

To show that a similar formula holds for larger powers of $\underline{A}_S$ suppose $k \in \mathscr{I}_S$. If $ik, kj \in \Omega_S$, then

$$\begin{aligned}(\underline{A}_S)_{ik}(\underline{A}_S)_{kj}(\mathbf{e}_S)_j &= \inf_{x\in\xi_{ik}} |f'(x)|^{-1} \inf_{x\in\xi_{kj}} |f'(x)|^{-1}\mu(\xi_j) \\ &\leq \mu\{x \in \xi_i : f_H(x) \in \xi_k, f_H^2(x) \in \xi_j\}.\end{aligned}$$

If either $ik \notin \Omega_S$ or $kj \notin \Omega_S$ then $(\underline{A}_S)_{ik}(\underline{A}_S)_{kj}(\mathbf{e}_S)_j = 0$.

Suppose $k \in M \setminus \mathscr{I}_S$. If $ikj \in \Omega_S$ then $ikj;1 \in M_S$ and

$$\begin{aligned}(\underline{A}_S)_{i,ikj;1}(\underline{A}_S)_{ikj;1,j}(\mathbf{e}_S)_j =& 1 \cdot \inf_{x\in\xi_{ikj}} |(f^2(x))'|^{-1}\mu(\xi_j) \\ \leq& \mu\{x \in \xi_i : f_H(x) \in \xi_k, f_H^2(x) \in \xi_j\}.\end{aligned}$$

If $ikj \notin \Omega_S$ then $ijk;1 \notin M_S$. Since

$$\begin{aligned}(\underline{A}_S^2)_{ij}(\mathbf{e}_S)_j &= \sum_{k\in M_S} (\underline{A}_S)_{ik}(\underline{A}_S)_{kj}\mu(\xi_j) \\ &= \sum_{ik,kj\in\Omega_S} (\underline{A}_S)_{ik}(\underline{A}_S)_{kj}\mu(\xi_j) + \sum_{ikj\in\Omega_S} (\underline{A}_S)_{i,ikj;1}(\underline{A}_S)_{ikj;1,j}\mu(\xi_j) \\ &\leq \sum_{k\in\mathscr{I}_S\cup(M\setminus\mathscr{I}_S)} \mu\{x \in \xi_i : f_H(x) \in \xi_k, f_H^2(x) \in \xi_j\} \\ &= \mu\{x \in \xi_i : f_H^2(x) \in \xi_j\}.\end{aligned}$$

Continuing in this manner it follows that

$$(\underline{A}_S^n)_{ij}(\mathbf{e}_S)_j \leq \mu\{x \in \xi_i : f_H^n(x) \in \xi_j\} \tag{7.10}$$

for $i \in M \setminus \mathscr{I}$, $j \in \mathscr{I}$, and $n \geq 1$. Since $(\mathbf{e}_S)_j = 0$ if $j \notin \mathscr{I}$ then for $n \geq 1$ equation (7.10) implies

$$\begin{aligned}\mathbf{1}_S\underline{A}_S^n\mathbf{e}_S = \sum_{i\in M}\sum_{j\in M_S} (\underline{A}_S^n)_{ij}(\mathbf{e}_S)_j &\leq \sum_{i\in M\setminus\mathscr{I}} \mu\{x \in \xi_i : f_H^n(x) \in H\} \\ &= \mu\{x \in I \setminus H : f_H^n(x) \in H\}.\end{aligned}$$

As $\mathbf{1}_S \underline{A}_S^0 \mathbf{e}_S = \mu(H)$ then $\underline{X}^n(Rf_S) \leq \mu(X^n(f_H)) \leq \overline{X}^n(Rf_S)$ for $n \geq 0$ where the second inequality follows by using the same argument with the matrix $\overline{A}_S$.

To show that $\underline{X}^n(f_H) \leq \underline{X}^n(Rf_S)$ we again suppose that $i \in M \setminus \mathscr{I}$ and $j \in \mathscr{I}$. Then we have

$$(\underline{A}_H)_{ij}(\mathbf{e}_H)_j = \begin{cases} \inf_{x \in \xi_{ij}} |f'(x)|^{-1}\mu(\xi_j) & \text{if } \xi_{ij} \neq \emptyset, \\ 0 & \text{otherwise} \end{cases} = (\underline{A}_S)_{ij}(\mathbf{e}_S)_j.$$

For larger matrix powers we have

$$\sum_{k \in M \setminus \mathscr{I}_S} (\underline{A}_H)_{ik}(\underline{A}_H)_{kj}\mu(\xi_j) = \sum_{k \in M \setminus \mathscr{I}_S} \inf_{x \in \xi_{ik}} |f'(x)|^{-1} \inf_{x \in \xi_{kj}} |f'(x)|^{-1}\mu(\xi_j)$$
$$\leq \sum_{k \in M \setminus \mathscr{I}_S} 1 \cdot \inf_{x \in \xi_{ikj}} |(f^2(x))'|^{-1}\mu(\xi_j) = \sum_{ikj \in \Omega_S} (\underline{A}_S)_{i,ikj;1}(\underline{A}_S)_{ikj;1,j}(\mathbf{e}_S)_j.$$

From this it follows that

$$\begin{aligned}(\underline{A}_H^2)_{ij}(\mathbf{e}_H)_j &= \sum_{k \in \mathscr{I}_S} (\underline{A}_H)_{ik}(\underline{A}_H)_{kj}\mu(\xi_j) + \sum_{k \in M \setminus \mathscr{I}_S} (\underline{A}_H)_{ik}(\underline{A}_H)_{kj}\mu(\xi_j) \\ &\leq \sum_{ik,kj \in \Omega_S} (\underline{A}_S)_{ik}(\underline{A}_S)_{kj}\mu(\xi_j) + \sum_{ikj \in \Omega_S} (\underline{A}_H)_{i,ikj;1}(\underline{A}_H)_{ikj;1,j}\mu(\xi_j) \\ &= (\underline{A}_S^2)_{ij}(\mathbf{e}_S)_j.\end{aligned}$$

Again, continuing in this manner we have $(\underline{A}_H^n)_{ij}(\mathbf{e}_H)_j \leq (\underline{A}_S^n)_{ij}(\mathbf{e}_S)_j$ for $i \in M \setminus \mathscr{I}$, $j \in \mathscr{I}$, and $n \geq 1$. As $\mathbf{1}_H \underline{A}_H^0 \mathbf{e}_H = \mu H = \mathbf{1}_S \underline{A}_S^0 \mathbf{e}_S$ then

$$\mathbf{1}_H \underline{A}_H^n \mathbf{e}_H = \sum_{i \in M} \sum_{j \in \mathscr{I}} (\underline{A}_H^n)_{ij}(\mathbf{e}_H)_j \leq \sum_{i \in M} \sum_{j \in M_S} (\underline{A}_S^n)_{ij}(\mathbf{e}_S)_j = \mathbf{1}_S \underline{A}_S^n \mathbf{e}_S$$

for $n \geq 0$. Hence, $\underline{X}^n(f_H) \leq \underline{X}^n(Rf_S)$.

By using the same argument with the matrix $\overline{A}_S$ we obtain the inequality $\underline{X}^n(Rf_S) \leq \underline{X}^n(f_H)$. The second set of inequalities in Theorem 7.3 then follow, which completes the proof. □

Example 7.5. We again consider the open system $g_H : I \to I$ from Example 7.4 where it was shown that $S = \{v_1, v_3, v_4\} \in \Gamma_H$. To compute the upper and lower transition matrices of Rg_H note that the system's admissible sequences are given by $\Omega_S = \{23, 24, 33, 34, 424, 423, 41\}$ implying

$$M_S = \{1, 2, 3, 4, 424; 1, 423; 1\}. \tag{7.11}$$

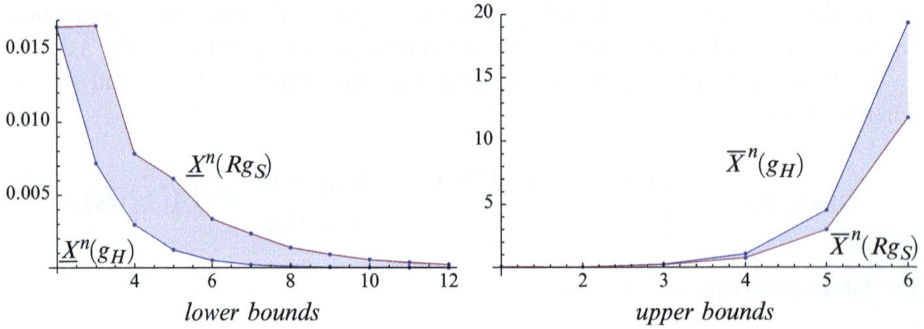

Fig. 7.6 Comparison between $\underline{X}^n(g_H)$ and $\underline{X}^n(Rg_H)$ (*left*) and $\overline{X}^n(g_H)$ and $\overline{X}^n(Rg_H)$ (*right*) from Examples 7.3 and 7.5

From Ω_S we compute that $\xi_{424} = (3/4, .85]$ and $\xi_{423} = (.85, .94]$. The other partition elements have been computed in Example 7.3. Using the order given in (7.11) we obtain

$$\underline{A}_S = \begin{bmatrix} 0 & 0 & 0 & 0 & 0 & 0 \\ 0 & 0 & .29 & 2/11 & 0 & 0 \\ 0 & 0 & .29 & 2/11 & 0 & 0 \\ 2/11 & 0 & 0 & 0 & 1 & 1 \\ 0 & 0 & 0 & .26 & 0 & 0 \\ 0 & .26 & 0 & 0 & 0 & 0 \end{bmatrix}, \quad \overline{A}_S = \begin{bmatrix} 0 & 0 & 0 & 0 & 0 & 0 \\ 0 & 0 & 4 & .29 & 0 & 0 \\ 0 & 0 & 4 & .29 & 0 & 0 \\ .29 & 0 & 0 & 0 & 1 & 1 \\ 0 & 0 & 0 & .73 & 0 & 0 \\ 0 & 1.18 & 0 & 0 & 0 & 0 \end{bmatrix}.$$

The vectors $\mathbf{1}_S$ and $\mathbf{e}_S$ are given by

$$\mathbf{1}_S = [1, 1, 1, 1, 0, 0] \text{ and } \mathbf{e}_S = [1/4, 0, 0, 0, 0, 0]^T.$$

Figure 7.6 shows that $\underline{X}^n(g_H) < \underline{X}^n(Rg_S)$ and $\overline{X}^n(Rg_S) < \overline{X}^n(g_H)$ for a number of $n-$values and indicates the extent to which using the delayed first return map Rg_H improves our estimates of $\mu(X^n(g_H))$. The shaded regions in these graphs represent the difference between these upper and lower estimates, respectively.

7.6 Conclusion

Our results demonstrate that the theory of isospectral transformations can be effectively applied to open dynamical systems to obtain sharper estimates of a system's survival probabilities. Previous applications of this theory have allowed for improvements in the classical estimates of matrix spectra and in obtaining stronger sufficient conditions for the global stability of dynamical networks. We do not doubt that this theory can be applied to other problems as well.

Concerning the results of the present paper it is easy to see that the improved estimates we obtained can also be found for a much broader class of open systems via the same technique. For instance, one could extend these techniques to higher-dimensional systems.

References

1. Afraimovich, V.S., Bunimovich, L.A.: Which hole is leaking the most: A topological approach to study open systems. Nonlinearity **23**, 644–657 (2010)
2. Afraimovich, V., Bunimovich, L.A.: Dynamical networks: Interplay of topology, interactions, and local dynamics. Nonlinearity **20**, 1761–1771 (2007)
3. Bakhtin, Y., Bunimovich, L.A.: The optimal sink and the best source in a Markov chain. J. Stat. Phys. **143**, 943–954 (2011)
4. Blank, M.L., Bunimovich, L.A.: Long range action in networks of chaotic elements. Nonlinearity **19**, 330–345 (2006)
5. Bunimovich, L.A.: Fair dice-like hyperbolic systems. Contemp. Math. **467**, 79–87 (2012)
6. Bunimovich, L.A., Webb, B.Z.: Isospectral graph transformations, spectral equivalence, and global stability of dynamical networks. Nonlinearity **25**, 211–254 (2012)
7. Bunimovich, L.A., Webb, B.Z.: Dynamical networks, isospectral graph reductions, and improved estimates of Matrices' spectra. Linear Algebra Appl. **437**(7), 1429–1457 (2012)
8. Bunimovich, L.A., Yurchenko, A.: Where to place a hole to achieve the fastest escape rate. Israel J. Math. **182**, 229–252 (2011)
9. Demers, M., Wright, P.: Behavior of the escape rate function in hyperbolic dynamical systems. Nonlinearity **25**, 2133–2150 (2012)
10. Varga, R.: Gershgorin and His Circles. Springer, Berlin (2004)

Concerning the results of the present paper, it is easy to see that the improved estimates we obtained can also be found for a much broader class of open systems via the same technique. For instance, one could extend these techniques to higher dimensional systems.

References

1. Afraimovich, V., Bunimovich, L.: [illegible] Nonlinearity 23, 643–[illegible] (2010)
[illegible]

Chapter 8
Dispersing Billiards with Small Holes

Mark F. Demers

Abstract We study several classes of dispersing billiards with holes, including both finite and infinite horizon Lorentz gases and tables with corner points. We allow holes in the form of arcs in the boundary and open sets in the interior of the table as well as generalized holes in which escape may depend on the angle of collision as well as the position. For a large class of initial distributions (including Lebesgue measure and the smooth invariant (SRB) measure for the billiard before the introduction of the hole), we prove the existence of a common escape rate and a limiting conditionally invariant distribution. The limiting distribution converges to the SRB measure for the billiard as the hole tends to zero. Finally, we are able to characterize the common escape rate via pressure on the survivor set.

8.1 Introduction

The study of deterministic dynamical systems with holes was introduced by Pianigiani and Yorke [32] who posed the following intuitive problem. Consider a point particle on a billiard table with chaotic dynamics. If a small hole is made in the table, what are the statistical properties of the trajectories of this system? At what rate does mass escape from the system with respect to a given reference measure? Given an initial distribution μ_0 and letting μ_n denote the normalized distribution at time n (assuming the particle has not escaped by time n), does μ_n converge to a limiting distribution μ? Such a limiting measure may constitute a *conditionally invariant measure* for the open system.

These initial questions in turn motivate many others. For example, does the conditionally invariant measure converge to an invariant measure for the closed

---

M.F. Demers (✉)
Fairfield University, 1073 N. Benson Road, Fairfield, CT 06824, USA
e-mail: mdemers@fairfield.edu

W. Bahsoun et al. (eds.), *Ergodic Theory, Open Dynamics, and Coherent Structures*, Springer Proceedings in Mathematics & Statistics 70, DOI 10.1007/978-1-4939-0419-8_8,
© Springer Science+Business Media New York 2014

system as the size of the hole tends to zero? Such a result views the open system as a perturbation of the closed system and examines the stability of invariant measures under this type of perturbation. Alternatively, one can consider the *survivor set*, the set of points which never escapes through the hole, and ask if there is a notion of pressure on the survivor set which characterizes the escape rate with respect to a certain class of initial distributions.

Such questions have been studied and affirmative or partial answers obtained for a wide variety of dynamical systems. The first results are for uniformly hyperbolic systems admitting a finite Markov partition: Expanding maps on $\mathbb{R}^n$ [16, 17, 32], Smale horseshoes [6, 7], billiards with convex scatterers satisfying a non-eclipsing condition [31, 33], and Anosov diffeomorphisms [9, 10, 14, 15]. Subsequent studies include piecewise expanding maps of the interval [2, 12, 18, 30] and certain unimodal maps [3, 19, 27]. Recently, such questions were answered in the context of the finite horizon periodic Lorentz gas [24, 25].

The purpose of this paper is to answer the questions posed above in the context of two-dimensional dispersing billiards. Our results apply to both finite and infinite horizon periodic Lorentz gases as well as billiard tables with corner points. In order to answer these questions, we will study the transfer operator associated with the billiard systems in question and use the recent Banach spaces constructed in [22, 23] on which these operators are known to admit a spectral gap. We show that the spectral gap persists for maps with small holes satisfying mild conditions. This unified approach allows us to study a wide variety of holes, including escape through partially absorbing boundaries, i.e., boundaries which may allow particles to escape from any position, but depending on the angle of collision. In addition to the new results in the case of the infinite horizon Lorentz gas and billiards with corner points, the current technique also strengthens results for the finite horizon case obtained in [24]. For example, left open in [24] is whether the escape rates with respect to Lebesgue measure and the smooth invariant measure (the SRB measure) for the closed system are equal. Additionally, does the push-forward of Lebesgue measure (renormalized to condition on non-escape) in the open system converge to a conditionally invariant measure? We answer both questions in the affirmative here.

In order to study the notion of pressure on the survivor set, we invoke another useful tool in the study of dynamical systems known as the Young tower, which is a type of Markov extension for the system. Young towers were introduced in [34] and constructed for piecewise hyperbolic systems (including billiards with corner points) under general conditions in [8]. Young towers have also been used to study open systems [3, 18, 24, 25], although in such cases, the towers must be constructed after the introduction of the hole due to the extra cutting created by the boundary of the hole. Young towers were used to prove a variational principle for the finite horizon Lorentz gas in [25]. We build on this work here to extend such results to more general dispersing billiards with holes and to complete the answers to the questions posed above.

8.1.1 Preliminaries

In this section, we recall some basic definitions that we shall use throughout this paper. Given a map $T : M \circlearrowleft$ and a hole $H \subset M$, we define $\mathring{M} = M \setminus H$ and $\mathring{M}^n = \cap_{i=0}^n T^{-i}(M \setminus H)$ to be the set of points in M that have not escaped by time n, $n \geq 1$.

We let $\mathring{T}^n = T^n|\mathring{M}^n$, for $n \geq 1$, denote the map with holes which loses track of points once they enter H. In this sense, we will refer to $\mathring{T}^n$ as the iterates of $\mathring{T}$ despite the fact that the domain of $\mathring{T}$ is not invariant.

8.1.1.1 Rates of Escape

Let μ be a measure on M (not necessarily invariant with respect to T). We define the *exponential rate of escape* with respect to μ to be $-\rho(\mu)$ where

$$\rho(\mu) = \lim_{n\to\infty} \frac{1}{n} \log \mu(\mathring{M}^n), \tag{8.1}$$

when the limit exists.

8.1.1.2 Conditionally Invariant Measures

Given a Borel probability measure μ on M, define $\mathring{T}_*\mu(A) = \mu(T^{-1}A \cap \mathring{M}^1)$ for any Borel $A \subset M$. We say μ is conditionally invariant for $\mathring{T}$ if

$$\frac{\mathring{T}_*\mu(A)}{\mathring{T}_*\mu(M)} = \mu(A) \quad \text{for all Borel } A \subset M. \tag{8.2}$$

The normalizing constant $\lambda = \mu(\mathring{M}^1)$ is often referred to as the eigenvalue of μ since iterating the above equation yields $\mathring{T}_*^n\mu(A) = \lambda^n\mu(A)$ for $n \in \mathbb{N}$. In particular, $\mu(\mathring{M}^n) = \mathring{T}_*^n\mu(M) = \lambda^n$ so that $-\log\lambda$ is the escape rate with respect to μ according to (8.1).

We remark that infinitely many conditionally invariant measures have been shown to exist under quite general conditions for any $0 \leq \lambda < 1$ [21] so that we are not interested in existence results for such measures. Rather, we will be interested in conditionally invariant measures with physical properties: measures that can be realized as the (renormalized) limit of Lebesgue measure or other physically relevant initial distributions. Such conditionally invariant measures will also describe a common rate of escape with respect to a large class of reference measures.

8.1.1.3 Pressure on the Survivor Set

The survivor set[1] $\mathring{M}^\infty := \cap_{i=-\infty}^{\infty} T^i(M \setminus H)$ is a $\mathring{T}$-invariant (and also T-invariant) set which supports all the invariant measures that persist after the introduction of the hole. We define the *pressure* on $\mathring{M}^\infty$ with respect to a class of invariant measures $\mathscr{C}$ to be

$$\mathscr{P}_{\mathscr{C}} = \sup_{\nu \in \mathscr{C}} P_\nu \quad \text{where} \quad P_\nu = h_\nu(T) - \int \chi^+(T)\, d\nu.$$

Here $h_\nu(T)$ denotes the Kolmogorov-Sinai entropy of T with respect to ν and $\chi^+(T)$ represents the sum of positive Lyapunov exponents, counted with multiplicity.

We say the open system satisfies a *variational principal* if $\rho(\mu) = \mathscr{P}_{\mathscr{C}}$ for some physically relevant reference measure μ and a class of invariant measures $\mathscr{C}$. If there is an invariant measure $\nu \in \mathscr{C}$ such that $\rho(\mu) = P_\nu$, we say that ν satisfies an *escape rate formula*.

Escape rate formulas have been proved for many of the uniformly hyperbolic systems described in the introduction; for such systems, variational principles are often formulated in terms of the associated symbolic dynamics. The recent reference [25] contains variational principles and inequalities for nonuniformly hyperbolic systems without appealing to symbolic dynamics.

8.2 Setting and Results

In this section, we describe the classes of billiards that we study in this paper and formulate precise conditions on the holes we introduce. We include a variety of examples of holes that meet these conditions.

8.2.1 Classes of Dispersing Billiards

We identify a domain $Q \subset \mathbb{R}^2$ or $\mathbb{T}^2$ (the 2-torus with Euclidean metric) as the billiard table and assume that ∂Q has d connected components, $\Gamma_1, \ldots, \Gamma_d$, each of which comprises a finite number of $\mathscr{C}^3$ smooth, compact arcs. The dynamics of the billiard flow on Q is induced by a particle traveling at unit speed and undergoing elastic collisions at the boundary. The phase space for the billiard flow is $\mathscr{M} = Q \times \mathbb{S}^1/\sim$ with the conventional identifications at collisions (see for example [11, Sect. 2.5]).

[1]We give the definition for invertible T. When T is not invertible, define $\mathring{M}^\infty = \cap_{i=0}^{\infty} T^{-i}(M \setminus H)$.

Define $M = \cup_{i=1}^{d}(\Gamma_i \times [-\pi/2, \pi/2])$ to be a union of cylinders. The billiard map $T : M \to M$ is the Poincaré map corresponding to collisions with the scatterers. We will denote coordinates on M by (r, φ), where $r \in \Gamma_i$ is parametrized by arclength (oriented positively in the usual sense) and φ is the angle that the (post-collision) velocity vector at r makes with the normal pointing into the domain Q. We shall denote normalized Lebesgue measure on M by m throughout. It is a standard fact that T preserves the smooth measure μ_{SRB} defined by $d\mu_{\text{SRB}} = \frac{\pi}{2} \cos\varphi \, dm$ [11, Sect. 2.12].

We include two types of dispersing billiard tables Q in the results of this paper.

8.2.1.1 Periodic Lorentz Gas

Let $\{B_i\}_{i=1}^{d}$ define a finite number of open convex regions in $\mathbb{T}^2$ such that each ∂B_i is $\mathscr{C}^3$ with strictly positive curvature. Then in this case, $Q = \mathbb{T}^2 \setminus (\cup_i B_i)$ and $\Gamma_i = \partial B_i$ for $i = 1, \ldots, d$. We refer to the billiard flow on such a domain as a periodic Lorentz gas.

For each $x = (r, \varphi) \in M$, define $\tau(x)$ to be the time of the first (nontangential) collision of the trajectory starting at x under the billiard flow. The billiard table Q is said to have finite horizon if τ is uniformly bounded above on M. Otherwise, Q is said to have infinite horizon. In this paper, we allow our Lorentz gas to have either finite or infinite horizon.

8.2.1.2 Billiards with Corner Points

Let $Q' \subset \mathbb{R}^2$ be a compact region whose boundary consists of finitely many $\mathscr{C}^3$ curves that are positioned convex inward to Q' with strictly positive curvature. In the interior of Q', we may also define a finite number of open convex regions $\{B_i\}_{i=1}^{d'}$ such that ∂B_i is also $\mathscr{C}^3$ with strictly positive curvature. These obstacles $\{B_i\}_{i=1}^{d'}$ are either pairwise connected or disjoint. We define the billiard table Q to be the compact region $Q := Q' \setminus (\cup_i B_i)$. We assume Q has a connected interior. Since the obstacles B_i may or may not overlap, the boundary of Q comprises a finite number $d \leq d' + 1$ of connected components, Γ_i. Each Γ_i consists of a finite number of smooth curves as described above; the endpoints of these smooth curves are called *corner points*. For such billiard tables, the horizon is always finite.

We make two additional assumptions in the case of billiards with corner points:

(C1) The intersections of the smooth curves comprising ∂Q are transverse, i.e., the angle at each corner point is positive.[2]

(C2) The hyperbolicity of the billiard map dominates the complexity in the sense of equation (8.10) of Sect. 8.3.4.

---

[2]In the presence of cusps (corner points whose angle is zero), it was proved in [CM07, CZ08] that the billiard map has polynomial decay of correlations and so in general will have polynomial rates of escape. Thus the present methods will not apply.

Assumption (C2) says that the number of singularity curves for T^{-n} which intersect at a single point cannot grow too quickly compared to the expansion of stable curves. Both assumptions are standard for billiards with corner points (see [4,5,8,23]).

8.2.1.3 Hyperbolicity and Singularities of Dispersing Billiards

Set $\mathcal{S}_0 = \{(r,\varphi) \in M : \varphi = \pm\pi/2\}$. Let $r_1, \ldots r_k$ denote the arclength coordinates of the corner points in ∂Q and set $P_0 = \{(r,\varphi) \in M : r = r_i, i = 1, \ldots k\}$ (for the periodic Lorentz gas, $P_0 = \emptyset$). The sets $\mathcal{S}_{\pm n} = \cup_{i=0}^{n} T^{\mp i}(\mathcal{S}_0 \cup P_0)$ are the singularity sets for $T^{\pm n}$, $n \geq 1$. The sets $\mathcal{S}_n$ comprise finitely many smooth compact curves in the finite horizon case and countably many in the infinite horizon case. Moreover, the curves $\mathcal{S}_n \setminus \mathcal{S}_0$ are decreasing for $n > 0$ and increasing for $n < 0$.

To control distortion near $\mathcal{S}_0$, we define the usual homogeneity strips following [4,5],

$$\mathbb{H}_k = \{(r,\varphi) \in M : \pi/2 - 1/k^2 \leq \varphi \leq \pi/2 - 1/(k+1)^2\},$$

for $k \geq k_0$ to be determined later. We define $\mathbb{H}_{-k}$ near $\varphi = -\pi/2$ similarly. By $\mathbb{H}_0$ we denote the complementary set $M \setminus (\cup_{|k| \geq k_0} \mathbb{H}_k)$.

The assumption of strict convexity of ∂Q for both classes of dispersing billiards above guarantees the hyperbolicity of T in the following sense: There exist invariant families of cones $C^s(x)$ (stable) and $C^u(x)$ (unstable), continuous on $M \setminus (\mathcal{S}_0 \cup P_0)$ and satisfying $DT^{-1}(x)C^s(x) \subset C^s(T^{-1}x)$ and $DT(x)C^u(x) \subset C^u(Tx)$ wherever T and T^{-1} are defined. Indeed, for all classes of billiards we consider here, the angle between $C^u(x)$ and $C^s(x)$ is uniformly bounded away from zero on M.

In terms of the global (r,φ) coordinates, the slopes of vectors in $C^s(x)$ are always negative, while those in $C^u(x)$ are always positive. Let $\mathcal{K}_{\min} > 0$ denote the minimum curvature of boundary curves in ∂Q. Although vectors in either cone can become arbitrarily close to vertical near corner points (but never at the same time), in all cases $d\varphi/dr \geq \mathcal{K}_{\min}$ for any $(dr, d\varphi) \in C^u(x)$ and $d\varphi/dr \leq -\mathcal{K}_{\min}$ for all $(dr, d\varphi) \in C^s(x)$ so that both cones are uniformly bounded away from the horizontal. We refer the interested reader to [11] for the Lorentz gas and to [8] for the case of billiards with corner points. See also [23, Sect. 6.1] for an explicit calculation of the upper boundaries of the cones during a corner series.

We say a curve $W \subset M$ is a *homogeneous stable curve* if W is contained in one homogeneity strip and the tangent vector to W at x belongs to $C^s(x)$ at every point x in W. Homogeneous unstable curves are defined similarly. It is a well-established fact that for all classes of billiards we consider, there exist families of stable and unstable curves $\mathcal{W}^s$ and $\mathcal{W}^u$ which are invariant in the following sense: For any $W \in \mathcal{W}^s$, $T^{-1}W$ consists of at most countably many connected components, each of which belong to $\mathcal{W}^s$. Indeed, we may choose $\mathcal{W}^s$ to consist of all homogeneous stable curves whose curvature is bounded above by a uniform constant $B_c > 0$ and whose length is no greater than some $\delta_0 > 0$ whose value we fix in Sect. 8.3.5, Eq. (8.11). $\mathcal{W}^u$ is similarly invariant under T after choosing B_c sufficiently large ([8, 11]).

8.2.2 *Admissible Holes*

We denote by $N_\varepsilon(A)$ the ε-neighborhood of a set A in M. In formulating our conditions below, we consider the maximum length scale δ_0 for stable curves, defined by (8.11), to be fixed before the introduction of the hole.

A hole $H \subset M$ is an open set with finitely many connected components whose boundary consists of finitely many compact smooth arcs. In addition, we require that:

(H1) *(Complexity)* There exists $B_0 > 0$ such that any stable curve of length less than δ_0 can be cut into at most B_0 pieces by ∂H;
(H2) *(Weak transversality)* There exists $C_0 > 0$ such that for any stable curve W, $m_W(N_\varepsilon(\partial H) \cap W) \le C_0 \varepsilon^{1/2}$ for all $\varepsilon > 0$ sufficiently small[3], where m_W denotes arclength measure on W.

Note that if ∂H is uniformly transverse to the stable cone, then (H2) is trivially satisfied. The weaker form of transversality defined above admits the "square-root type" tangencies that appear between singularity curves and stable and unstable curves in billiards with corner points.

Although formally we consider holes in M satisfying the above assumptions, we are especially interested in holes which are actually made in the configuration space Q and in turn induce holes in M that satisfy (H1) and (H2). Such holes are more physically relevant from the point of view of the billiard flow. Below we list several examples of holes in Q which induce holes in M satisfying (H1) and (H2). We follow [24] in our exposition of the first two types of holes.

8.2.2.1 I. Openings in the Boundary of Q

Let ω denote an open arc in one of the Γ_i comprising ∂Q. If we consider ω as an absorbing boundary, then ω induces a hole H_ω in M which is simply a vertical rectangle $H_\omega = (a, b) \times [-\pi/2, \pi/2]$, where (a, b) is the arclength interval corresponding to ω. Such holes are called holes of Type I in [24].

---

[3]According to [23, Sect. 2.1], we could take $m_W(N_\varepsilon(\partial H) \cap W) \le C_0 \varepsilon^{t_0}$ for any exponent $t_0 > 0$. We choose $\varepsilon_0 = 1/2$ here to simplify the exposition and choice of constants in the norms and because this already contains an interesting class of examples (see also (A3)(2) of Sect. 8.3.3).

It is clear that (H1) is satisfied with $B_0 = 3$. If ω is a positive distance from a corner point, (H2) is also satisfied since stable cones have strictly negative slopes bounded away from the vertical direction outside any neighborhood of the corner points so that ∂H is uniformly transverse to $C^s(x)$. If ω contains a corner point, (H2) is also satisfied due to the square-root type tangencies that appear between stable curves and the vertical lines in P_0 at corner points [8, Sect. 9].

8.2.2.2 II. Holes in the Interior of Q

Let ω be an open connected set in the interior of Q. ω does not immediately correspond to a subset of M since it is a positive distance from ∂Q. Thus we have a choice in declaring what we consider the induced hole H_ω to be either the backward shadow of the hole, i.e., the set of all points in $(r, \varphi) \in M$ whose forward trajectories under the flow enter ω before they reach ∂Q again, or the forward shadow of the hole, i.e., the set of all points $(r, \varphi) \in M$ whose backward trajectories under the flow enter ω before reaching ∂Q. We choose to define H_ω as the latter, the forward shadow of the hole. Thus those trajectories that are about to enter the hole in forward time are still considered in the system, while those which would have passed through the hole on their way to their current collision are considered out. Such holes are called holes of Type II in [24].

We remark that with this definition it does not matter whether ω is convex or not. The induced hole H_ω is always the forward shadow of the convex hull of ω. We thus adopt the convention that ω of this type are always convex. Also, a single ω of this type induces $H_\omega \subset M$ with multiple connected components since each Γ_i with a line of sight to ω contains a component of H_ω.

One can easily obtain geometric properties of ∂H_ω by considering ω as a convex scatterer in Q. Then ∂H_ω will correspond to the forward images of the trajectories that meet $\partial\omega$ tangentially, i.e., the forward image of the set $\partial\omega \times [-\pi/2, \pi/2]$ if ω were considered to be a scatterer. Other components of ∂H_ω may comprise curves in $\mathcal{S}_0$ and $T(\mathcal{S}_0)$. Curves in $T(\mathcal{S}_0)$ and in the forward image of $\partial\omega \times [-\pi/2, \pi/2]$ have positive slopes, while curves in $\mathcal{S}_0$ are horizontal. All such curves are uniformly transverse to the stable cone which has vectors with strictly negative slopes bounded away from 0; thus (H2) is satisfied. See Fig. 8.1 for an example of the geometry of H_ω induced by a hole of Type II. If ω does not lie in an infinite horizon corridor, due to the uniform transversality and the fact that the connected components of H_ω are a positive distance apart, (H1) is also satisfied and one has $B_0 = 3$ if δ_0 is sufficiently small. In any case, $B_0 \leq 2k + 1$ where k is the maximal number of connected components of H_ω on a single scatterer.

8.2.2.3 III. Corner and Side Pockets

We can combine holes of the first two types described above by placing an open hole ω which intersects ∂Q. Indeed, we even allow ω to contain a corner point.

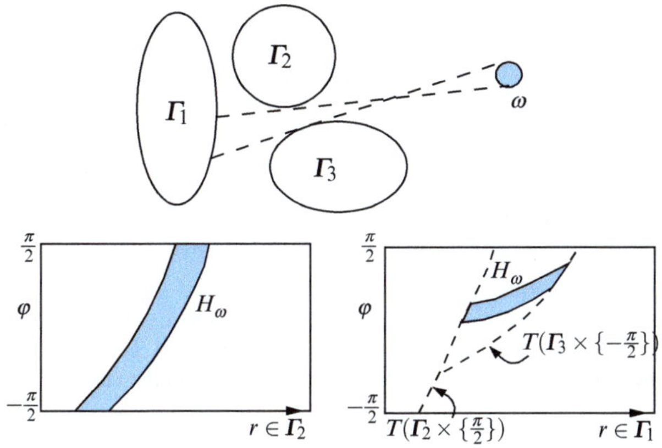

Fig. 8.1 *Top:* A hole ω of Type II on the billiard table is shown with a line of sight between two scatterers. *Bottom left:* The induced hole H_ω in the component of the phase space M corresponding to Γ_2. *Bottom right:* The induced hole H_ω in the component of the phase space M corresponding to Γ_1. Notice that ∂H_ω is comprised of three types of curves: curves corresponding to $\partial\omega \times \{\pm\frac{\pi}{2}\}$ if ω is considered as a scatterer, curves in $\mathscr{S}_0$, and singularity curves in $T(\mathscr{S}_0)$. In all cases, ∂H_ω is uniformly transverse to the stable cone

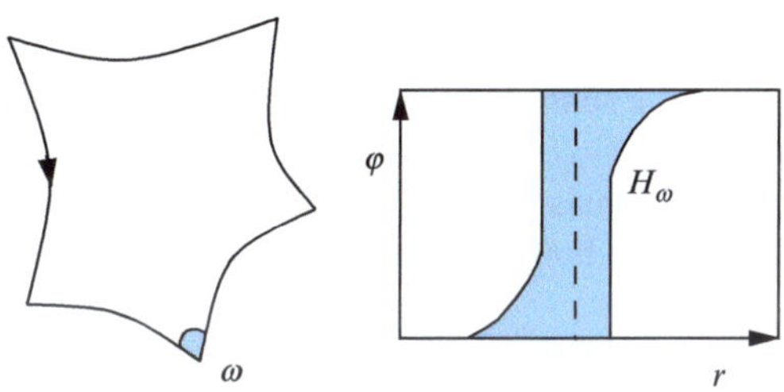

Fig. 8.2 *Left:* A hole ω positioned as a corner pocket in a dispersing billiard table. *Right:* The induced hole H_ω in the phase space M. The *dotted line* corresponds to the component of P_0 created by the corner point. The portion of ∂H_ω formed by the forward image of $\partial\omega \times \{\pm\frac{\pi}{2}\}$ has positive slopes, similar to a hole of Type II. Such boundary curves as well as components of ∂H_ω formed by $\mathscr{S}_0$ are uniformly transverse to the stable cone. The vertical segments formed by the rectangle above $\omega \cap \partial Q$ are not uniformly transverse to the stable cone as the diameter of ω shrinks; however, stable curves do satisfy a weaker "square-root type" tangency with components of P_0 so that (H2) is satisfied

Now $\omega \cap \partial Q$ corresponds to a vertical rectangle in M as described for Type I holes above, while taking the forward image of $\partial\omega \setminus \partial Q$ induces boundaries which are curves with positive slopes in M. These two sets overlap, forming H_ω. See Fig. 8.2 for an example of a hole in the form of a corner pocket.

(H1) is clearly satisfied with $B_0 = 3$ again. Near corner points, the stable cones are not necessarily bounded away from the vertical and so are not uniformly transverse to ∂H_ω. However, they do satisfy the square-root type tangency condition required by (H2) (see [8, Sect. 9]).

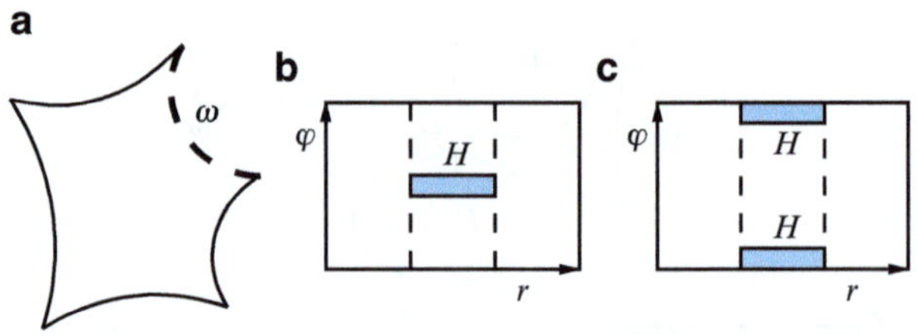

Fig. 8.3 (**a**) A partially absorbing hole ω (*dashed curve*) in a dispersing billiard table. (**b**) The induced hole H in M if reflections close to normal are absorbed. (**c**) The two components of the induced hole H in M if reflections close to tangential are absorbed. The *dotted vertical lines* in (**b**) and (**c**) refer to singularity lines corresponding to corner points in the table

8.2.2.4 IV. Generalized Holes and Partially Absorbing Walls

A generalized hole is a subset ω of configuration space through which trajectories escape if some additional condition is met. As a simple example, we consider ω to be an arc along the boundary and allow trajectories to be absorbed depending on the angle of collision. In other words, if (a,b) denotes the arclength interval corresponding to ω, we choose $\phi_1, \phi_2 \in [-\pi/2, \pi/2]$ and declare $H_\omega = (a,b) \times (\phi_1, \phi_2)$. We may even allow ω to be the entire boundary ∂Q. For large ω, the smallness of the hole, in the sense we shall make precise later, can be guaranteed by choosing $|\phi_1 - \phi_2|$ small.

In any case, since ∂H_ω consists of vertical and horizontal line segments, it is easy to see that (H1) and (H2) are both satisfied. See Fig. 8.3 for an example of partially absorbing walls.

There are many more interesting examples of generalized holes one can construct in a similar manner. For example, one can allow the restriction on the angle to vary depending on the position as long as the resulting holes in M satisfy (H1) and (H2). We leave the generation of such additional examples to the interested reader.

8.2.3 *Transfer Operator*

If ψ is a smooth test function, then $\psi \circ T$ is only piecewise smooth due to the singularities of T. For this reason, we introduce scales of spaces, defined using the invariant family of homogeneous stable curves $\mathcal{W}^s$ introduced in Sect. 8.2.1.3 (and also described in property (A4) in Sect. 8.3.3) on which to describe the action of the *transfer operator* $\mathcal{L} = \mathcal{L}_T$ associated with T.

Define $T^{-n}\mathcal{W}^s$ to be the set of homogeneous stable curves W such that T^n is smooth on W and $T^i W \in \mathcal{W}^s$ for $0 \le i \le n$. Then $T^{-n}\mathcal{W}^s \subset \mathcal{W}^s$ and it follows from the definition of $\mathcal{W}^s$ that the connected components of $T^{-n}W$ belong to $\mathcal{W}^s$ whenever $W \in \mathcal{W}^s$ (up to subdividing long pieces).

For $W \in T^{-n}\mathscr{W}^s$, a complex-valued test function $\psi : M \to \mathbb{C}$ and $0 < p \leq 1$ define $H^p_W(\psi)$ to be the Hölder constant of ψ on W with exponent p measured in the metric d_W. Define $H^p_n(\psi) = \sup_{W \in T^{-n}\mathscr{W}^s} H^p_W(\psi)$ and let $\tilde{\mathscr{C}}^p(T^{-n}\mathscr{W}^s) = \{\psi : M \to \mathbb{C} \mid |\psi|_\infty + H^p_n(\psi) < \infty\}$, denote the set of bounded complex-valued functions which are Hölder continuous on elements of $T^{-n}\mathscr{W}^s$. The set $\tilde{\mathscr{C}}^p(T^{-n}\mathscr{W}^s)$ equipped with the norm $|\psi|_{\mathscr{C}^p(T^{-n}\mathscr{W}^s)} = |\psi|_\infty + H^p_n(\psi)$ is a Banach space. We define $\mathscr{C}^p(T^{-n}\mathscr{W}^s)$ to be the closure of $\tilde{\mathscr{C}}^1(T^{-n}\mathscr{W}^s)$ in $\tilde{\mathscr{C}}^p(T^{-n}\mathscr{W}^s, \mathbb{C})$.[4]

It follows from the hyperbolicity of T (see (A2) from Sect. 8.3.3) that

$$H^p_{W_i}(\psi \circ T) \leq C H^p_W(\psi),$$

for each connected component $W_i \subset T^{-1}W$ and a uniform constant C depending only on T. Thus if $\psi \in \tilde{\mathscr{C}}^p(T^{-(n-1)}\mathscr{W}^s)$, then $\psi \circ T \in \tilde{\mathscr{C}}^p(T^{-n}\mathscr{W}^s)$. Similarly, if $\zeta \in \tilde{\mathscr{C}}^1(T^{-(n-1)}\mathscr{W}^s)$, then $\zeta \circ T \in \tilde{\mathscr{C}}^1(T^{-n}\mathscr{W}^s)$. These two facts together imply that for $p < 1$, if $\psi \in \mathscr{C}^p(T^{-(n-1)}\mathscr{W}^s)$, then $\psi \circ T \in \mathscr{C}^p(T^{-n}\mathscr{W}^s)$.

If $f \in (\mathscr{C}^p(T^{-n}\mathscr{W}^s))'$ is an element of the dual of $\mathscr{C}^p(T^{-n}\mathscr{W}^s)$, then $\mathscr{L} : (\mathscr{C}^p(T^{-n}\mathscr{W}^s))' \to (\mathscr{C}^p(T^{-(n-1)}\mathscr{W}^s))'$ acts on f by

$$\mathscr{L} f(\psi) = f(\psi \circ T) \quad \forall \psi \in \mathscr{C}^p(T^{-(n-1)}\mathscr{W}^s).$$

Recall that m denotes (normalized) Lebesgue measure on M. If $f \in L^1(m)$, then f is canonically identified with a signed measure absolutely continuous with respect to Lebesgue, which we shall also call f, i.e.,

$$f(\psi) = \int_M \psi f \, dm.$$

With the above identification, we write $L^1(M, m) \subset (\mathscr{C}^p(T^{-n}\mathscr{W}^s))'$ for each $n \in \mathbb{N}$. Then restricted to $L^1(M, m)$, $\mathscr{L}$ acts according to the familiar expression

$$\mathscr{L}^n f = f \circ T^{-n} \, |\det DT^n(T^{-n})|^{-1} \quad \text{for any } n \geq 0 \text{ and any } f \in L^1(M, m).$$

When we wish to be explicit about the dependence of $\mathscr{L}$ on a map T, we will use the notation $\mathscr{L}_T$.

When we introduce a hole $H \subset M$, the transfer operator corresponding to $\mathring{T}$, which we shall denote by $\mathring{\mathscr{L}}$, is defined by the equivalent expressions,

$$\mathring{\mathscr{L}} = \boldsymbol{I}_{\mathring{M}} \mathscr{L} \boldsymbol{I}_{\mathring{M}} = \mathscr{L} \boldsymbol{I}_{\mathring{M}^1}, \tag{8.3}$$

[4] Here by $\tilde{\mathscr{C}}^1(\mathscr{W}^s)$, we mean to indicate $\tilde{\mathscr{C}}^p(\mathscr{W}^s)$ with $p = 1$, i.e., functions which are Lipschitz on elements of $\mathscr{W}^s$.

where $\boldsymbol{I}_A$ indicates the indicator function for the set A. Thus for any test function $\psi \in \mathscr{C}^p(\mathscr{W}^s)$ and $f \in (\mathscr{C}^p(T^{-1}\mathscr{W}^s))'$, we have

$$\mathring{\mathscr{L}} f(\psi) = \boldsymbol{I}_{\mathring{M}} \mathscr{L}(\boldsymbol{I}_{\mathring{M}} f)(\psi) = \mathscr{L}(\boldsymbol{I}_{\mathring{M}} f)(\psi \boldsymbol{I}_{\mathring{M}}) = f(\psi \circ T \cdot \boldsymbol{I}_{\mathring{M}^1}),$$

since $\boldsymbol{I}_{\mathring{M}} \cdot \boldsymbol{I}_{\mathring{M}} \circ T = \boldsymbol{I}_{\mathring{M}^1}$. Iterating this expression we obtain

$$\mathring{\mathscr{L}}^n f(\psi) = f(\psi \circ T^n \cdot \boldsymbol{I}_{\mathring{M}^n}), \quad \text{for each } n \in \mathbb{N}.$$

When we wish to be explicit about the dependence of $\mathring{\mathscr{L}}$ on a hole H, we will use the notation $\mathring{\mathscr{L}}_H$.

8.2.4 Main Results

In Sect. 8.3.2 we shall define the Banach spaces $(\mathscr{B}, \|\cdot\|_{\mathscr{B}})$ and $(\mathscr{B}_w, |\cdot|_w)$ used in [23]. It was proved there that $\mathscr{L}_T$ is quasi-compact on $(\mathscr{B}, \|\cdot\|_{\mathscr{B}})$ under certain abstract assumptions on the map T. We postpone the definition of these norms and the abstract assumptions on the map (which will be satisfied by all classes of billiards we consider here) and first state our results.

In order to obtain information about the spectrum of $\mathring{\mathscr{L}}$ from the spectrum of $\mathscr{L}$, we will use the perturbative framework of Keller and Liverani [29]. This framework requires two ingredients: (1) uniform Lasota-Yorke inequalities along a sequence of holes and (2) smallness of the perturbation in the following norm:

$$|||\mathscr{L}||| := \{|\mathscr{L} f|_w : \|f\|_{\mathscr{B}} \le 1\}. \tag{8.4}$$

The first two propositions establish these ingredients.

Let $\mathscr{H}(B_0, C_0)$ denote a family of holes in M satisfying (H1) and (H2) with uniform constants B_0 and C_0. Throughout this section, T is assumed to be a billiard map corresponding to either a periodic Lorentz gas or a bounded domain with corner points as described in Sect. 8.2.1.

Proposition 8.1. *Fix $B_0, C_0 > 0$ and let $\mathscr{H}(B_0, C_0)$ denote the corresponding family of holes satisfying (H1) and (H2). Then there exist constants $C > 0$, $\sigma < 1$ depending only on T, B_0, and C_0 such that for all $H \in \mathscr{H}(B_0, C_0)$ and $n \in \mathbb{N}$,*

$$\|\mathring{\mathscr{L}}_H^n f\|_{\mathscr{B}} \le C\sigma^n \|f\|_{\mathscr{B}} + C|f|_w \quad \textit{for all } f \in \mathscr{B}; \tag{8.5}$$

$$\|\mathscr{L}^n f\|_{\mathscr{B}} \le C\sigma^n \|f\|_{\mathscr{B}} + C|f|_w \quad \textit{for all } f \in \mathscr{B}; \tag{8.6}$$

$$|\mathring{\mathscr{L}}_H^n f|_w \le C|f|_w \quad \textit{and} \quad |\mathscr{L}^n f|_w \le C|f|_w \quad \textit{for all } f \in \mathscr{B}_w. \tag{8.7}$$

For a hole $H \subset M$, we define $\mathrm{diam}_s(H) = \sup_{W \in \mathscr{W}^s} |W \cap H|$ and refer to this quantity as the *stable diameter* of H. We define the unstable diameter $\mathrm{diam}_u(H)$ similarly.

Proposition 8.2. *Suppose H is a hole satisfying (H1) and (H2) and let $h = \mathrm{diam}_s(H)$. Then there exists $C > 0$, depending only on C_0, B_0, and T, such that*

$$|||\mathscr{L} - \mathring{\mathscr{L}}_H||| \leq C h^{\alpha-\gamma},$$

where $0 < \gamma < \alpha$ are from the norms, Sect. 8.3.2.

Theorem 8.1. *Fix $B_0, C_0 > 0$. Then for all $H \in \mathscr{H}(B_0, C_0)$ with $\mathrm{diam}_s(H)$ sufficiently small, $\mathring{\mathscr{L}}_H$ has a spectral gap. Its eigenvalue of maximum modulus λ_H is real and its associated eigenvector μ_H is a conditionally invariant measure for $\mathring{T}$ which is singular with respect to Lebesgue measure.*

Moreover, for any probability measure $\mu \in \mathscr{B}$ such that $\lim_{n\to\infty} \lambda_H^{-n} \mathring{\mathscr{L}}_H^n \mu \neq 0$, we have

(i) $\rho(\mu) = \log \lambda_H$;

(ii) $\left\| \dfrac{\mathring{\mathscr{L}}_H^n \mu}{|\mathring{\mathscr{L}}_H^n \mu|} - \mu_H \right\|_{\mathscr{B}} \leq C\sigma_1^n$, *for some $C > 0$, $\sigma_1 < 1$.*

In particular, both Lebesgue measure and the smooth invariant measure μ_{SRB} for T have the same escape rate and converge to μ_H under the normalized action of $\mathring{T}$.

Theorem 8.2. *Let H_ε be a sequence of holes in $\mathscr{H}(B_0, C_0)$ such that $\mathrm{diam}_s(H_\varepsilon) \leq \varepsilon$. Let μ_ε denote the conditionally invariant measures corresponding to λ_ε from Theorem 8.1. Then*

$$\lim_{\varepsilon\to 0} |\mu_\varepsilon - \mu_{SRB}|_w = 0,$$

and $\lambda_\varepsilon \to 1$ as $\varepsilon \to 0$.

We remark that convergence in the weak norm $|\cdot|_w$ implies the weak convergence of measures.

Next we proceed to study the connection between escape rate and pressure on the survivor set, $\mathring{M}^\infty$. Let $\mathscr{M}_{\mathring{T}}$ denote the set of ergodic, $\mathring{T}$-invariant probability measures supported on $\mathring{M}^\infty$. Following [25], we define a class of invariant measures by

$$\mathscr{G}_H = \{\nu \in \mathscr{M}_{\mathring{T}} : \exists C, r > 0 \text{ such that } \forall \varepsilon > 0, \nu(N_\varepsilon(\mathscr{S}_0 \cup P_0 \cup \partial H)) \leq C\varepsilon^r\}. \tag{8.8}$$

If we omit ∂H, the condition on $N_\varepsilon(\mathscr{S}_0 \cup P_0)$ is the same as that used in [28] to ensure the existence of Lyapunov exponents and stable and unstable manifolds for ν-a.e. point. Thus this restriction, or something like it, on the class of invariant measures is necessary for maps with singularities.

Theorem 8.3. *Suppose $H \in \mathcal{H}(B_0, C_0)$ satisfies the assumptions of Theorem 8.1. If* $\mathrm{diam}_u(H)$ *is sufficiently small, then*

$$\rho(m) = \log \lambda_H = \sup_{\nu \in \mathcal{G}_H} \{h_\nu(T) - \chi_\nu^+(T)\}.$$

Moreover, we may define a measure ν_H via the limit,

$$\nu_H(\psi) = \lim_{n \to \infty} \lambda_H^{-n} \mu_H(\psi) \quad \textit{for all } \psi \in \mathcal{C}^0(M),$$

and ν_H is an invariant probability measure for $\mathring{T}$ belonging to $\mathcal{G}_H$ that achieves the supremum in the variational principle above, i.e., $\rho(m) = h_{\nu_H}(T) - \chi_{\nu_H}^+(T)$.

8.3 Analytical Framework

In this section we introduce the necessary definitions and abstract assumptions on the class of maps studied in [23]. We will then show in Sect. 8.4 that an iterate of our map T with the expanded singularity set induced by ∂H satisfies these abstract conditions.

8.3.1 Representation of Admissible Stable Curves

Recall from Sect. 8.2.1.3 that the stable cone $C^s(x)$ is bounded away from the horizontal direction in all cases we consider. Thus, any curve $W \in \mathcal{W}^s$ can be viewed as the graph of a function $r_W(\varphi)$ of the angular coordinate φ with derivative uniformly bounded above. For each homogeneous stable curve W, let I_W denote the φ-interval on which r_W is defined and define $G_W(\varphi) = (r_W(\varphi), \varphi)$ so that $W = \{G_W(\varphi) : \varphi \in I_W\}$.[5]

With this view of stable curves, we may redefine $\mathcal{W}^s$ to be the set of homogeneous stable curves satisfying $|W| \le \delta_0$ and $|\frac{d^2 r_W}{d\varphi^2}| \le B_c$ for some $\delta_0, B_c > 0$. $\mathcal{W}^s$ is invariant under T^{-1} in the sense described in Sect. 8.2.1.3 as long as B_c is chosen sufficiently large [8]. From this point forward, we fix such a choice of B_c once and for all.

The family $\mathcal{W}^u$ of unstable curves has an analogous characterization.

[5] Our treatment of stable curves here differs from that in [23]. In that abstract setting, stable curves are defined via graphs in charts of the given manifold. In the present more concrete setting, we dispense with charts and use the global (r, φ) coordinates.

We define a distance in $\mathcal{W}^s$ as follows. Let $W_i = G_{W_i}(I_i)$, $i = 1, 2$ be two curves in $\mathcal{W}^s$ with defining functions r_{W_i}. We denote by $\ell(I_1 \triangle I_2)$ the length of the symmetric difference of the φ-intervals on which they are defined. Then the distance between W_1 and W_2 is defined as

$$d_{\mathcal{W}^s}(W_1, W_2) = \eta(W_1, W_2) + \ell(I_1 \triangle I_2) + |r_{W_1} - r_{W_2}|_{\mathcal{C}^1(I_1 \cap I_2)},$$

where $\eta(W_1, W_2) = 0$ if W_1 and W_2 lie in the same homogeneity strip in the same component of M and $\eta = \infty$ otherwise.

For two functions $\psi_i \in \mathcal{C}^p(W_i)$, we define the distance between them to be

$$d_q(\psi_1, \psi_2) = |\psi_1 \circ G_{W_1} - \psi_2 \circ G_{W_2}|_{\mathcal{C}^q(I_1 \cap I_2)},$$

where $q < 1$ is from the definition of the strong stable norm in Sect. 8.3.2.

8.3.2 Norms

Given a curve $W \in \mathcal{W}^s$ and $0 \le p \le 1$, we define $\tilde{\mathcal{C}}^p(W)$ to be the set of complex-valued Hölder continuous functions on W with exponent p, with distance measured in the Euclidean metric along W. We set $\mathcal{C}^p(W)$ to be the closure of $\tilde{\mathcal{C}}^1(W)$ in the $\tilde{\mathcal{C}}^p$-norm: $|\psi|_{\mathcal{C}^p(W)} = |\psi|_{\mathcal{C}^0(W)} + H^p_W(\psi)$, where $H^p_W(\psi)$ denotes the Hölder constant of ψ on W as in Sect. 8.2.3. $\tilde{\mathcal{C}}^p(M)$ and $\mathcal{C}^p(M)$ are defined similarly.

For $\alpha, p \ge 0$, define the following norms for test functions

$$|\psi|_{W,\alpha,p} := |W|^\alpha \cdot \cos W \cdot |\psi|_{\mathcal{C}^p(W)},$$

where $\cos W$ denotes the average value of $\cos\varphi$ along W, with respect to arclength.

We choose constants to define our norms as follows. Choose $\alpha, \gamma > 0$ such that $\gamma < \alpha < \frac{1}{3}$. Next choose $p, q > 0$ such that $q < p < \gamma$ and note that $p < \frac{1}{3}$ necessarily by the restriction on γ. Finally, choose $\beta > 0$ such that $\beta < \min\{(\alpha - \gamma)/6, p - q, \frac{1}{3} - \alpha\}$.

Given a function $f \in \mathcal{C}^1(M)$, define the *weak norm* of f by

$$|f|_w := \sup_{W \in \mathcal{W}^s} \sup_{\substack{\psi \in \mathcal{C}^p(W) \\ |\psi|_{W,\gamma,p} \le 1}} \int_W f\psi \, dm_W.$$

We define the *strong stable norm* of f by

$$\|f\|_s := \sup_{W \in \mathcal{W}^s} \sup_{\substack{\psi \in \mathcal{C}^q(W) \\ |\psi|_{W,\alpha,q} \le 1}} \int_W f\psi \, dm_W,$$

and the *strong unstable norm* of f by

$$\|f\|_u := \sup_{\varepsilon\le\varepsilon_0} \sup_{\substack{W_1,W_2\in\mathscr{W}^s\\ d_{\mathscr{W}^s}(W_1,W_2)<\varepsilon}} \sup_{\substack{\psi_i\in\mathscr{C}^p(W_i)\\ |\psi_i|_{W_i,\gamma,p}\le 1\\ d_q(\psi_1,\psi_2)<\varepsilon}} \varepsilon^{-\beta}\left|\int_{W_1} f\psi_1\,dm_W - \int_{W_2} f\psi_2\,dm_W\right|,$$

where $\varepsilon_0 > 0$ is chosen less than δ_0, the maximum length of $W \in \mathscr{W}^s$. We then define the *strong norm* of f by

$$\|f\|_{\mathscr{B}} = \|f\|_s + b\|f\|_u,$$

where b is a small constant chosen in (8.12).

We define $\mathscr{B}$ to be the completion of $\mathscr{C}^1(M)$ in the strong norm and $\mathscr{B}_w$ to be the completion of $\mathscr{C}^1(M)$ in the weak norm.

8.3.3 Uniform Properties of T

In this section we recall the uniform properties (A1)–(A5) for a hyperbolic map T used in [23] to prove the required Lasota-Yorke inequalities for $\mathscr{L}_T$. The following properties refer to the map before the introduction of a hole. Rather than recall the properties in the abstract setting used in [23], we translate them into the concrete setting of dispersing billiards which we have adopted here. This will make the properties easier to verify and will avoid cumbersome additional notation which serves no purpose here. Translated into this setting, most of the abstract assumptions read as well-known facts about dispersing billiards.

(A1) Jacobian. $|DT(x)| := |\det DT(x)| = \cos\varphi(x)/\cos\varphi(Tx)$ wherever $DT(x)$ exists.

(A2) Hyperbolicity. The set $\mathscr{S}_0 \cup P_0$ consists of finitely many curves, although $\mathscr{S}_{\pm n}$, $n \ge 1$, may be finite or countable. There exist families of stable and unstable cones, continuous on the closure of each component of $M \setminus (\mathscr{S}_0 \cup P_0)$, such that the angle between $C^s(x)$ and $C^u(x)$ is uniformly bounded away from 0 on M. Furthermore, there exist constants $C > 0$, $\Lambda > 1$ such that the following hold

(1) $DT(C^u(x)) \subset C^u(Tx)$ and $DT^{-1}(C^s(x)) \subset C^s(T^{-1}x)$ whenever DT and DT^{-1} exist.
(2) $\|DT(x)v\|_* \ge \Lambda\|v\|_*, \forall v \in C^u(x)$ and $\|DT^{-1}(x)v\|_* \ge \Lambda\|v\|_*, \forall v \in C^s(x)$, where $\|\cdot\|_*$ is an adapted norm, uniformly equivalent to the Euclidean norm, $\|\cdot\|$.

(A3) Structure of Singularities.

(1) There exists $C_1 > 0$ such that for all $x \in M$,

$$C_1 \frac{\tau(T^{-1}x)}{\cos\varphi(T^{-1}x)} \le \frac{\|DT^{-1}(x)v\|}{\|v\|} \le C_1^{-1}\frac{\tau(T^{-1}x)}{\cos\varphi(T^{-1}x)}, \quad \forall v \in C^s(x).$$

Let $\exp_x$ denote the exponential map from the tangent space $\mathscr{T}_x M$ to M. Then,

$$\|D^2 T^{-1}(x)v\| \le C_1^{-1}\tau^2(T^{-1}x)(\cos\varphi(T^{-1}x))^{-3},$$

for all $v \in \mathscr{T}_x M$ such that $T^{-1}(\exp_x(v))$ and $T^{-1}x$ lie in the same homogeneity strip.

In the infinite horizon case, let x_∞ denote one of the finitely many infinite horizon points: points in $\mathscr{S}_0$ which are the accumulation points of a sequence of curves $S_n \subset \mathscr{S}_{-1}$. Let $D_{n,k}$ denote the set of points between S_n and S_{n+1} and whose image under T^{-1} lies in $\mathbb{H}_k$. Then there exists a constant $c_s > 0$ such that $k \ge c_s n^{1/4}$.

(2) There exists $C_2 > 0$ such that for any stable curve $W \in \mathscr{W}^s$ and any smooth curve $S \subset \mathscr{S}_{-n}$, we have $m_W(N_\varepsilon(S) \cap W) \le C_2 \varepsilon^{1/2}$ for all $\varepsilon > 0$ sufficiently small.

(3) $\partial\mathbb{H}_k$ are uniformly transverse to the stable cones.

(4) $\exists C > 0$ such that for all $k > k_0$, if $W \in \mathscr{W}^s$ and $W \subset \mathbb{H}_k$, then $|W| \le Ck^{-3}$.

(5) The sum $\sum_{k \ge k_0} \cos(\mathbb{H}_k) < \infty$, where $\cos(\mathbb{H}_k)$ is the average value of $\cos\varphi$ on $\mathbb{H}_k$.

(A4) Invariant Families of Stable and Unstable Curves. There are invariant families of curves $\mathscr{W}^s$ and $\mathscr{W}^u$ with the properties described in Sects. 8.2.1.3 and 8.3.1. Moreover, we require the following distortion bounds along stable curves.

There exists $C_d > 0$ such that for any $W \in \mathscr{W}^s$ with $T^i W \in \mathscr{W}^s$ for $i = 0, 1, \ldots, n$, and any $x, y \in W$,

$$\left|\frac{J_W T^n(x)}{J_W T^n(y)} - 1\right| \le C_d d_W(x,y)^{1/3} \quad \text{and} \quad \left|\frac{|DT^n(x)|}{|DT^n(y)|} - 1\right| \le C_d d_W(x,y)^{1/3},$$

where $J_W T(x)$ denotes the Jacobian of T along W.

We also require an analogous distortion bound along unstable curves. If $T^i W \in \mathscr{W}^u$ for $0 \le i \le n$, then for any $x, y \in W$,

$$\left|\frac{|DT^n(x)|}{|DT^n(y)|} - 1\right| \le C_d d_W(T^n x, T^n y)^{1/3}.$$

(A5) One-Step Expansion. Let $W \in \mathscr{W}^s$ and partition the connected components of $T^{-1}W$ into maximal pieces V_i such that each V_i is a homogeneous stable curve (not necessarily of length at most δ_0). Let $|J_{V_i} T|_*$ denote the minimum contraction on V_i under T in the metric induced by the adapted norm from (A2)(2). There exists a choice of k_0 for the homogeneity strips such that

$$\lim_{\delta \to 0} \sup_{\substack{W \in \mathscr{W}^s \\ |W| < \delta}} \sum_i |J_{V_i} T|_* < 1. \tag{8.9}$$

8.3.4 Verifying (A1)–(A5) for Our Classes of Maps

For both the finite and infinite horizon periodic Lorentz gas, (A1)–(A5) are well-established properties of the billiard map T (see [11]). In this concrete setting, (A3)(3) is due to the fact that stable cones are bounded away from the horizontal, (A3)(4) follows directly from (A3)(3), and the series in (A3)(5) is majorized by the series C/k^2 for some $C > 0$.

For billiards with corner points, the uniform transversality between P_0 and the stable cones fails. However, the tangencies between stable curves and vertical lines in P_0 are at worst of a square-root type so that (A3)(2) holds [8, Sect. 9].

In addition, the uniform expansion (A2)(2) and the one-step expansion (A5) fail. In order to regain uniform hyperbolicity, the usual solution is to consider a higher iterate of T, $T_1 = T^{n_1}$ and prove the above properties hold for T_1. This is done in the present specific setting in [23, Sect. 6] based on the facts established in [8, Sect. 9]. Indeed, in this case one may take the adapted norm $\|\cdot\|_*$ to be simply the Euclidean norm. Below, we formulate precisely the complexity condition referred to in condition (C2) in Sect. 8.2.1 for billiards with corner points which allows us to regain uniform hyperbolicity for this higher iterate of T.

A consequence of (C1) is that there is a uniform upper bound on the number of consecutive collisions near a corner point. More precisely, let $\phi_0 > 0$ be the minimum angle of intersection of ∂Q at the corner points and define $m_0 = [\pi/\phi_0] + 1$. Then there exists a constant $\tau_c > 0$ such that for each $x \in M$, there is an $i \in \{0, \ldots, m_0 - 1\}$ such that $\tau(T^i x) \geq \tau_c$ [4, 8]. Define

$$\Lambda_0 := (1 + \tau_c \mathscr{K}_{\min})^{1/m_0} > 1 \quad \text{and} \quad n_0 := \left\lceil \frac{\ln(1 + \mathscr{K}_{\min}^{-2})}{2 \ln \Lambda_0} \right\rceil.$$

Let K_n denote the maximum number of smooth curves in $\cup_{i=0}^{n} T^i(\mathscr{S}_0 \cup V_0)$ that intersect or terminate at any one point in M. We assume there exists $n_2 \geq 1$ such that

$$K_{n_2} < \Lambda_0^{n_2}. \tag{8.10}$$

This inequality can be iterated since $K_{\ell n_2} \leq (K_{n_2})^\ell$ for each $\ell \in \mathbb{N}$. Let $s = K_{n_2} \Lambda_0^{-n_2} < 1$. Choose ℓ_1 so that $n_2 \ell_1 > m_0 + n_0$ and $\max\{s^{\ell_1}, \Lambda_0^{-\ell_1 n_2}\} < \frac{1}{2} \Lambda_0^{-m_0 - n_0}$. We define $n_1 = n_2 \ell_1$ and estimate

$$K_{n_1} \leq s^{\ell_1} \Lambda_0^{n_1} < \Lambda_0^{n_1} (\Lambda_0^{-m_0 - n_0} - \Lambda_0^{-n_1}) = \Lambda_0^{n_1 - m_0 - n_0} - 1.$$

Now defining $T_1 = T^{n_1}$, it is shown explicitly in [23, Sect. 6] that T_1 satisfies (A2)(2) with respect to the Euclidean metric with $\Lambda := \Lambda_0^{n_1 - m_0 - n_0} > 1$. Also, since $T_1^{-1} W$ can be cut into at most $K_{n_1} + 1 < \Lambda$ pieces by $\mathscr{S}_{-n_1}$, it follows that the one-step expansion (8.9) of (A5) holds by choosing k_0 large enough.

8.3.5 Properties of the Banach Spaces

Properties (A1)–(A5) are sufficient to prove the following set of Lasota-Yorke inequalities, with constants depending only on the quantities appearing in (A1)–(A5). We have one more constant to fix.

In light of (A5), we fix $\delta_0 > 0$ in the definition of $\mathcal{W}^s$ sufficiently small so that

$$\sup_{W \in \mathcal{W}^s} \sum_i |J_{V_i} T|_* =: \theta_* < 1. \tag{8.11}$$

The following proposition is [23, Proposition 2.3].

Proposition 8.3. *Let T satisfy properties (A1)–(A5). There exists $C > 0$, depending only on the quantities in (A1)–(A5) such that for all $n \in \mathbb{N}$ and all $f \in \mathcal{B}$,*

$$|\mathcal{L}^n f|_w \leq C|h|_w$$
$$\|\mathcal{L}^n f\|_s \leq C(\theta_*^{n(1-\alpha)} + \Lambda^{-qn})\|f\|_s + C\delta_0^{\gamma-\alpha}|f|_w$$
$$\|\mathcal{L}^n f|_u \leq C\Lambda^{-\beta n}\|h\|_u + C\,n\,\|h\|_s,$$

where θ_* is from (8.11).

The proposition is enough to conclude the required Lasota-Yorke inequality in (8.6) since if we choose $1 > \sigma > \max\{\Lambda^{-\beta}, \theta_*^{1-\alpha}, \Lambda^{-q}\}$, then there exists $N \geq 0$ such that

$$\begin{aligned}\|\mathcal{L}^N h\|_{\mathcal{B}} = \|\mathcal{L}^N h\|_s + b\|\mathcal{L}^N h\|_u &\leq \frac{\sigma^N}{2}\|h\|_s + C\delta_0^{\gamma-\alpha}|h|_w + b\sigma^N\|h\|_u + bCN\|h\|_s \\ &\leq \sigma^N\|h\|_{\mathcal{B}} + C\delta_0^{\gamma-\alpha}|h|_w,\end{aligned} \tag{8.12}$$

provided b is chosen small enough with respect to N.

It is proved in [23, Lemma 3.9] that the unit ball of $\mathcal{B}$ is compactly embedded in $\mathcal{B}_w$. Thus it follows from standard arguments (see [1,26]) that the essential spectral radius of $\mathcal{L}$ on $\mathcal{B}$ is bounded by σ, while the spectral radius of $\mathcal{L}$ is at most 1. The fact that the spectral radius actually equals 1 is proved in [23, Lemma 5.1].

Next we state two lemmas proved in [23] that we shall need.

Lemma 8.1 ([23, Lemma 3.5]). *Let $\mathcal{P}$ be a (mod 0) countable partition of M into open, simply connected sets such that (1) for each $k \in \mathbb{N}$, there is an $N_k < \infty$ such that at most N_k elements $P \in \mathcal{P}$ intersect $\mathbb{H}_k$; (2) there are constants $K, C_3 > 0$ such that for each $P \in \mathcal{P}$ and $W \in \mathcal{W}^s$, $P \cap W$ comprises at most K connected components and for any $\varepsilon > 0$, $m_W(N_\varepsilon(\partial P) \cap W) \leq C_3\varepsilon^{1/2}$.*

Let $s > \beta/(1-\beta)$. If $f \in \mathcal{C}^s(P)$ for each $P \in \mathcal{P}$ and $\sup_{P \in \mathcal{P}} |f|_{\mathcal{C}^s(P)} < \infty$, then $f \in \mathcal{B}$. In particular, $\mathcal{C}^s(M) \subset \mathcal{B}$ and both Lebesgue measure and the smooth SRB measure for T are in $\mathcal{B}$.

Lemma 8.2 ([23, Lemma 5.3]). *If* $f \in \mathcal{B}$ *and* $\psi \in \mathcal{C}^s(M)$, $s > \max\{\beta/(1-\beta), p\}$, *then* $\psi f \in \mathcal{B}$. *Moreover,* $\|\psi f\|_{\mathcal{B}} \leq C\|f\|_{\mathcal{B}}|\psi|_{\mathcal{C}^p(M)}$ *for some* $C > 0$ *independent of* ψ *and* f.

Other properties of $\mathcal{B}$ and $\mathcal{B}_w$ proved in [23] include:

- $\mathcal{L}$ is well defined as a continuous linear operator on both $\mathcal{B}$ and $\mathcal{B}_w$. Moreover, there is a sequence of continuous and injective embeddings $\mathcal{C}^s(M) \hookrightarrow \mathcal{B} \hookrightarrow \mathcal{B}_w \hookrightarrow (\mathcal{C}^p(M))'$ for all $s > \beta/(1-\beta)$.
- The elements of the peripheral spectrum of $\mathcal{L}$ on $\mathcal{B}$ are measures and all physical measures belong to $\mathcal{B}$.[6]
- If T is mixing, then $\mathcal{L}$ has a spectral gap, i.e., 1 is a simple eigenvalue and all other eigenvalues have modulus strictly less than 1.

8.3.5.1 Working with Higher Iterates of *T*

As mentioned in Sect. 8.3.4, (A2)(2) and (A5) fail for billiards with corner points so that we must work with a higher iterate of T, $T_1 = T^{n_1}$. In this context, the spectral gap is originally proved for $\mathcal{L}_1 = \mathcal{L}^{n_1}$. In order to conclude that $\mathcal{L}$ also has a spectral gap, one must show that $\|\mathcal{L}\|_{\mathcal{B}}$ is finite. This is done for billiards with corner points in [23, Sect. 6.4].

The proof there relies on the fact that although (A4) for T fails in the sense that the sum in (8.9) cannot be made less than 1 (so θ_* would be greater than 1 in Proposition 8.3), the sum in (8.9) does remain finite. Also, in (A2)(2), Λ is not strictly greater than 1. Yet the rest of the Lasota-Yorke inequalities go through without any problem for T so that one obtains the inequalities of the same form as Proposition 8.3, but without contraction. This suffices to conclude that $\mathcal{L}$ is bounded on $\mathcal{B}$.

We will make a similar argument for our map with holes since the added discontinuities coming from ∂H will cause (8.9) to fail, but due to the finiteness assumption (H1), the sum will still converge, keeping $\mathring{\mathcal{L}}_H$ bounded. In order to regain contraction for the open system, we will work with a higher iterate of T even in the case of the Lorentz gas.

[6]Recall that a physical measure for T is an ergodic, invariant probability measure μ for which there exists a positive Lebesgue measure set B_μ, with $\mu(B_\mu) = 1$, such that $\lim_{n\to\infty} \frac{1}{n}\sum_{i=0}^{n-1} \psi(T^i x) = \mu(\psi)$ for all $x \in B_\mu$ and all continuous functions ψ.

8.4 Extension to Open Systems

In this section we will prove Propositions 8.1 and 8.2 and Theorems 8.1 and 8.2. Proposition 8.3 yields the required inequalities for Proposition 8.1 for $\mathcal{L}$. We begin by explaining how to extend those inequalities to $\mathring{\mathcal{L}}_H$.

In this section, we *assume* that we have a map T (without holes) satisfying (A1)–(A5) (in the case of corner points this is already an iterate $T_1 = T^{n_1}$) and then show how to choose a higher iterate of T once we introduce the additional cuts made by ∂H so that (A1)–(A5) are still satisfied.

We fix $B_0, C_0 > 0$ and the set of holes $\mathscr{H}(B_0, C_0)$ satisfying (H1) and (H2). We choose $H \in \mathscr{H}(B_0, C_0)$ and define $\mathring{T}$ as in Sect. 8.1.1.

We want to think of ∂H as an extended singularity set for T. To this end, we define a map $\hat{T}$ which is equal to T everywhere, except $\hat{T}$ has the expanded singularity set induced by $\mathscr{S}_0 \cup P_0 \cup \partial H$. Thus when iterating $\hat{T}$, we introduce artificial cuts according to ∂H. When we want to consider the map with a hole, we simply drop the pieces that would have entered H.

Note that by (H2) and (A3)(2), ∂H has the same properties as $\mathscr{S}_0 \cup P_0$. Since $\hat{T}$ and T are the same map everywhere, properties (A1)–(A4) hold for $\hat{T}$ with essentially the same constants as for T (we may have to replace C_2 by C_0 in (A3)(2), but taking C_2' to be the larger of these two numbers, we note that both maps satisfy (A3)(2) with respect to C_2').

Thus the only thing which we need to address is (A5) and in particular (8.9) which may fail for $\hat{T}$ due to the additional cuts. Note that since ∂H increases the sums in (A5) by at most a factor of B_0, both sums are still finite. This is sufficient to ruin contraction in the Lasota-Yorke inequalities, but still yields a finite bound on $\|\mathcal{L}_{\hat{T}}\|$ according to the discussion in Sect. 8.3.5.1, where $\mathcal{L}_{\hat{T}}$ denotes the transfer operator corresponding to $\hat{T}$.

8.4.1 Complexity Bound and Proof of Proposition 8.1

In this section, we prove the following lemmas, which will allow us to regain (A5) for a higher iterate of $\hat{T}$. This in turn will allow us to prove Proposition 8.1.

Lemma 8.3. *There exists a sequence $\delta_n \downarrow 0$ such that*

$$\sup_{\substack{W \in \mathscr{W}^s \\ |W| \le \delta_n}} \sum_i |J_{V_i^n} T^n|_* \le \theta_*^n, \tag{8.13}$$

where V_i^n denote the maximal homogeneous stable curves in $T^{-n}W$ on which T^n is smooth.

Proof. We prove the lemma by induction on n. The case $n = 1$ follows from (A5) and (8.11) by taking $\delta_1 = \delta_0$.

Now assume (8.13) holds for all $0 \le k \le n$. In order to extend this inequality to $n + 1$, we claim that $\delta_{n+1} \le \delta_n$ can be chosen so small that $|V_i^n| \le \delta_0$ whenever $|W| \le \delta_{n+1}$. In this way, V_i^n will belong to $\mathcal{W}^s$ and we may apply (A5) to each such curve without additional artificial subdivisions. Let $A(V_i^n)$ comprise those indices j such that $TV_j^{n+1} \subset V_i^n$. Then grouping V_j^{n+1} according to the sets $A(V_i^n)$, we have

$$\sum_j |J_{V_j^{n+1}} T^{n+1}|_* \le \sum_i \sum_{j \in A(V_i^n)} |J_{V_j^{n+1}} T|_* |J_{V_i^n} T^n|_* \le \sum_i |J_{V_i^n} T^n|_* \theta_* \le \theta_*^{n+1},$$

as required. It remains to prove the claim.

The claim follows from the fact that if T^{-1} is smooth on a stable curve W, then there exists a uniform constant C, depending only on T, such that $|T^{-1}W| \le C|W|^{1/2}$ in the finite horizon case and $|T^{-1}W| \le C|W|^{1/3}$ in the infinite horizon case [11, Sect. 4.9].[7] Thus the lemma follows if we inductively choose $\delta_{n+1} = \delta_n^2$ in the finite horizon case and $\delta_{n+1} = \delta_n^3$ in the infinite horizon case. □

For $W \in \mathcal{W}^s$, Let $\hat{V}_i^n$ denote the maximal homogeneous stable curves in $\hat{T}^{-n}W$ on which $\hat{T}^n$ is smooth.

Lemma 8.4. *For $n \in \mathbb{N}$, let δ_n be from Lemma 8.3. Then*

$$\sup_{\substack{W \in \mathcal{W}^s \\ |W| \le \delta_n}} \sum_i |J_{\hat{V}_i^n} \hat{T}^n|_* \le (1 + n(B_0 - 1))\theta_*^n.$$

Proof. Fix $W \in \mathcal{W}^s$ with $|W| \le \delta_n$. Each V_i^n comprises one or more $\hat{V}_j^n$ due to the expanded singularity set for $\hat{T}$. For a fixed V_i^n, we must estimate the cardinality of the curves $\hat{V}_j^n \subset V_i^n$.

Let $U_i^n = T^n V_i^n$ and $\hat{U}_j^n = T^n \hat{V}_j^n$ for each i and j. Note that if $\hat{V}_{j_1}^n$ and $\hat{V}_{j_2}^n$ belong to the same curve V_i^n, then in fact $T^{-k}\hat{U}_{j_1}^n$ and $T^{-k}\hat{U}_{j_2}^n$ belong to the same smooth curve $T^{-k}U_i^n$ for each $0 \le k \le n$ since the additional cuts due to $\hat{T}$ are artificial and do not change the orbits of points. Also, $|T^{-k}U_i^n| \le \delta_0$ for each $k \le n - 1$ by choice of δ_n from the proof of Lemma 8.3.

Applying (H2) to $T^{-k+1}U_i^n$, the total number of new cuts in $T^{-k}U_i^n$ compared to $T^{-k+1}U_i^n$ can be no more than $B_0 - 1$. Inductively, the total number of cuts introduced into V_i^n by time n can be no more than $n(B_0 - 1)$. Thus the cardinality of the set of j such that $\hat{V}_j^n \subset V_i^n$ is at most $1 + n(B_0 - 1)$. This plus the fact that $|J_{\hat{V}_j^n} \hat{T}^n|_* \le |J_{V_i^n} T^n|_*$ whenever $\hat{V}_j^n \subset V_i^n$ completes the proof of the lemma. □

[7] Indeed, [11, Sect. 4.9] does not address the infinite horizon case explicitly, but a quick calculation shows that an exponent of $1/3$ is in fact needed.

Proof (Proof of Proposition 8.1). Now we choose n_0 such that $(1+n_0(B_0-1))\theta_*^{n_0} = \theta_0 < 1$. Then setting $\hat{T}_0 = \hat{T}^{n_0}$, and choosing δ_{n_0} from Lemma 8.3 to be the maximum length scale of curves in $\mathscr{W}^s$, we have (A1)–(A5) satisfied for $\hat{T}_0$. Thus the results of [23] imply the uniform Lasota-Yorke inequalities for $\mathscr{L}_{\hat{T}_0}$ given by Proposition 8.3 with δ_{n_0} in place of δ_0 and θ_0 in place of θ_* with the same choices of constants in the norms.

Notice that we need not change the definition of the Banach spaces $\mathscr{B}$ and $\mathscr{B}_w$. This is because once we have the uniform Lasota-Yorke inequalities measured on $|W| \le \delta_{n_0}$, we can quickly extend them to $|W| \le \delta_0$ by subdividing such curves into at most $[\delta_0/\delta_{n_0}]+1$ pieces of length at most δ_{n_0} and then applying the estimates on the shorter pieces. This has the effect of multiplying all the inequalities in Proposition 8.3 by the factor $[\delta_0/\delta_{n_0}]+1$ which affects neither the essential spectral radius nor the spectral radius of $\mathscr{L}_{\hat{T}_0}$.

Since $\mathscr{L}_{\hat{T}}$ is bounded as an operator on $\mathscr{B}$ as mentioned previously, this implies that $\mathscr{L}_{\hat{T}}$ also satisfies a uniform set of Lasota-Yorke inequalities with the contraction factor in (8.12) weakened to σ^{1/n_0}.

Now the transfer operator $\mathring{\mathscr{L}}_{\mathring{T}}$ corresponding to the map with a hole satisfies the same Lasota-Yorke inequalities as $\mathscr{L}_{\hat{T}}$ with the same constants since the pieces we must sum over are fewer (we drop the pieces that pass through H), but the estimates on each surviving piece remain the same (the maps T, $\mathring{T}$, and $\hat{T}$ are all the same on such pieces). The equalities of Proposition 8.1 now follow with constants depending only on (A1)–(A5), B_0, and C_0, as required. □

8.4.2 Proof of Proposition 8.2

Fix $H \in \mathscr{H}(B_0, C_0)$ with $\operatorname{diam}_s(H) \le h$. We must estimate $|||\mathscr{L} - \mathring{\mathscr{L}}|||$. We do this estimate for T directly rather than some power of T that we have been working with in previous sections. This is because we do not need contraction for the present estimate, but only use the smallness of the hole. We assume only that T satisfies (A1)–(A5) with $\Lambda = 1$ and the sum in (A5) finite, but not necessarily contracting. As noted previously, these weaker conditions are satisfied for all our classes of billiard maps (see Sects. 8.3.4 and 8.3.5.1).

We choose $f \in \mathscr{B}$ and recalling (8.3), we estimate

$$\begin{aligned} |\mathscr{L}f - \boldsymbol{I}_{\mathring{M}}\mathscr{L}(\boldsymbol{I}_{\mathring{M}}f)|_w &\le |\mathscr{L}f - \boldsymbol{I}_{\mathring{M}}\mathscr{L}f|_w + |\boldsymbol{I}_{\mathring{M}}\mathscr{L}f - \boldsymbol{I}_{\mathring{M}}\mathscr{L}(\boldsymbol{I}_{\mathring{M}}f)|_w \\ &\le |\boldsymbol{I}_H\mathscr{L}f|_w + |\boldsymbol{I}_{\mathring{M}}\mathscr{L}(\boldsymbol{I}_H f)|_w, \end{aligned} \tag{8.14}$$

by linearity since $H = M \setminus \mathring{M}$.

In order to estimate the two terms in (8.14), we will need the following two lemmas. The first lemma shows that the indicator functions $\boldsymbol{I}_H$ and $\boldsymbol{I}_{\mathring{M}}$ are bounded multipliers in both spaces $\mathcal{B}$ and $\mathcal{B}_w$. The second lemma shows that the hole is a small perturbation in the sense of $|||\cdot|||$.

Lemma 8.5. *Suppose $f \in \mathcal{B}$ and $H \in \mathcal{H}(B_0, C_0)$. There exists $C > 0$, depending only on B_0 and C_0 and (A1)–(A5), such that $\|\boldsymbol{I}_H f\|_{\mathcal{B}} \le C\|f\|_{\mathcal{B}}$ and $|\boldsymbol{I}_H f|_w \le C|f|_w$. Similar bounds hold for $\boldsymbol{I}_{\mathring{M}} f$.*

Lemma 8.6. *If $f \in \mathcal{B}$ and $H \in \mathcal{H}(B_0, C_0)$, then $|\boldsymbol{I}_H f|_w \le Ch^{\alpha-\gamma}\|f\|_s$, where $h = \mathrm{diam}_s(H)$ and $C > 0$ depends only on B_0.*

Postponing the proofs of these two lemmas, we first show how they complete the proof of Proposition 8.2. Let $C_w = \sup\{|\mathcal{L}f|_w : f \in \mathcal{B}_w, |f|_w \le 1\}$ and $C_B = \sup\{\|\mathcal{L}f\|_{\mathcal{B}} : f \in \mathcal{B}, \|f\|_{\mathcal{B}} \le 1\}$ denote the norm of $\mathcal{L}$ in the two spaces $\mathcal{B}_w$ and $\mathcal{B}$, respectively. Then using the two lemmas together with (8.14), we have

$$\begin{aligned}|\mathcal{L}f - \boldsymbol{I}_{\mathring{M}}\mathcal{L}(\boldsymbol{I}_{\mathring{M}} f)|_w &\le |\boldsymbol{I}_H \mathcal{L}f|_w + |\boldsymbol{I}_{\mathring{M}}\mathcal{L}(\boldsymbol{I}_H f)|_w \\ &\le Ch^{\alpha-\gamma}\|\mathcal{L}f\|_{\mathcal{B}} + CC_w|1_H f|_w \\ &\le CC_B h^{\alpha-\gamma}\|f\|_{\mathcal{B}} + CC_w h^{\alpha-\gamma}\|f\|_{\mathcal{B}}.\end{aligned}$$

Now taking the supremum over $f \in \mathcal{B}$, $\|f\|_{\mathcal{B}} \le 1$, completes the proof of the proposition.

8.4.2.1 Proofs of Lemmas 8.5 and 8.6

We will use below that there exists a constant $C_c > 0$ such that

$$C_c^{-1} \le \frac{\cos\varphi(x)}{\cos\varphi(y)} \le C_c \tag{8.15}$$

whenever x and y lie in the same homogeneity strip.

Proof (Proof of Lemma 8.5). We begin by checking that the partition $\mathcal{P}$ formed by the open, connected components of H and $M \setminus H$ satisfies assumptions (1) and (2) of Lemma 8.1. Assumption (1) holds with N_k uniformly bounded in k due to the fact that H has finitely many connected components. Also, assumption (2) of Lemma 8.1 is satisfied due to (H2) and (A3)(2) with $C_3 = \max\{C_0, C_2\}$ and $K = B_0$.

By density, it suffices to prove the lemma for $f \in \mathcal{C}^1(M)$. We fix $f \in \mathcal{C}^1(M)$ and note that $\boldsymbol{I}_H f$ and $\boldsymbol{I}_{\mathring{M}} f$ have the type of singularity admitted in Lemma 8.1 so that $\boldsymbol{I}_H f, \boldsymbol{I}_{\mathring{M}} f \in \mathcal{B}$.

We now proceed to estimate $\|\boldsymbol{I}_H f\|_{\mathcal{B}} \le C\|f\|_{\mathcal{B}}$. The estimate for the weak norm is similar to that for the strong stable norm and is omitted. From these, the estimates on $\boldsymbol{I}_{\mathring{M}} f$ follow by linearity since $\boldsymbol{I}_{\mathring{M}} = \boldsymbol{I}_M - \boldsymbol{I}_H$.

To estimate the strong stable norm of $\boldsymbol{I}_H f$, let $W \in \mathcal{W}^s$ and $\psi \in \mathcal{C}^q(W)$ with $|\psi|_{W,\alpha,q} \le 1$. Note that $|\psi|_{\mathcal{C}^q(W)} \le (\cos W)^{-1}|W|^{-\alpha}$. Then since $W \cap H$ comprises at most B_0 curves $W_i \in \mathcal{W}^s$, we have

$$\int_W \boldsymbol{I}_H f\, \psi\, dm_W = \sum_i \int_{W_i} f\, \psi\, dm_W \le \|f\|_s \sum_i |W_i|^\alpha \cos W_i\, |\psi|_{\mathscr{C}^q(W)}$$
$$\le \|f\|_s \sum_i \frac{|W_i|^\alpha}{|W|^\alpha} \frac{\cos W_i}{\cos W} \le \|f\|_s C_c B_0,$$

since $\cos W_i / \cos W$ is uniformly bounded by (8.15) and $|W_i| \le |W|$. Taking the supremum over $W \in \mathscr{W}^s$ and $\psi \in \mathscr{C}^q(W)$ yields $\|\boldsymbol{I}_H f\|_s \le C_c B_0 \|f\|_s$.

Next we estimate the strong unstable norm of $\boldsymbol{I}_H f$. Let $\varepsilon \le \varepsilon_0$ and choose $W^1, W^2 \in \mathscr{W}^s$ with $d_{\mathscr{W}^s}(W^1, W^2) \le \varepsilon$. For $\ell = 1, 2$, let $\psi_\ell \in W^\ell$ with $|\psi_\ell|_{W^\ell,\gamma,p} \le 1$ and $d_q(\psi_1, \psi_2) \le \varepsilon$. We must estimate

$$\int_{W^1} \boldsymbol{I}_H f\, \psi_1\, dm_W - \int_{W^2} \boldsymbol{I}_H f\, \psi_2\, dm_W.$$

Recalling the notation of Sect. 8.3.1, we consider W^ℓ as graphs of functions of their angular coordinates, $r_{W^\ell}(\varphi)$, and write $W^\ell = G_{W^\ell}(I_{W^\ell})$, $\ell = 1, 2$. We subdivide $W^1 \cap H$ and $W^2 \cap H$ into matched pieces U_j^1 and U_j^2 and unmatched pieces V_k^1 and V_k^2, respectively, using a foliation of horizontal line segments in M. Thus U_j^1 and U_j^2 are matched if both are defined over the same φ-interval I_j.

Due to (H1), there are at most B_0 matched pieces and $2B_0 + 2$ unmatched pieces V_k^ℓ created by ∂H and near the endpoints of W^ℓ. Note that due to (H2) and (A3)(2), we have $|V_k^\ell| \le C_2' \varepsilon^{1/2}$, where $C_2' = \max\{C_0, C_2\}$, since $d_{\mathscr{W}^s}(W_1, W_2) \le \varepsilon$. We split the estimate into matched and unmatched pieces

$$\int_{W^1} \boldsymbol{I}_H f\, \psi_1\, dm_W - \int_{W^2} \boldsymbol{I}_H f\, \psi_2\, dm_W = \sum_j \int_{U_j^1} f\, \psi_1\, dm_W - \int_{U_j^2} f\, \psi_2\, dm_W$$
$$+ \sum_{\ell,k} \int_{V_k^\ell} f\, \psi_\ell\, dm_W.$$

We estimate the integrals on unmatched pieces first,

$$\int_{V_k^\ell} f\, \psi_\ell\, dm_W \le \|f\|_s |V_k^\ell|^\alpha \cos V_k^\ell |\psi_\ell|_{\mathscr{C}^q(W_\ell)} \le \|f\|_s \frac{|V_k^\ell|^\alpha \cos V_k^\ell}{|W^\ell|^\gamma \cos W^\ell}$$
$$\le C_c C_2' \|f\|_s \varepsilon^{(\alpha-\gamma)/2}, \tag{8.16}$$

where we have used (8.15) and $|V_k^\ell| \le |W^\ell|$ in the last step.

To estimate the integrals on matched pieces, note that

$$d_{\mathscr{W}^s}(U_j^1, U_j^2) \le d_{\mathscr{W}^s}(W^1, W^2) \le \varepsilon.$$

Also,

$$|\psi_1 \circ G_{U_j^1} - \psi_2 \circ G_{U_j^2}|_{\mathscr{C}^q(I_j)} \le |\psi_1 \circ G_{W^1} - \psi_2 \circ G_{W^2}|_{\mathscr{C}^q(I_{W^1} \cap I_{W^2})} \le \varepsilon,$$

since $r_{U_j^1}$ and $r_{U_j^2}$ are simply the restrictions of r_{W^1} and r_{W^2} to I_j, respectively. Thus,

$$\left| \int_{U_j^1} f\,\psi_1\,dm_W - \int_{U_j^2} f\,\psi_2\,dm_W \right| \le \|h\|_u \varepsilon^\beta.$$

Putting this estimate together with (8.16) and using the fact that the number of matched and unmatched pieces is finite as mentioned earlier, we obtain

$$\left| \int_{W^1} \boldsymbol{I}_H f\,\psi_1\,dm_W - \int_{W^2} \boldsymbol{I}_H f\,\psi_2\,dm_W \right| \le C_c C_2'(2B_0+2)\|h\|_s \varepsilon^{(\alpha-\gamma)/2} + B_0\|h\|_u \varepsilon^\beta,$$

which, since $\beta \le (\alpha-\gamma)/2$, means we may divide through by ε^β to complete the estimate on the strong unstable norm and the proof of the lemma. □

Proof (Proof of Lemma 8.6). Again, by density, it suffices to do this estimate for $f \in \mathscr{C}^1(M)$.

Let $f \in \mathscr{C}^1(M)$ and $W \in \mathscr{W}^s$. Take $\psi \in \mathscr{C}^p(W)$ with $|\psi|_{W,\gamma,p} \le 1$. Let W_i denote the at most B_0 connected components of $W \cap H$. Then each W_i belongs to $\mathscr{W}^s$ and $|W_i| \le h$ by definition of the stable diameter. We thus estimate

$$\int_W \boldsymbol{I}_H f\,\psi\,dm_W = \sum_i \int_{W_i} f\psi\,dm_W \le \sum_i \|f\|_s |W_i|^\alpha \cos W_i |\psi|_{\mathscr{C}^q(W_i^1)}$$

$$\le \sum_i \|f\|_s h^{\alpha-\gamma} \frac{|W_i|^\gamma \cos W_i}{|W|^\gamma \cos W} \le B_0 C_c \|f\|_s h^{\alpha-\gamma}$$

where we have used (8.15) in the last line.

Taking the supremum over $W \in \mathscr{W}^s$ and $\psi \in \mathscr{C}^p(W)$, the lemma is proved. □

8.4.3 Proof of Theorems 8.1 and 8.2

Proof (Proof of Theorem 8.1). Propositions 8.1 and 8.2 allow us to apply the perturbative framework of Keller and Liverani [29] in the following way. We fix B_0, C_0 and consider the family of holes $\mathscr{H}(B_0, C_0)$. Proposition 8.1 guarantees uniform Lasota-Yorke inequalities for $\mathring{\mathscr{L}}_H$ for all $H \in \mathscr{H}(B_0, C_0)$. Then for $H \in \mathscr{H}(B_0, C_0)$ with $\text{diam}_s(H)$ sufficiently small, Proposition 8.2 and [29, Corollary 1] imply that the spectra outside the disk of radius $\sigma < 1$ and the corresponding spectral projectors of $\mathring{\mathscr{L}}_H$ move Hölder continuously for $H \in \mathscr{H}(B_0, C_0)$. Thus for $\mathrm{diam}_s(H)$ sufficiently small, $\mathring{\mathcal{L}}_H$ has a spectral gap. We prove the remainder of the theorem assuming that $\mathring{\mathcal{L}}_H$ has a spectral gap in this context.

Since $\mathring{\mathcal{L}}_H$ is real, its eigenvalue of maximum modulus, λ_H, must persist in being real and positive. We claim that its corresponding eigenvector $\mu_H \in \mathcal{B}$ is a measure. It follows from the spectral decomposition of $\mathring{\mathcal{L}}_H$ that for each $f \in \mathcal{B}$, there exists a constant c_f such that

$$\lim_{n\to\infty} \lambda_H^{-n} \mathring{\mathcal{L}}_H^n f(\psi) = c_f \mu_H(\psi), \quad \forall \psi \in \mathcal{C}^p(M). \tag{8.17}$$

Indeed, the above limit defines the spectral projector Π_{λ_H} onto the eigenspace corresponding to λ_H for $\mathring{\mathcal{L}}_H$. Letting Π_1 denote the eigenprojector onto the eigenspace corresponding to eigenvalue 1 for $\mathcal{L}$, we know that these projectors vary Hölder continuously in the $|||\cdot|||$-norm from (8.4) according to [29, Corollary 2]. Recall that μ_{SRB} denotes the smooth invariant measure for T before the introduction of the hole (see Sect. 8.2.1). Then since $\Pi_1 m(1) = \mu_{\text{SRB}}(1) = 1$, where m denotes Lebesgue measure, we must have $\Pi_{\lambda_H} m(1) = c_m \mu_H(1) > 0$. Indeed, the positivity of $\mathring{\mathcal{L}}$ requires both $c_m > 0$ and $\mu_H(1) > 0$.

Now (8.17) with 1 (the density of m) in place of f implies $|\mu_H(\psi)| \le |\psi|_\infty |\mu_H(1)|$, which implies that μ_H is a measure. Since $\mu_H(1) > 0$ by the positivity of $\mathring{\mathcal{L}}_H$, we may normalize μ_H to be a probability measure, $\mu_H(1) = 1$. It is now clear that μ_H is a conditionally invariant measure for $\mathring{T}$.

From (8.2) and the injectivity of T, it follows that μ_H cannot be supported on the set $\cup_{i\ge 0} T^i(H)$. Since this set has full Lebesgue measure by the ergodicity of μ_{SRB}, μ_H must be singular with respect to Lebesgue.

Now suppose that $\mu \in \mathcal{B}$ is a probability measure such that $c_\mu > 0$. Then by (8.17),

$$c_\mu \mu_H(1) = \lim_{n\to\infty} \lambda_H^{-n} \mathring{\mathcal{L}}_H^n \mu(1) = \lim_{n\to\infty} \lambda_H^{-n} \mu(\mathring{M}^n),$$

so that the escape rates with respect to μ_H and μ are equal, i.e., $\rho(\mu) = \log \lambda_H$. Moreover,

$$\frac{\mathring{\mathcal{L}}_H^n \mu}{|\mathring{\mathcal{L}}_H^n \mu|} = \frac{\mathring{\mathcal{L}}_H^n \mu}{\lambda_H^n} \frac{\lambda_H^n}{\mathring{\mathcal{L}}_H^n \mu(1)} = c_\mu \mu_H \cdot \frac{1}{c_\mu \mu_H(1)} = \mu_H,$$

and the convergence is at an exponential rate in $\mathcal{B}$ due to the spectral decomposition of $\mathring{\mathcal{L}}_H$.

We complete the proof by remarking that $c_m, c_{\mu_{\text{SRB}}} > 0$ by continuity of the spectral projectors so that Lebesgue and the smooth SRB measure for T are both included in this class of measures in $\mathcal{B}$. □

Proof (Proof of Theorem 8.2). With Propositions 8.1 and 8.2 established, the convergence of μ_H to μ_{SRB} and λ_H to 1 follows immediately from the continuity of the spectral projectors corresponding to $\mathring{\mathscr{L}}_H$ given by [29, Corollary 2] as long as we take a sequence of holes in $\mathscr{H}(B_0, C_0)$ with B_0 and C_0 fixed. □

8.5 Variational Principle

For all the maps and holes that we consider, we have $\rho(m) \geq \mathscr{P}_{\mathscr{G}_H}$, where $\mathscr{G}_H$ is from (8.8) by [25, Theorem C]. (Indeed, the setting in terms of both the maps and the permissible holes considered in [25] is much more general than the classes of billiard maps we consider here.) Thus, in order to show that a variational principle holds, we need to find $\nu_H \in \mathscr{G}_H$ such that $\rho(m) = P_{\nu_H}$. We proceed to construct such a measure.

8.5.1 Definition of ν_H

Let $s > \max\{\beta/(1-\beta), p\}$. We define a linear functional on $\mathscr{C}^s(M)$ by

$$\nu_H(\psi) = \lim_{n\to\infty} \lambda_H^{-n} \mu_H(\boldsymbol{I}_{\mathring{M}^n} \psi), \qquad \forall \psi \in \mathscr{C}^s(M). \tag{8.18}$$

The limit is well defined by (8.17) since $\mu_H(\boldsymbol{I}_{\mathring{M}^n}\psi) = \mathring{\mathscr{L}}^n(\psi\mu_H)(1)$ and $\psi\mu_H \in \mathscr{B}$ by Lemma 8.2. Indeed, $\nu_H(\psi) = c_{\psi\mu_H}$ in the notation of (8.17).

Since $|\nu_H(\psi)| \leq |\psi|_\infty$, ν_H extends as a bounded linear functional to $\mathscr{C}^0(M)$ so that by the Riesz representation theorem, ν_H is a measure. Also, the fact that $\nu_H(1) = 1$ and the positivity of the limit in the definition of ν_H implies that indeed, ν_H is a probability measure supported on the survivor set, $\mathring{M}^\infty$.

Now $\boldsymbol{I}_{\mathring{M}^{n-1}} \circ \mathring{T} = \boldsymbol{I}_{\mathring{M}^n}$ due to the nested nature of the sequence $\mathring{M}^n$. We use this to write

$$\begin{aligned}\nu_H(\psi \circ \mathring{T}) &= \lim_{n\to\infty} \lambda_H^{-n} \mu_H(\boldsymbol{I}_{\mathring{M}^n} \psi \circ \mathring{T}) = \lim_{n\to\infty} \lambda_H^{-n} \mathring{\mathscr{L}}_H \mu_H(\boldsymbol{I}_{\mathring{M}^{n-1}} \psi) \\ &= \lim_{n\to\infty} \lambda_H^{-(n-1)} \mu_H(\boldsymbol{I}_{\mathring{M}^{n-1}} \psi) = \nu_H(\psi),\end{aligned}$$

so that ν_H is an invariant measure for $\mathring{T}$ and also for T since $\mathring{T} = T$ on $\mathring{M}^\infty$.

In the subsequent sections, we will show that ν_H defined in this way belongs to $\mathscr{G}_H$ and that $\rho(m) = P_{\nu_H}$. To do this, we will use a Young tower constructed for the open system. This approach combines the work of [8, 34] to define and construct the tower under general assumptions on the map and singularities (all of which are satisfied in the present setting) and then applies the results of [25] to conclude the variational principle for the open system.

8.5.2 Review: Young Towers with Holes

In this section, we recall some of the important definitions regarding Young towers for piecewise hyperbolic maps. We refer the reader to [34] for details.

8.5.2.1 Generalized Horseshoes

Given a piecewise $C^{1+\epsilon}$ diffeomorphism $T : M \circlearrowleft$ of a Riemannian manifold M, the tower is built on a compact set X with a hyperbolic product structure: $X = (\cup \Gamma^u) \cap (\cup \Gamma^s)$ where Γ^u and Γ^s are continuous families of local unstable and stable manifolds, respectively, with $m_\omega(\omega \cap \Delta_0) > 0$ for every $\omega \in \Gamma^u$, where m_ω is the Riemannian volume on ω. We define an s-subset of X to be a set $X^s = (\cup \Gamma^u) \cap (\cup \tilde{\Gamma}^s)$ for some $\tilde{\Gamma}^s \subset \Gamma^s$; u-subsets are defined similarly.

Modulo a set of m_ω-measure zero, X is a countable disjoint union of closed s-subsets X_i with the property that for each i, there exists $R_i \in \mathbb{Z}^+$ such that $T^{R_i}(X_i)$ is a u-subset of X. The function $R : X \to \mathbb{Z}^+$ is called the return time function to X. We say the inducing scheme (T, X, R) has *exponential tail* if there exist $C > 0$, $\vartheta < 1$ such that $m_\omega(R > n) \leq C\vartheta^n$ for all $\omega \in \Gamma^u$.

In [34], the return map $T^R : X \circlearrowleft$ satisfies certain uniform hyperbolic properties listed as (P1)–(P5) in that paper. We omit the formal statements of those properties and instead focus on the properties we shall need in what follows. Essentially, the structure of (T^R, X) is that of a generalized horseshoe with countably many branches and variable return times.

For $x \in X$, let $\omega^s(x)$ denote the element of Γ^s containing x. An important quantity in the construction of the horseshoe is the *separation time* $s : X \times X \to \mathbb{Z}^+$ with the following properties: (i) $s(x, y) = s(x', y')$ for $x' \in \omega^s(x)$, $y' \in \omega^s(y)$; (ii) for $x, y \in X_i$, $s(x, y) \geq R_i$; (iii) for $x \in X_i$, $y \in X_j$, $i \neq j$, we have $s(x, y) \leq \min(R_i, R_j)$. The separation time is used in the construction of the horseshoe to determine when the derivative along orbits starting in X cease to be comparable, either due to discontinuities or subdivisions to control distortion.

8.5.2.2 Induced Markov Extensions

To the inducing scheme (T, X, R) described above, it is shown in [34] that one can associate a Markov extension $F : \Delta \circlearrowleft$ which inherits the uniform hyperbolicity of T^R. The Young tower $\Delta := \cup_{\ell \geq 0} \Delta_\ell$ is the disjoint union of sets $\Delta_\ell := \{(x, \ell) : x \in X, R(x) > \ell\}$. The tower map F is defined by $F(x, \ell) = (x, \ell + 1)$ for $\ell < R(x) - 1$ and $F(x, R(x) - 1) = (T^R x, 0)$. Thus, F maps x up the levels of the tower until the return time is reached. Identifying Δ_0 with X, we have a uniquely defined projection $\pi : \Delta \to M$ such that $T \circ \pi = \pi \circ F$.

The separation time $s(\cdot,\cdot)$ defines a natural countable Markov partition $\{\Delta_{\ell,j}\}$ on Δ: for $x, y \in \Delta_0$, $s(x,y) = \inf\{n > 0 : F^n x, F^n y \text{ lie in different } \Delta_{\ell,j}\}$. We call the Young tower *mixing* if g.c.d.$\{R\} = 1$.

8.5.2.3 Towers with Holes

Given a hole $H \subset M$, we say that a constructed tower (F, Δ) *respects the hole* if the following two conditions are met.

(R1) $\pi^{-1} H$ is the union of countably many elements of $\{\Delta_{\ell,j}\}$.
(R2) $\pi(\Delta_0) \subset M \setminus H$ and there exist $\delta > 0$, $\xi_1 > 1$ such that all $x \in \Delta_0$ satisfy $d(T^n x, \mathscr{S} \cup \partial H) \geq \delta \xi_1^{-n}$ for all $n \geq 0$, where $\mathscr{S}$ is the singularity set for T.

The first condition above guarantees that the tower map with a hole $\mathring{F}$ is still a Markov map on Δ. The second is a controlled approach condition to ∂H similar to that required for the singularity set for the map. Both conditions have appeared in several previous works, [20, 24, 25].

In this setting, the following theorem is proved in [25]. In the setting of [25], T is a $C^{1+\epsilon}$ piecewise smooth diffeomorphism of a manifold M satisfying the Katok-Strelcyn conditions [28], which are more general than our assumptions (A1)–(A5) (and indeed include a wide variety of billiard maps) so that we may apply the results of [25] in the present setting.

Theorem 8.4. *([25, Theorem D]) Suppose T satisfies (A1)–(A4) and that $(T, M; H)$ admits a mixing Young tower $(F, \Delta; \tilde{H})$ with an exponential tail such that (F, Δ) respects the hole. Let μ_{SRB} denote the unique invariant SRB measure of T supported on $\pi(\Delta)$. If the transfer operator on the tower with a hole has a spectral gap with leading eigenvalue λ_Δ, then*

(a) *$\rho(\mu_{SRB})$ is well defined and equals* $\log \lambda_\Delta$*;*
(b) *$\mathring{T}^n_* \mu_{SRB} / |\mathring{T}^n_* \mu_{SRB}|$ converges weakly to a conditionally invariant measure $\tilde{\mu}_H$ with eigenvalue λ_Δ;*
(c) *There exists $\tilde{\nu}_H \in \mathscr{G}_H$ such that*

$$\rho(\mu_{SRB}) = P_{\tilde{\nu}_H} := h_{\tilde{\nu}_H}(T) - \chi^+_{\tilde{\nu}_H}(T)\,.$$

(d) *$\tilde{\nu}_H$ is defined by*

$$\tilde{\nu}_H(\psi) = \lim_{n\to\infty} \lambda_\Delta^{-n} \int_{\mathring{M}^n} \psi \, d\tilde{\mu}_H \qquad \textit{for all } \psi \in \mathscr{C}^0(M).$$

In addition, $\tilde{\nu}_H$ enjoys exponential decay of correlations on Hölder observables.

8.5.3 Proof of Theorem 8.3 Assuming a Young Tower Respecting H

We proceed to prove Theorem 8.3 under the assumption that the tower construction necessary to invoke Theorem 8.4 holds for some iterate $T_0 = T^{n_0}$, where T is one of the billiard maps considered here. More precisely, we assume that T_0 satisfies (A1)–(A5) and that T_0 admits a mixing Young tower with an exponential tail which respects the hole H. Moreover, we assume the hole is sufficiently small that the transfer operator $\mathring{\mathscr{L}}_T$ has a spectral gap by Theorem 8.1 and also the transfer operator on the tower for $(\mathring{F}, \Delta; \tilde{H})$ has a spectral gap.

In this context, Theorem 8.4 gives information about objects with respect to T_0. So for example, we distinguish between the escape rates with respect to T and T_0 via the notation, $\rho(\mu_{\text{SRB}}; T)$ and $\rho(\mu_{\text{SRB}}; T_0)$. Obviously, $\rho(\mu_{\text{SRB}}; T_0) = n_0 \rho(\mu_{\text{SRB}}; T)$.

Since $\mu_{\text{SRB}} = \frac{\pi}{2} \cos\varphi \, m \in \mathscr{B}$ by Lemma 8.1, we have $\mathring{\mathscr{L}}_T^n \mu_{\text{SRB}} / |\mathring{\mathscr{L}}_T^n \mu_{\text{SRB}}| \to \mu_H$ and $\mu_H \in \mathscr{B}$ is a conditionally invariant measure for $\mathring{T}$ with eigenvalue λ_H by Theorem 8.1. Thus, applying Theorem 8.4(b) to T_0, we must have $\tilde{\mu}_H = \mu_H$ and $\lambda_\Delta = \lambda_H^{n_0}$, since λ_Δ is the eigenvalue of μ_H with respect to T_0.

Since $\tilde{\mu}_H = \mu_H$, comparing the definitions of ν_H in (8.18) and $\tilde{\nu}_H$ in Theorem 8.4(d), we must have $\tilde{\nu}_H = \nu_H$ as well. Thus by Theorem 8.4(c), we have $\rho(\mu_{\text{SRB}}; T_0) = h_{\nu_H}(T_0) - \chi_{\nu_H}^+(T_0)$. But since $h_{\nu_H}(T_0) = n_0 h_{\nu_H}(T)$ and $\chi_{\nu_H}^+(T_0) = n_0 \chi_{\nu_H}^+(T)$, this implies

$$\rho(\mu_{\text{SRB}}; T) = h_{\nu_H}(T) - \chi_{\nu_H}^+(T).$$

Now since $\rho(m; T) = \rho(\mu_{\text{SRB}}; T) = \log \lambda_H$ by Theorem 8.1, this completes the proof of Theorem 8.3.

8.5.4 Existence of a Young Tower Respecting the Hole

In this section we prove that a Young tower satisfying the assumptions of Theorem 8.4 can be constructed for some iterate of our map with holes. Recall the definition of $\hat{T}_0 = \hat{T}^{n_0}$ from Sect. 8.4.1, where $\hat{T}$ is the same as T, but with expanded singularity set due to ∂H.

We showed in Sect. 8.4.1 that $\hat{T}_0$ satisfies (A1)–(A5) so that the one-step expansion condition (8.9) is recovered via Lemma 8.4 even in the presence of the additional cuts due to ∂H.

In this setting, $\hat{T}_0$ satisfies the abstract conditions used in [8] to construct Young towers for a general class of hyperbolic maps with singularities. Below, we recall the simplified two-dimensional version of these properties used in [13, Sect. 4] (see also [8, Sect. 2] for the more general version).

(1) *Smoothness.* Assume that M is a smooth, two-dimensional compact Riemannian manifold and that $\mathscr{F}$ is a C^2 diffeomorphism of $M \setminus D$ onto $\mathscr{F}(M \setminus D)$, where D is a closed set of Lebesgue measure zero that is referred to as the singularity set of $\mathscr{F}$.
(2) *Hyperbolicity.* There exist two families of cones $C^s(x)$ and $C^u(x)$ satisfying our assumptions (A2)(1) and (2). The singularity sets D are transverse to stable and unstable cones as in our assumption (A2)(4) and (5). Defining $D_{\pm n} = \cup_{i=1}^{n} \mathscr{F}^{\mp i}$, the tangent vectors to $D_n \setminus D$ are in stable cones for $n > 0$ and in unstable cones for $n < 0$.
(3) *SRB measure.* $\mathscr{F}$ preserves a mixing SRB measure.
(4) *Distortion bounds.* Our condition (A3), but for the corresponding unstable Jacobian of $\mathscr{F}$. To obtain this, D necessarily includes the boundaries of homogeneity strips.
(5) *Bounded curvature.* The same as what we require in (A3), but for unstable manifolds.
(6) *Absolute continuity.* The holonomy map between unstable manifolds is assumed to be absolutely continuous wherever it is defined.
(7) *One-step expansion condition.* This is a slightly more complicated version of our one-step expansion (8.9), but for unstable manifolds.

All these conditions are verified for the periodic Lorentz gas and dispersing billiards with corner points in [8, Sects. 7–9]. The only way in which our map $\hat{T}_0$ differs from $\mathscr{F}$ above is in the one-step expansion condition due to our expanded singularity set. Although the formulation of this condition in [8] has a more complicated form, it is proved in [13, Lemma 8] that our one-step expansion condition (8.9) implies that the more complicated form holds for the billiard classes considered here due to the controlled accumulation of singularity curves described in our (A3)(1). This control is automatic for finite horizon billiards since the singularity sets comprise finitely many smooth compact curves; it is proved via simple geometrical estimates in the case of the infinite horizon Lorentz gas.

Thus our map $\hat{T}_0$ satisfies the assumptions of [13, Theorem 10] which yields a mixing tower with exponential tails for $\hat{T}_0$. We proceed to check that this tower respects the hole.

The fact that the singularity set for $\hat{T}_0$ includes ∂H implies that for each stable rectangle X_i in the horseshoe X and each $n \le R(X_i)$, we have either $\hat{T}_0^n(X_i) \subset H$ or $\hat{T}_0^n(X_i) \cap H = \emptyset$. This guarantees condition (R1).

Define for $\xi_1 > 1$ and $\delta > 0$,

$$M^{\pm}_{\xi_1,\delta}\{x \in M : \mathrm{dist}(\mathscr{F}^{\pm n}, D \cup \partial M) > \delta \xi_1^{-n}, \quad \forall n \ge 0\}.$$

The reference set X with hyperbolic product structure for the generalized horseshoe constructed in [8] is defined on the positive measure Cantor set given by $M^{+}_{\xi_1,\delta} \cap M^{-}_{\xi_1,\delta}$ for fixed $\xi_1 > 1$ and $\delta > 0$ chosen sufficiently small. Since ∂H is included in the singular set for $\hat{T}_0$, this guarantees the slow approach condition (R2). Thus the constructed tower respects the hole.

The final point to check is that the hole on the tower $\pi^{-1}H$ can be made sufficiently small by making H sufficiently small in M. This follows from the slow approach condition (R2) (this argument also appears in [20, 24]). If $\mathrm{diam}_u(H)$ denotes the unstable diameter of H, then the slow approach condition (R2) implies that $\pi^{-1}H$ cannot appear in the tower below level $\ell \sim \log \mathrm{diam}_u(H)$. Thus by decreasing the unstable diameter of H, we can make $\pi^{-1}H$ sufficiently small as long as we remain in the class $\mathscr{H}(B_0, C_0)$. This argument relies on the fact that the tail bounds on the tower depend only on the quantities introduced in items 1–7 above, which depend only on the map and the added complexity due to the hole. Thus keeping B_0 and C_0 fixed, we retain uniform tail estimates constructed for the towers respecting holes in $\mathscr{H}(B_0, C_0)$.

Now for $\pi^{-1}H$ sufficiently small in Δ, it follows from [3, Proposition 2.4] (see also [24, Theorem 4.4]) that the transfer operator on the tower with a hole has a spectral gap. This completes the verification of the assumptions of Theorem 8.4, so we may apply the conclusions of that theorem to our map $\hat{T}_0$ as we did in Sect. 8.5.3.

Acknowledgements This research is partially supported by NSF grant DMS-1101572.

References

1. Baladi, V.: Positive transfer operators and decay of correlations. Advanced Series in Nonlinear Dynamics, vol.16, World Scientific, Singapore (2000)
2. Baladi, V., Keller, G.: Zeta functions and transfer operators for piecewise monotonic transformations. Comm. Math. Phys. **127**, 459–477 (1990)
3. Bruin, H., Demers, M., Melbourne, I.: Existence and convergence properties of physical measures for certain dynamical systems with holes. Ergod. Th. Dynam. Sys. **30**, 687–728 (2010)
4. Bunimovich, L.A., Sinai, Ya. G., Chernov, N.: Markov partitions for two-dimensional hyperbolic billiards. Russian Math. Surveys **45**, 105–152 (1990)
5. Bunimovich, L.A., Sinai, Ya. G., Chernov, N.: Statistical properties of two-dimensional hyperbolic billiards. Russian Math. Surv. **46**, 47–106 (1991)
6. Čencova, N.N.: A natural invariant measure on Smale's horseshoe. Soviet Math. Dokl. **23**, 87–91 (1981)
7. Čencova, N.N.: Statistical properties of smooth Smale horseshoes. In: Dobrushin, R.L. (ed.) Mathematical Problems of Statistical Mechanics and Dynamics, pp. 199–256. Reidel, Dordrecht (1986)
8. Chernov, N.: Decay of correlations and dispersing billiards. J. Stat. Phys. **94**, 513–556 (1999)
9. Chernov, N., Markarian, R.: Ergodic properties of Anosov maps with rectangular holes. Bol. Soc. Bras. Mat. **28**, 271–314 (1997)
10. Chernov, N., Markarian, R.: Anosov maps with rectangular holes. Nonergodic cases. Bol. Soc. Bras. Mat. **28**, 315–342 (1997)
11. Chernov, N., Markarian, R.: Chaotic Billiards. Mathematical Surveys and Monographs, vol. 127, AMS, Providence, RI (2006)
12. Chernov, N., van den Bedem, H.: Expanding maps of an interval with holes. Ergod. Th. Dynam. Sys. **22**, 637–654 (2002)
13. Chernov, N., Zhang, H.-K.: Billiards with polynomial mixing rates. Nonlinearity **18**, 1527–1553 (2005)

14. Chernov, N., Markarian, R., Troubetskoy, S.: Conditionally invariant measures for Anosov maps with small holes. Ergod. Th. Dynam. Sys. **18**, 1049–1073 (1998)
15. Chernov, N., Markarian, R., Troubetskoy, S.: Invariant measures for Anosov maps with small holes. Ergod. Th. Dynam. Sys. **20**, 1007–1044 (2000)
16. Collet, P., Martínez, S., Schmitt, B.: The Yorke-Pianigiani measure and the asymptotic law on the limit Cantor set of expanding systems. Nonlinearity **7**, 1437–1443 (1994)
17. Collet, P., Martínez, S., Schmitt, B.: Quasi-stationary distribution and Gibbs measure of expanding systems. In: Tirapegui, E., Zeller, W. (eds.) Instabilities and Nonequilibrium Structures V, pp. 205–19. Kluwer, Dordrecht (1996)
18. Demers, M.F.: Markov Extensions for Dynamical Systems with Holes: An Application to Expanding Maps of the Interval. Israel J. Math. **146**, 189–221 (2005)
19. Demers, M.F.: Markov Extensions and Conditionally Invariant Measures for Certain Logistic Maps with Small Holes. Ergod. Th. Dynam. Sys. **25**(4), 1139–1171 (2005)
20. Demers, M.F., Wright, P.: Behavior of the escape rate function in hyperbolic dynamical systems. Nonlinearity **25**, 2133–2150 (2012)
21. Demers, M.F., Young, L.-S.: Escape rates and conditionally invariant measures. Nonlinearity **19**, 377–397 (2006)
22. Demers, M.F., Zhang, H.-K.: Spectral analysis of the transfer operator for the Lorentz gas. J. Mod. Dynam. **5**(4), 665–709 (2011)
23. Demers, M.F., Zhang, H.-K.: Spectral analysis for hyperbolic systems with singularities. Nonlinearity **27**, 379–433 (2014)
24. Demers, M.F., Wright, P., Young, L.-S.: Escape rates and physically relevant measures for billiards with small holes. Commun. Math. Phys. **294**(2), 353–388 (2010)
25. Demers, M.F., Wright, P., Young, L.-S.: Entropy, Lyapunov exponents and escape rates in open systems. Ergod. Th. Dynam. Sys. **32**(4), 1270–1301 (2012)
26. Hennion, H., Hevré, L.: Limit theorems for markov chains and stochastic properties of dynamical systems by quasi-Compactness. Lecture Notes in Mathematics, vol. 1766, Springer, Berlin (2001)
27. Homburg, A., Young, T.: Intermittency in families of unimodal maps. Ergod. Th. Dynam. Sys. **22**(1), 203–225 (2002)
28. Katok, A., Strelcyn, J.-M.: with the collaboration of F. Ledrappier and F. Przytycki, Invariant Manifolds, Entropy and Billiards; Smooth Maps with Singularities. Lecture Notes in Mathematics, vol. 1222, 283 pp. Springer, Berlin (1986)
29. Keller, G., Liverani, G.: Stability of the spectrum for transfer operators. Ann. Scuola Norm. Sup. Pisa Cl. Sci. **28**(4), 141–152 (1999)
30. Liverani, C., Maume-Deschamps, V.: Lasota-Yorke maps with holes: conditionally invariant probability measures and invariant probability measures on the survivor set. Annales de l'Institut Henri Poincaré Probab. Stat. **39**, 385–412 (2003)
31. Lopes, A., Markarian, R.: Open Billiards: Cantor sets, invariant and conditionally invariant probabilities. SIAM J. Appl. Math. **56**, 651–680 (1996)
32. Pianigiani, G., Yorke, J.: Expanding maps on sets which are almost invariant: decay and chaos. Trans. Amer. Math. Soc. **252**, 351–366 (1979)
33. Richardson, Jr., P.A.: Natural measures on the unstable and invariant manifolds of open billiard dynamical systems. Doctoral Dissertation, Department of Mathematics, University of North Texas, (1999)
34. Young, L.-S.: Statistical properties of dynamical systems with some hyperbolicity. Ann. Math. **147**, 585–650 (1998)

Chapter 9
Almost-Invariant and Finite-Time Coherent Sets: Directionality, Duration, and Diffusion

Gary Froyland and Kathrin Padberg-Gehle

Abstract Regions in the phase space of a dynamical system that resist mixing over a finite-time duration are known as almost-invariant sets (for autonomous dynamics) or coherent sets (for nonautonomous or time-dependent dynamics). These regions provide valuable information for transport and mixing processes; almost-invariant sets mitigate transport between their interior and the rest of phase space, and coherent sets are good transporters of 'mass' precisely because they move about with minimal dispersion (e.g. oceanic eddies are good transporters of water that is warmer/cooler/saltier than the surrounding water). The most efficient approach to date for the identification of almost-invariant and coherent sets is via transfer operators. In this chapter we describe a unified setting for optimal almost-invariant and coherent set constructions and introduce a new coherent set construction that is suited to tracking coherent sets over several finite-time intervals. Under this unified treatment we are able to clearly explain the fundamental differences in the aims of the techniques and describe the differences and similarities in the mathematical and numerical constructions. We explore the role of diffusion, the influence of the finite-time duration, and discuss the relationship of time directionality with hyperbolic dynamics. All of these issues are elucidated in detailed case studies of two well-known systems.

G. Froyland (✉)
School of Mathematics and Statistics, University of New South Wales,
Sydney NSW 2052, Australia
e-mail: g.froyland@unsw.edu.au

K. Padberg-Gehle
Institute of Scientific Computing, Technische Universität Dresden,
D-01069 Dresden, Germany
e-mail: kathrin.padberg@tu-dresden.de

W. Bahsoun et al. (eds.), *Ergodic Theory, Open Dynamics, and Coherent Structures*, Springer Proceedings in Mathematics & Statistics 70, DOI 10.1007/978-1-4939-0419-8_9,
© Springer Science+Business Media New York 2014

9.1 Introduction

The mathematical description of transport and mixing processes in dynamical systems has been the subject of intense research over the last two decades. Relevant applications include astrodynamics, molecular dynamics, geophysical flows, and biological systems; see, e.g. [2, 34, 50, 51] for discussions and reviews of transport and mixing phenomena.

Much research has focussed on the detection, approximation, and analysis of the *geometrical* structures that may explain transport barriers and the underlying transport mechanisms in autonomous and nonautonomous dynamical systems. These methods include topics such as (time-dependent) invariant manifolds, lobe dynamics, and finite-time Lyapunov exponents and related local extremisers of stretching and shearing; see, e.g. [24–26, 41, 42, 45, 50, 51] and references therein. A recent *topological* approach [1] attempts to find transport boundaries in two-dimensional forced systems by inferring growth rates of loops from particle trajectories.

Probabilistic approaches provide a macroscopic view of the dynamics, studying the global evolution of densities. These techniques can be used to detect regions in phase space that remain *coherent* under the action of the dynamical system. In autonomous systems such regions are termed *almost-invariant* or metastable sets, introduced in the past 15 years in the context of dynamical systems [8, 10] and time-symmetric Markov processes [11, 27, 44]. These concepts rely on the Perron-Frobenius (or transfer) operator, a linear Markov operator. Subdominant eigenfunctions of this operator are heuristically used to estimate almost-invariant sets; see [12, 13, 15] for further extensions to this approach. Studies on the connections between eigenmodes of evolution operators and slow mixing in fluid flow can be found in [33, 39, 40, 47], numerical investigations of stochastically perturbed transfer operators include [3, 5], and a related series of work beginning with [35, 36] decomposes the phase space into ergodic components. A numerical study of two-dimensional periodically forced flows [48] connects transfer operator results with topological braiding approaches [21], reminiscent of the almost-cyclic behaviour in [10].

Building on the transfer operator framework, a mathematical definition and the corresponding numerical treatment of coherent sets in nonautonomous systems has only recently been proposed in the time-asymptotic [17, 18] and finite-time [14, 19] settings. The mathematical concepts introduced in [12] for the autonomous case and [19] for the finite-time case deal with finite-state Markov chains, i.e. discretised transfer operators, and are thus purely finitary. Froyland [14] recently proposed a transfer operator-based framework for identifying finite-time coherent sets, generalising the matrix-based approach of [19]. A transfer operator is defined via an appropriately chosen stochastic kernel and is shown to be a compact L^2-operator with suitable spectral properties. Of particular focus in [14] is a Perron-Frobenius operator pre- and post-composed with ϵ-diffusion; the influence of noise on the spectrum and singular vectors is studied. While [14] focusses on developing an analytic framework for the matrix set-up used in [19], we will show in this paper that the construction of [14] is rather generally applicable. We will adapt it to verify the assumptions underlying the finitary almost-invariant sets framework in [12] and to make further theoretical extensions to finite-time coherent sets concepts. We also demonstrate how the duration of the finite-time interval under consideration and the strength of diffusion influence the structure of the regions in phase space that most resist mixing. Furthermore, we show how different concepts give rise to, or suppress, the impact of time directionality on almost-invariant sets and finite-time coherent sets in strongly hyperbolic conditions.

The paper is organised as follows. Section 9.2 begins by giving brief background information on the Perron-Frobenius operator and is followed by three introductory sections that describe the three dynamical settings considered in this chapter and set up the associated problems of finding optimal almost-invariant and coherent sets. Section 9.3 introduces the two key tools we will use: firstly, a variant of the Perron-Frobenius operator developed in [14] that is adapted to the finite-time setting and secondly, some background on the minimax properties of eigenvectors of compact, self-adjoint operators on Hilbert space. Section 9.4 contains the mathematical set-ups to handle the three problems described in Sect. 9.2. Section 9.5 discusses the differences and similarities of single- and bidirectional coherence, describes how one can create a sequence of finite-time coherent sets, and summarises some further mathematical properties of the framework. Section 9.6 details how one can numerically implement the theory in Sect. 9.4 in each of the three dynamical settings, and Sect. 9.7 contains the two main case studies, in which we explain the similarities and differences of the three dynamical set-ups and demonstrate the influence of finite-time duration, diffusion amplitude, and connections with time directions.

9.2 Transfer Operators and Three Transport Problems

Let $M \subset \mathbb{R}^d$ be compact, ℓ denote Lebesgue (or volume) measure on M, and consider a map $T : M \to M$, which is non-singular[1]. The map T may describe discrete time dynamics or may be a time-t map of a continuous time flow. The *Perron-Frobenius operator* for T, denoted $\mathbf{P} : L^1(M,\ell) \to L^1(M,\ell)$, describes the evolution of densities under T. Its action on an $f \in L^1(M,\ell)$ is defined by insisting that $\int_A \mathbf{P}f\, d\ell = \int_{T^{-1}A} f\, d\ell$ for all (Borel-)measurable $A \subset M$. In the situation where T is differentiable, one has the equivalent definition of $\mathbf{P}$: $\mathbf{P}f(y) = \sum_{x\in T^{-1}y} f(x)/|\det DT(x)|$, where $DT(x)$ is the spatial linearisation of T at $x \in M$. If T is differentiable and invertible, as in the situation where T

[1]A map T is non-singular if $\ell(T^{-1}A) = 0$ when $\ell(A) = 0$; thus volume cannot be created from nothing by pulling back with T or alternatively cannot be completely destroyed by pushing forward with T.

arises as a time-t map of a smooth flow, then the expression above simplifies to $\mathbf{P}$: $\mathbf{P}f(y) = f(T^{-1}y)/|\det DT(T^{-1}y)|$. The operator $\mathbf{P}$ is bounded; in fact $\|\mathbf{P}\|_1 = 1$, and preserves nonnegative functions; that is, $\mathbf{P}f \geq 0$ if $f \geq 0$. A density is invariant under T if the density is a fixed point of $\mathbf{P}$. Further details on the Perron-Frobenius operator may be found in [30].

In the following, we will introduce linear operators that are built around the Perron-Frobenius operator. We will introduce some noise or diffusion to the purely deterministic action of T, perform some 'change of measure space' transformations, and combine these constructions with their duals in order to solve specific dynamical transport problems. Before getting into details, we briefly describe the problems we are interested in solving and prototype operator constructions for solving them.

An overarching goal is to detect and locate slow mixing dynamical structures. These structures should be macroscopic in size and by 'slow mixing' we have in mind a geometric mixing rate that is slower than $1/\Lambda$, where Λ is the largest[2] positive Lyapunov multiplier. Thus, such 'slow mixing' cannot be explained by *local* stretching, but is instead due to the way in which the dynamics acts *globally*.

9.2.1 Autonomous, Time-Independent, or Periodically Forced Dynamics

In this setting we have a single map $T : X \to X$, where $X \subset M$ is the compact region in which we will search for a single almost-invariant partition; repeated iteration of T defines the dynamics. The map T may arise as a time-t map of an autonomous ODE $\dot{x} = F(x)$. Alternatively, if the dynamics is periodic with period p, then in discrete time one may study $T = T_p \circ \cdots \circ T_2 \circ T_1$ (or cyclic permutations) as a single map, or in continuous time, $\dot{x} = F(x,t)$, where $F(x,t) = F(x,t+p)$ for all $x \in X, t \in \mathbb{R}$, one may set T to the time-p map of the ODE.

We assume that there is some T-invariant[3] probability measure μ and that one is interested in tracking the transport with respect to this measure. For example, if T is volume-preserving, then a natural choice for μ is Lebesgue measure.

As the dynamics is fixed in time, the global slowly mixing structures we seek are also fixed sets. If a set $A \subset X$ satisfies $A \approx T^{-1}A$, then the set of points that are *currently in A and will remain in A after one time step*, namely $A \cap T^{-1}A$, is large relative to A in the sense of μ-measure, i.e. $\frac{\mu(A\cap T^{-1}A)}{\mu(A)} \approx 1$. Thus, the probability to leave the set A in one time step is low; because of this property and the approximate invariance[4] property, these structures are known in the literature as *almost-invariant*

[2]As we are considering structures of full dimension, these structures will be stretched and mixed according to the largest positive Lyapunov exponent.

[3]A probability measure μ is called T-invariant if $\mu \circ T^{-1} = \mu$.

[4]A set A is called *invariant* if $A = T^{-1}A$.

sets or *metastable sets*. We will be interested in optimal almost-invariant sets; optimal in the sense that the conditional probability to leave the set in one time step is minimal.

In terms of operators, in the following sections we will work with an operator $\mathbf{L}$ that is dynamically similar to $\mathbf{P}$, but with the property that $\mathbf{L}\mathbf{1}_X = \mathbf{1}_X$, where $\mathbf{1}_X$ denotes the characteristic function of X. If T is volume-preserving, then $\mathbf{L} = \mathbf{P}$. The invariance condition $A \approx T^{-1}A$ can be translated into the operator equation $\mathbf{L}\mathbf{1}_A \approx \mathbf{1}_A$, where we think of $\mathbf{1}_A$ as a functional representation of the set A. For technical reasons discussed later (in addition to a formal definition of $\mathbf{L}$), solving the problem of finding *optimal* almost-invariant sets is helped by forming a symmetrised operator $(\mathbf{L} + \mathbf{L}^*)/2$, where $\mathbf{L}^*$ is the dual of $\mathbf{L}$. While the action of $\mathbf{L}$ can be interpreted as a push-forward under the dynamics T, the action of $\mathbf{L}^*$ can be interpreted as a pull-back (applying T^{-1}). Thus, our symmetrised operator is effectively checking how mass is transported in forward and backward time. Because our measure μ is T-invariant, it does not matter in which time direction we check for mass loss from a set A; we do however benefit from optimality properties of self-adjoint operators, so the symmetrisation construction is important.

9.2.2 Nonautonomous, Time-Dependent, or Aperiodically Forced Dynamics: Single Time Direction

In contrast to the single map setting above, we now have a sequence of maps $T_{t+\tau} \circ \cdots \circ T_{t+1} \circ T_t$ over a duration of τ time steps. Alternatively, in continuous time, one studies the flow of an ODE $\dot{x} = F(x,t)$ from some initial time t to some final time $t + \tau$. In both cases, we construct a single-map $T : X \to T(X)$, which in the discrete time setting is simply $T = T_{t+\tau} \circ \cdots \circ T_{t+1} \circ T_t$ and in the continuous time setting is the flow map of the ODE from time t to time $t + \tau$.

Superficially, we may now appear to be back in the autonomous setting discussed in Sect. 9.2.1; however, there are two crucial differences. Firstly, in the autonomous setting, our symmetrisation construction implicitly assumed that the dynamics in backward time is T^{-1}. However, in a general nonautonomous setting, the backward-time dynamics is $T_{t-\tau}^{-1} \circ \cdots \circ T_{t-2}^{-1} \circ T_{t-1}^{-1}$, which is different to $T^{-1} = T_t^{-1} \circ T_{t+\tau-1}^{-1} \circ T_{t+\tau}^{-1}$. Secondly, because of the general time-dependence, the slowly mixing spatial structures we are interested in may be *time-varying*, rather than fixed in space; for example, consider oceanic eddies or atmospheric vortices, both of which move over time. Because of this general time-dependence, we do *not* insist on tracking the finite-time transport with respect to an *invariant* measure, but rather begin with a user-specified probability measure μ at time t. This measure describes a mass distribution that we are interested in transporting; by the final time $t+\tau$, μ will have transformed into another probability measure ν via the transformation[5] $\nu = \mu \circ T^{-1}$.

[5]Note that $\nu = \mu \circ T^{-1}$ is the natural push-forward of μ under T. For example, if μ is supported on a set $A \subset X$, then $\nu(T(A)) = \mu \circ T^{-1}T(A) \geq \mu(A)$ shows that ν is supported on $T(A)$.

As our slowly mixing spatial structures are time-varying, we call them *coherent sets* to distinguish them from the spatially fixed almost-invariant sets. If the sets $A_t, A_{t+\tau}$ satisfy $A_t \approx T^{-1}A_{t+\tau}$, then the set of points that are *currently in A_t and will be carried to $A_{t+\tau}$ after τ time steps*, namely $A_t \cap T^{-1}A_{t+\tau}$, is large relative to A_t, i.e. $\frac{\mu(A_t \cap T^{-1}A_{t+\tau})}{\mu(A_t)} \approx 1$. Thus, the conditional probability to *not* be carried from A_t to $A_{t+\tau}$ is low, and we say that such $A_t, A_{t+\tau}$ constitute a pair of coherent sets. We will be interested in optimal coherent sets; optimal in the sense that the conditional probability to not be carried from A_t to $A_{t+\tau}$ is minimal.

In terms of operators, we again make use of an operator $\mathbf{L}$ that is dynamically similar to $\mathbf{P}$, where $\mathbf{P}$ is the Perron-Frobenius operator for the map T which governs the dynamics from time t to $t+\tau$; roughly speaking, $\mathbf{L}$ will have the property that $\mathbf{L}\mathbf{1}_X = \mathbf{1}_{T(X)}$. The equivariance condition $A_t \approx T^{-1}A_{t+\tau}$ can be translated into the operator equation $\mathbf{L}\mathbf{1}_{A_t} \approx \mathbf{1}_{A_{t+\tau}}$. Clearly, this operator equation cannot be directly solved as an eigenequation as $\mathbf{1}_{A_t}$ and $\mathbf{1}_{A_{t+\tau}}$ may be very different functions. Instead, our approach will be to push forward $\mathbf{1}_{A_t}$ with $\mathbf{L}$, to obtain something close to $\mathbf{1}_{A_{t+\tau}}$, and then pull back with $\mathbf{L}^*$ to return to something close to $\mathbf{1}_{A_t}$. The operators $\mathbf{L}$ and $\mathbf{L}^*$ will include a small amount of diffusive dynamics, and at the operator level, the way in which mass is not carried from A_t to $A_{t+\tau}$ is by mass diffusing from both A_t and $A_{t+\tau}$. Leaving technical details for later, we consider eigenfunctions[6] of $\mathbf{L}^*\mathbf{L}$ (push forward from t to $t+\tau$, then pull back from $t+\tau$ to t) to determine the set A_t and eigenfunctions of $\mathbf{L}\mathbf{L}^*$ (pull back from $t+\tau$ to t then push forward from t to $t+\tau$) to determine the set $A_{t+\tau}$. Both $\mathbf{L}^*\mathbf{L}$ and $\mathbf{L}\mathbf{L}^*$ are self-adjoint in appropriate Hilbert spaces and we make use of associated optimality properties to find the *optimal* pair of coherent sets.

9.2.3 Nonautonomous, Time-Dependent, or Aperiodically Forced Dynamics: Both Time Directions

The dynamical setting is almost the same as in Sect. 9.2.2, except that we consider two segments of time. Firstly, we consider dynamics from time t to $t-\tau_1$, governed by a map T_-, and secondly from t to $t+\tau_2$, governed by a map T_+. The maps T_- and T_+ are generated in the same way as in Sect. 9.2.2. Our focus is on finding a coherent set at the intermediate time t, one that is coherent in *both*[7] forward and backward time. Thus, we wish to select a triple of sets $A_{t-\tau_1}, A_t, A_{t+\tau_2}$ with the property that $A_{t-\tau_1} \approx T_-A_t$ *and* $A_{t+\tau_2} \approx T_+A_t$. As in Sect. 9.2.2, the user prescribes a probability measure μ at time t that represents a mass distribution we are interested in transporting.

[6]In numerical computations it is cheaper to compute singular vectors of $\mathbf{L}$, than to form $\mathbf{L}^*\mathbf{L}$ and compute eigenvectors (see Algorithm 2 in this chapter and Sect. 6 [14]).

[7]A similar result could also be achieved by a more direct application of the construction in the previous section; we discuss this further in Sect. 9.5.2.

In terms of operators, we again make use of operators $\mathbf{L}_-$ and $\mathbf{L}_+$ that are dynamically similar to $\mathbf{P}_{T_-}$ and $\mathbf{P}_{T_+}$, where $\mathbf{P}_{T_\pm}$ is the Perron-Frobenius operator for the map $T_\pm$. The equivariance conditions $A_{t-\tau_1} \approx T_- A_t$ and $A_t \approx T_+^{-1} A_{t+\tau_2}$ can be translated into the operator equations $\mathbf{L}_- \mathbf{1}_{A_t} \approx \mathbf{1}_{A_{t-\tau_1}}$ and $\mathbf{L}_+ \mathbf{1}_{A_t} \approx \mathbf{1}_{A_{t+\tau_2}}$, respectively. As in Sect. 9.2.2, these operator equations cannot be directly solved as an eigenequation. In analogy to the single-direction case, our approach will be to push forward $\mathbf{1}_{A_t}$ with $\mathbf{L}_+$, to obtain something close to $\mathbf{1}_{A_{t+\tau_2}}$, and then pull back with $\mathbf{L}_+^*$, to return to something close to $\mathbf{1}_{A_t}$; this utilises the dynamics on the interval $[t, t+\tau_2]$. Similarly, we also push forward $\mathbf{1}_{A_t}$ (under backward-time dynamics) with $\mathbf{L}_-$ and then pull back with $\mathbf{L}_-^*$ to again return to something close to $\mathbf{1}_{A_t}$, however, this time utilising the dynamics from t to $t-\tau_1$. Both $\mathbf{L}_+^*\mathbf{L}_+$ and $\mathbf{L}_-^*\mathbf{L}_-$ are self-adjoint in the same Hilbert space (anchored at time t). We may now average the effect of the dynamics over the intervals $[t-\tau_1, t]$ and $[t, t+\tau_2]$ by averaging the operators to form $(\mathbf{L}_-^*\mathbf{L}_- + \mathbf{L}_+^*\mathbf{L}_+)/2$, again obtaining a self-adjoint operator, which has the necessary optimality properties to find the *optimal* triple of coherent sets.

9.3 Two Key Tools

In this section we introduce two key objects for the analysis that follows. The first is the operator $\mathbf{L}$, which acts as a building block for purpose-built operators for each of the three dynamical settings we consider. The second is a class of self-adjoint operators, which we use heavily to obtain optimality results.

9.3.1 A Building Block Operator

Our dynamical system is acting on a subset M of $\mathbb{R}^d$ and our transport analysis will be confined to a compact subset X of M and neighbourhoods of X and $T(X)$, denoted X_ϵ and Y_ϵ, respectively, where ϵ is a diffusion parameter related to the magnitude of diffusion. A specific type of diffusion is discussed at the end of Sect. 9.3.1.

We construct our basic building block operator $\mathbf{L}$ from the Perron-Frobenius operator for T. The constructions here are common to the three settings we consider and may be found in greater detail in [14]. Specialisations for the autonomous case will be pointed out after the main construction. We endow X with a probability measure μ, which we assume is absolutely continuous with positive density h_μ.

We begin building the operator $\mathbf{L}$ by pre- and post-applying diffusion to $\mathbf{P}$. We define a diffusion operator on X, $\mathbf{D}_{X,\epsilon} : L^1(X,\ell) \to L^1(X_\epsilon,\ell)$ by $\mathbf{D}_{X,\epsilon} f(y) = \int_X \alpha_{X,\epsilon}(y-x) f(x)dx$, where $\alpha_{X,\epsilon} : X_\epsilon \to \mathbb{R}_+$ is bounded and satisfies $\int_{X_\epsilon} \alpha_{X,\epsilon}(y-x)\, dy = 1$ for all $x \in X$. The operator $D_{X,\epsilon}$ acts as convolution with a stochastic kernel; see, e.g. §10.5 [30]. Similarly we define $\mathbf{D}_{Y'_\epsilon,\epsilon} : L^1(Y'_\epsilon, \ell) \to L^1(Y_\epsilon, \ell)$ by $\mathbf{D}_{Y'_\epsilon,\epsilon} f(y) = \int_{Y'_\epsilon} \alpha_{Y,\epsilon}(y-x) f(x)dx$, where $\alpha_{Y,\epsilon}$ is bounded and satisfies $\int_{Y_\epsilon} \alpha_{Y,\epsilon}(y-x)\, dy = 1$ for $x \in Y'_\epsilon$. We think of $X_\epsilon = \text{supp}(\mathbf{D}_{X,\epsilon}\mathbf{1}_X)$, $Y'_\epsilon = T(X_\epsilon)$ and $Y_\epsilon = \text{supp}(\mathbf{D}_{Y'_\epsilon,\epsilon}\mathbf{1}_{Y'_\epsilon})$. In terms of function spaces we have

$$L^1(X,\ell) \xrightarrow{\mathbf{D}_{X,\epsilon}} L^1(X_\epsilon,\ell) \xrightarrow{\mathbf{P}} L^1(Y'_\epsilon,\ell) \xrightarrow{\mathbf{D}_{Y'_\epsilon,\epsilon}} L^1(Y_\epsilon,\ell) \tag{9.1}$$

As an intermediate operator, we define $\mathbf{P}_\epsilon = \mathbf{D}_{Y'_\epsilon,\epsilon}\mathbf{P}\mathbf{D}_{X,\epsilon}$, our pre- and post-diffused Perron-Frobenius operator. In general $\mathbf{P}_\epsilon \mathbf{1}_X \neq \mathbf{1}_{Y_\epsilon}$; to obtain this property, we perform a 'change of measure' transformation and define

$$\mathbf{L}_\epsilon f = \mathbf{P}_\epsilon(f \cdot h_\mu)/\mathbf{P}_\epsilon(h_\mu). \tag{9.2}$$

Lemma 9.1. *One has* $\mathbf{L}_\epsilon \mathbf{1}_X = \mathbf{1}_{Y_\epsilon}$ *and* $\mathbf{L}_\epsilon^* \mathbf{1}_{Y_\epsilon} = \mathbf{1}_X$.

We refer the reader to Sect. 4.2 [14] for the necessary calculations to prove Lemma 9.1. The set-up described above covers both time-dependent situations under consideration; the map T can control the dynamics over the interval $[t, t+\tau]$ as in Sect. 9.2.2 or over one of the intervals $[t, t+\tau_2]$, $[t-\tau_1, t]$ as in Sect. 9.2.3. In the autonomous setting, there is one additional hypothesis, namely that h_μ is fixed by the advective and diffusive dynamics, that is[8], $\mathbf{P}_\epsilon h_\mu = h_\mu$. It follows that $X = Y_\epsilon$; examples of such a situation include:

(i) X is boundaryless (e.g. a solid torus represented by a solid cube in $\mathbb{R}^3$, identifying opposing faces),
(ii) X is an attractor,
(iii) X has boundaries; however, $\alpha_{X,\epsilon}, \alpha_{Y,\epsilon}$ are chosen in such a way that the diffusion does not perturb points across the boundary of X.

From Sect. 9.4 onwards, we will consider the situations where one chooses $\alpha_{X,\epsilon} = \alpha_{Y,\epsilon} = (1/\ell(B_\epsilon(0)))\mathbf{1}_{B_\epsilon(0)}$. Dynamically, this means that one diffuses uniformly over an ϵ-ball, then applies the deterministic dynamics T, and then diffuses uniformly over an ϵ-ball again. We have chosen such $\alpha_{X,\epsilon}, \alpha_{Y,\epsilon}$ as these are very natural choices of small random perturbations of the dynamics with bounded support. An important property of our building block operator is compactness.

Theorem 9.1 ([14]). *If* Y_ϵ *has finite Lebesgue measure and one chooses* $\alpha_{X,\epsilon} = \alpha_{Y,\epsilon} = (1/\ell(B_\epsilon(0)))\mathbf{1}_{B_\epsilon(0)}$, *then* $\mathbf{L}_\epsilon : L^2(X,\mu) \to L^2(Y_\epsilon, \nu_\epsilon)$ *is compact operator.*

Moreover, this choice of $\alpha_{X,\epsilon}, \alpha_{Y,\epsilon}$ is numerically accessible by representing the ϵ-ball diffusion via a fine grid of points uniformly distributed over $B_\epsilon(0)$; see Sect 9.6.2 for further details.

[8] In general, $\mathbf{P}_\epsilon h_\mu = h_{\nu_\epsilon}$, where h_{ν_ϵ} is a density of the push-forward of h_μ by the dynamics.

9.3.2 Optimality Properties of Compact Self-Adjoint Operators on Hilbert Space

We recall the minimax principle for compact self-adjoint operators, which forms a key part of our approach. Let $\mathbf{Q} : \mathscr{H} \to \mathscr{H}$ be a compact, self-adjoint operator on a Hilbert space $\mathscr{H}$. Then $\mathbf{Q}$ has only a countable number of distinct eigenvalues. We enumerate the positive eigenvalues of $\mathbf{Q}$, $\lambda_1^+ \geq \lambda_2^+ \geq \cdots$ and the negative eigenvalues, $\lambda_1^- \leq \lambda_2^- \leq \cdots$, where the number of occurrences equals the multiplicity of the eigenvalue. We choose orthonormal bases of eigenvectors $u_l^\pm$, enumerated so that $\mathbf{Q}u_l^\pm = \lambda_l^\pm u_l^\pm$. We may write

$$\mathbf{Q} = \sum_{l=1}^{M^-} \lambda_l^- \langle \cdot, u_l^- \rangle u_l^- + \sum_{l=1}^{M^+} \lambda_l^+ \langle \cdot, u_l^+ \rangle u_l^+, \tag{9.3}$$

where $M^\pm$ may be finite or infinite (see, e.g. Theorem II.5.1 [7]).

If $\mathbf{Q}$ is also positive[9], then

$$\mathbf{Q} = \sum_{l=1}^{M^+} \lambda_l^+ \langle \cdot, u_l^+ \rangle u_l^+, \tag{9.4}$$

where M^+ may be finite or infinite. All eigenvalues of $\mathbf{Q}$ are nonnegative and are denoted $\lambda_1^+ \geq \lambda_2^+ \geq \cdots$; see, e.g., Proposition II.7.14 [7].

One has the following minimax principle (see, e.g. Theorem 9.2.4, p212 [4]):

Theorem 9.2. *Let $\mathbf{Q} : \mathscr{H} \to \mathscr{H}$ be compact and self-adjoint and V denote a subspace of $\mathscr{H}$. Then*

$$\lambda_l^+ = \min_{V:\operatorname{codim} V \leq l-1 < M^+} \max_{0 \neq f \in V} \frac{\langle \mathbf{Q}f, f \rangle}{\langle f, f \rangle}, l = 1, \ldots, M^+ \tag{9.5}$$

and

$$\lambda_l^- = - \min_{V:\operatorname{codim} V \leq l-1 < M^-} \max_{0 \neq f \in V} \frac{-\langle \mathbf{Q}f, f \rangle}{\langle f, f \rangle}, l = 1, \ldots, M^-. \tag{9.6}$$

Furthermore, the maximising fs are u_l^+ and u_l^-, respectively.

The eigenvalue we are particularly interested in is λ_2^+, as this eigenvalue will correspond to the subspace of functions with the slowest nontrivial decay[10].

---

[9] An operator $\mathbf{Q}$ on a Hilbert space $\mathscr{H}$ is called positive if $\langle \mathbf{Q}x, x \rangle \geq 0$ for all $x \in \mathscr{H}$.

[10] In our dynamical application, all eigenvalues of $\mathbf{Q}$ will have magnitude less than or equal to 1, and those with magnitude strictly smaller than 1 correspond to decay. The corresponding eigenfunctions will ultimately help us to find the most almost-invariant or coherent sets.

A natural Hilbert space to consider is $\mathscr{H} = L^2(X,\mu) \subset L^1(X,\mu)$. With $\langle f, g\rangle_\mu = \int f \cdot g \, d\mu$, one has

Corollary 9.1. *If λ_1^+ is simple, then the V in (9.5) is $\mathrm{sp}\{u_1^+\}^\perp$, one has*

$$\lambda_2^+ = \max_{f \in L^2(X,\mu)} \left\{ \frac{\langle \mathbf{Q} f, f\rangle_\mu}{\langle f, f\rangle_\mu} : \langle f, u_1^+\rangle_\mu = 0 \right\}. \tag{9.7}$$

and the maximising f is u_2^+.

In the following we will only be interested in the large magnitude eigenvalues in the positive part of the spectrum, and we henceforth drop the '+' superscripts from these eigenvalues and eigenvectors.

9.4 Main Constructions and Results

Using our building block operator $\mathbf{L}$, in each of the three dynamical settings we construct a suitable operator-based optimisation problem whose solution yields optimal almost-invariant or coherent sets. We then form a relaxation of this problem and show that at its core is a self-adjoint operator. Finally, we utilise the optimality properties of this self-adjoint operator to exactly solve this relaxed problem via eigenvectors and then use these eigenvectors to construct almost-invariant or coherent sets.

9.4.1 *Autonomous Dynamics*

We consider the operator $\mathbf{L}_\epsilon : L^2(X,\mu) \to L^2(X,\mu)$, defined as in (9.2) with $\mathbf{P}$ the Perron-Frobenius operator for T. We wish to measurably partition $X = A \cup A^c$ such that $\mathbf{L}_\epsilon \mathbf{1}_A \approx \mathbf{1}_A$, $\mathbf{L}_\epsilon \mathbf{1}_{A^c} \approx \mathbf{1}_{A^c}$, and $\mu(A) \approx \mu(A^c)$; the latter condition avoids highly unbalanced partitions where one of the sets contains almost all of the measure. This can be achieved by considering

$$\rho(A) = \frac{\langle \mathbf{L}_\epsilon \mathbf{1}_A, \mathbf{1}_A\rangle_\mu}{\mu(A)} + \frac{\langle \mathbf{L}_\epsilon \mathbf{1}_{A^c}, \mathbf{1}_{A^c}\rangle_\mu}{\mu(A^c)}. \tag{9.8}$$

The above expression can be interpreted as follows. Supposing for the moment that $\mathbf{L}_\epsilon = \mathbf{L}_0$ defined as $\mathbf{L}_0 f = \mathbf{P}(f \cdot h_\mu)/h_\mu$; this is an 'advection only' version of $\mathbf{L}_\epsilon$. Then (using the duality property of $\mathbf{P}$ with respect to ℓ: $\langle \mathbf{P} f, g\rangle_\ell = \langle f, g \circ T\rangle_\ell$),

$$\begin{aligned}\frac{\langle \mathbf{L}_0 \mathbf{1}_A, \mathbf{1}_A\rangle_\mu}{\mu(A)} &= \frac{\langle \mathbf{P}(\mathbf{1}_A \cdot h_\mu)/h_\mu, \mathbf{1}_A\rangle_\mu}{\mu(A)} \\ &= \frac{\langle \mathbf{P}(\mathbf{1}_A \cdot h_\mu), \mathbf{1}_A\rangle_\ell}{\mu(A)} = \frac{\langle \mathbf{1}_A \cdot h_\mu, \mathbf{1}_A \circ T\rangle_\ell}{\mu(A)} = \frac{\langle \mathbf{1}_A, \mathbf{1}_A \circ T\rangle_\mu}{\mu(A)} = \frac{\mu(A \cap T^{-1}A)}{\mu(A)}.\end{aligned} \tag{9.9}$$

Similarly, $\langle \mathbf{L}_0 \mathbf{1}_{A^c}, \mathbf{1}_{A^c}\rangle_\mu/\mu(A^c) = \mu(A^c \cap T^{-1}A^c)/\mu(A^c)$, thus, in the zero diffusion setting

$$\rho_0(A) = \frac{\mu(A \cap T^{-1}A)}{\mu(A)} + \frac{\mu(A^c \cap T^{-1}A^c)}{\mu(A^c)},$$

which is a natural expression to optimise over A to find the optimal almost-invariant set A. Maximising ρ_0 over all measurable sets is not a well-posed optimisation problem so we use the addition of a small amount of diffusion via the operators $\mathbf{D}_{X,\epsilon}$ to regularise this problem, while not changing greatly the value of the expression for $\rho(A)$.

As $\langle \mathbf{L}_\epsilon \mathbf{1}_A, \mathbf{1}_A\rangle_\mu = \overline{\langle \mathbf{L}_\epsilon \mathbf{1}_A, \mathbf{1}_A\rangle_\mu} = \langle \mathbf{L}_\epsilon^* \mathbf{1}_A, \mathbf{1}_A\rangle_\mu$ one has

$$\rho(A) = \frac{\langle \mathbf{L}_\epsilon \mathbf{1}_A, \mathbf{1}_A\rangle_\mu}{\mu(A)} + \frac{\langle \mathbf{L}_\epsilon \mathbf{1}_{A^c}, \mathbf{1}_{A^c}\rangle_\mu}{\mu(A^c)} = \frac{\langle \mathbf{Q}_\epsilon \mathbf{1}_A, \mathbf{1}_A\rangle_\mu}{\mu(A)} + \frac{\langle \mathbf{Q}_\epsilon \mathbf{1}_{A^c}, \mathbf{1}_{A^c}\rangle_\mu}{\mu(A^c)}, \tag{9.10}$$

where $\mathbf{Q}_\epsilon = (\mathbf{L}_\epsilon + \mathbf{L}_\epsilon^*)/2$. The advantage of $\mathbf{Q}_\epsilon$ is that it is self-adjoint in $L^2(X,\mu)$. Now one has

$$\begin{aligned}
&\rho(A) - 1\\
&= \frac{\langle \mathbf{Q}_\epsilon \mathbf{1}_A, \mathbf{1}_A\rangle_\mu}{\mu(A)} + \frac{\langle \mathbf{Q}_\epsilon \mathbf{1}_{A^c}, \mathbf{1}_{A^c}\rangle_\mu}{\mu(A^c)} - 1\\
&= \left(\frac{\mu(A^c)}{\mu(A)} + 1\right)\langle \mathbf{Q}_\epsilon \mathbf{1}_A, \mathbf{1}_A\rangle_\mu + \left(\frac{\mu(A)}{\mu(A^c)} + 1\right)\langle \mathbf{Q}_\epsilon \mathbf{1}_{A^c}, \mathbf{1}_{A^c}\rangle_\mu - (\langle \mathbf{Q}_\epsilon \mathbf{1}_A, \mathbf{1}_X\rangle_\mu + \langle \mathbf{Q}_\epsilon \mathbf{1}_{A^c}, \mathbf{1}_X\rangle_\mu)\\
&= \left(\frac{\mu(A^c)}{\mu(A)}\langle \mathbf{Q}_\epsilon \mathbf{1}_A, \mathbf{1}_A\rangle_\mu + \frac{\mu(A)}{\mu(A^c)}\langle \mathbf{Q}_\epsilon \mathbf{1}_{A^c}, \mathbf{1}_{A^c}\rangle_\mu\right) - (\langle \mathbf{Q}_\epsilon \mathbf{1}_A, \mathbf{1}_X - \mathbf{1}_A\rangle_\mu + \langle \mathbf{Q}_\epsilon \mathbf{1}_{A^c}, \mathbf{1}_X - \mathbf{1}_{A^c}\rangle_\mu)\\
&= \left(\frac{\mu(A^c)}{\mu(A)}\int \mathbf{Q}_\epsilon \mathbf{1}_A \cdot \mathbf{1}_A\, d\mu + \frac{\mu(A)}{\mu(A^c)}\int \mathbf{Q}_\epsilon \mathbf{1}_{A^c} \cdot \mathbf{1}_{A^c}\, d\mu\right) - \left(\int \mathbf{Q}_\epsilon \mathbf{1}_A \cdot \mathbf{1}_{A^c}\, d\mu + \int \mathbf{Q}_\epsilon \mathbf{1}_{A^c} \cdot \mathbf{1}_A\, d\mu\right)\\
&= \left\langle \mathbf{Q}_\epsilon \left(\sqrt{\frac{\mu(A^c)}{\mu(A)}}\mathbf{1}_A - \sqrt{\frac{\mu(A)}{\mu(A^c)}}\mathbf{1}_{A^c}\right), \left(\sqrt{\frac{\mu(A^c)}{\mu(A)}}\mathbf{1}_A - \sqrt{\frac{\mu(A)}{\mu(A^c)}}\mathbf{1}_{A^c}\right)\right\rangle_\mu
\end{aligned}$$

Thus, finding $A \subset X$ that maximises $\rho(A)$ is equivalent to the problem

$$\max_{A \subset X}\left\langle \mathbf{Q}_\epsilon \left(\sqrt{\frac{\mu(A^c)}{\mu(A)}}\mathbf{1}_A - \sqrt{\frac{\mu(A)}{\mu(A^c)}}\mathbf{1}_{A^c}\right), \left(\sqrt{\frac{\mu(A^c)}{\mu(A)}}\mathbf{1}_A - \sqrt{\frac{\mu(A)}{\mu(A^c)}}\mathbf{1}_{A^c}\right)\right\rangle_\mu. \tag{9.11}$$

Note that $\left\langle \sqrt{\frac{\mu(A^c)}{\mu(A)}}\mathbf{1}_A - \sqrt{\frac{\mu(A)}{\mu(A^c)}}\mathbf{1}_{A^c}, \mathbf{1}_X\right\rangle_\mu = 0$ for any choice of A. Thus, a relaxed form of (9.11), where we remove the restriction that the argument of $\mathbf{Q}_\epsilon$ is a difference of characteristic functions, but retain the orthogonality property of this ansatz, is

$$\max_{f \in L^2(X,\mu)}\left\{\frac{\langle \mathbf{Q}_\epsilon f, f\rangle_\mu}{\langle f, f, \rangle_\mu} : \langle f, \mathbf{1}_X\rangle_\mu = 0\right\}. \tag{9.12}$$

Recall that $\mathbf{Q}_\epsilon \mathbf{1}_X = \mathbf{1}_X$ and that $\mathbf{Q}_\epsilon$ is self-adjoint and compact (by Theorem 9.1 setting $Y_\epsilon = X$ and $\nu_\epsilon = \mu$), but not positive, in general. Furthermore (see discussion in Sect. 9.5.3), $\lambda_1 = 1$ is simple. We may therefore apply Corollary 9.1 to see that the maximum value of (9.12) is λ_2, the second largest eigenvalue of $\mathbf{Q}_\epsilon$, and that the maximising f is u_2, the corresponding eigenvector of $\mathbf{Q}_\epsilon$. As (9.12) is a relaxation of (9.11), we immediately see that $\rho(A) \leq 1 + \lambda_2$ for all measurable $A \subset X$. One can also obtain an *a priori* lower bound for $\rho(A)$:

Theorem 9.3.

$$2 - 2\sqrt{2(1-\lambda_2)} \leq \sup_{A \subset X} \rho(A) \leq 1 + \lambda_2.$$

Proof. See the Appendix.

The above result is strongly related to classic conductance bounds in the reversible Markov chain literature [31,46] and has been discussed in regard to almost-invariant sets in a matrix setting [12]. An *a posteriori* lower bound (relying on eigenvector computations) has been proposed by Huisinga and Schmidt [27].

Finally, to construct a partition A, A^c from the solution to (9.12), we set $A = A_\beta := \{x \in X : u_2(x) > \beta\}$ and $A^c = A^c_\beta := \{x \in X : u_2(x) \leq \beta\}$, where β is chosen to maximise $\rho(A_\beta)$. Intuitively, one may think of u_2 as a signed fuzzy partition, and the thresholding procedure makes a 'hard', non-fuzzy choice of partition into sets. In numerical computations, β is obtained via a line search, as described in Sect. 9.6.3.

9.4.2 Nonautonomous or Time-Dependent Dynamics: Single Time Direction

The constructions in this section may be found in greater detail in [14]; we recall only the main points here. We consider the operator $\mathbf{L}_\epsilon : L^2(X,\mu) \to L^2(Y_\epsilon, \nu_\epsilon)$ defined as in (9.2) using $\mathbf{P}$ the Perron-Frobenius operator for T, where T represents the finite-time dynamics[11] from some time t to $t + \tau$. We wish to measurably partition $X = A \cup A^c$ and $Y_\epsilon = B \cup B^c$ such that:

[11] We have argued earlier it is natural to look for almost-invariant sets in autonomous systems and coherent sets in nonautonomous systems; however, there may be situations in which one is interested in finding coherent sets in autonomous systems. The coherent set framework described here and in [14] can be seamlessly applied to the situation where the map T arises from autonomous dynamics.

(i) $\mathbf{L}_\epsilon \mathbf{1}_A \approx \mathbf{1}_B$ and $\mathbf{L}_\epsilon \mathbf{1}_{A^c} \approx \mathbf{1}_{B^c}$,
(ii) $\mu(A) = \nu_\epsilon(B)$ and $\mu(A^c) = \nu_\epsilon(B^c)$.

We would also like $\mu(A) \approx \mu(A^c)$ and $\nu_\epsilon(B) \approx \nu_\epsilon(B^c)$ to avoid highly unbalanced partitions where one of the sets contains almost all of the measure.

This can be achieved by considering

$$\rho(A,B) = \frac{\langle \mathbf{L}_\epsilon \mathbf{1}_A, \mathbf{1}_B \rangle_{\nu_\epsilon}}{\mu(A)} + \frac{\langle \mathbf{L}_\epsilon \mathbf{1}_{A^c}, \mathbf{1}_{B^c} \rangle_{\nu_\epsilon}}{\mu(A^c)}. \tag{9.13}$$

The two terms in (9.13) are directly related to the two properties in point (i) above. Using arguments identical to the autonomous case above, one can show that in the 'advection only' setting, one has

$$\rho_0(A,B) = \frac{\mu(A \cap T^{-1}B)}{\mu(A)} + \frac{\mu(A^c \cap T^{-1}B^c)}{\mu(A^c)}.$$

Thus, it is now clear that to maximise $\rho(A,B)$ one should choose A so that it is close to the preimage of B (and likewise for the complements). Of course, one could choose $A = T^{-1}B$ to obtain $\rho(A,B) = 2$ and so there is a problem of nonuniqueness in solutions to an optimisation of $\rho(A,B)$ over A and B. As in the autonomous setting we require the small amount of diffusion via $\mathbf{D}_{X,\epsilon}, \mathbf{D}_{Y'_\epsilon,\epsilon}$ to regularise the optimisation of $\rho(A,B)$ over A and B; this also (generically) produces a unique optimum.

Consider the optimisation problem

$$\sup_{A \subset X, B \subset Y_\epsilon} \{\rho(A,B) - 1 : \mu(A) = \nu_\epsilon(B)\}, \tag{9.14}$$

where we have subtracted a constant 1 for convenience. One has (see [14] for details)

$$\rho(A,B) - 1 \tag{9.15}$$

$$= \frac{\langle \mathbf{L}_\epsilon \mathbf{1}_A, \mathbf{1}_B \rangle_{\nu_\epsilon}}{\mu(A)} + \frac{\langle \mathbf{L}_\epsilon \mathbf{1}_{A^c}, \mathbf{1}_{B^c} \rangle_{\nu_\epsilon}}{\mu(A^c)} - 1 \tag{9.16}$$

$$= \left\langle \mathbf{L}_\epsilon \left(\sqrt{\frac{\mu(A^c)}{\mu(A)}} \mathbf{1}_A - \sqrt{\frac{\mu(A)}{\mu(A^c)}} \mathbf{1}_{A^c} \right), \left(\sqrt{\frac{\nu_\epsilon(B^c)}{\nu_\epsilon(B)}} \mathbf{1}_B - \sqrt{\frac{\nu_\epsilon(B)}{\nu_\epsilon(B^c)}} \mathbf{1}_{B^c} \right) \right\rangle_{\nu_\epsilon} \tag{9.17}$$

We note that $\left\langle \sqrt{\frac{\mu(A^c)}{\mu(A)}} \mathbf{1}_A - \sqrt{\frac{\mu(A)}{\mu(A^c)}} \mathbf{1}_{A^c}, \mathbf{1}_X \right\rangle_\mu = 0$ and $\left\langle \sqrt{\frac{\nu_\epsilon(B^c)}{\nu_\epsilon(B)}} \mathbf{1}_B - \sqrt{\frac{\nu_\epsilon(B)}{\nu_\epsilon(B^c)}} \mathbf{1}_{B^c}, \mathbf{1}_{Y_\epsilon} \right\rangle_{\nu_\epsilon} = 0$ for any choice of A, B. Thus, a relaxed form of (9.14) where we remove the restriction that the argument of $\mathbf{L}_\epsilon$ is a difference of characteristic functions, but retain the orthogonality property of this ansatz, is

$$\max_{f\in L^2(X,\mu),g\in L^2(Y_\epsilon,\nu_\epsilon)}\left\{\frac{\langle \mathbf{L}_\epsilon f, g\rangle_{\nu_\epsilon}}{\|f\|_\mu\|g\|_{\nu_\epsilon}} : \langle f, \mathbf{1}_X\rangle_\mu = \langle g, \mathbf{1}_{Y_\epsilon}\rangle_{\nu_\epsilon} = 0,\right\}. \tag{9.18}$$

Proposition 1 [14] shows that the value of (9.18) is $\lambda_2^{1/2}$, the square root of the second largest eigenvalue of $\mathbf{Q}_\epsilon := \mathbf{L}_\epsilon^*\mathbf{L}_\epsilon$, and the maximising f (resp. g) is u_2 (resp. v_2), the corresponding eigenvector of $\mathbf{Q}_\epsilon$ (resp. $\mathbf{Q}_\epsilon^*$). In the language of singular values and singular vectors, $\lambda_2^{1/2}$ is the second largest singular value of $\mathbf{L}_\epsilon$ and u_2 (resp. v_2) is the corresponding left (resp. right) singular vector. As (9.18) is a relaxation of (9.14), using (9.15)–(9.17) one obtains

Theorem 9.4 (Froyland [14]).

$$\max_{A\subset X, B\subset Y_\epsilon}\{\rho(A,B) : \mu(A) = \nu_\epsilon(B)\} \le 1 + \lambda_2^{1/2}.$$

To construct a partition A, A^c from the solution to (9.18), we set $A = A_\beta := \{x \in X : u_2(x) > \beta\}$, $A^c = A_\beta^c := \{x \in X : u_2(x) \le \beta\}$, $B = B_{\beta'} := \{y \in Y_\epsilon : v_2(y) > \beta'\}$, $B^c = B_{\beta'}^c := \{y \in Y_\epsilon : v_2(y) \le \beta'\}$, where β is chosen to maximise $\rho(A_\beta, B_{\beta'})$ and $\beta' = \beta'(\beta)$ is chosen so that $\mu(A_\beta) = \nu_\epsilon(B_{\beta'})$. In practice, the optimal value of β is found via a line search; see Sect. 9.6.4 for details.

9.4.3 *Nonautonomous or Time-Dependent Dynamics: Both Time Directions*

Recall that T_+ represents the finite-time dynamics from some time t to $t + \tau_2$ and that T_- represents the finite-time dynamics from some time t to $t - \tau_1$. We consider the operators $\mathbf{L}_{\epsilon,+} : L^2(X,\mu) \to L^2(Y_\epsilon, \nu_\epsilon)$ and $\mathbf{L}_{\epsilon,-} : L^2(X,\mu) \to L^2(Z_\epsilon, \eta_\epsilon)$. The operator $\mathbf{L}_{\epsilon,+}$ is defined by substituting $\mathbf{P}_{T_+}$ (the Perron-Frobenius operator for T_+) for $\mathbf{P}$ in the expression (9.2) for $\mathbf{L}$ in Sect. 9.3.1. The operator $\mathbf{L}_{\epsilon,-}$ is defined by substituting $\mathbf{P}_{T_-}$ (the Perron-Frobenius operator for T_-) for $\mathbf{P}$ in (9.2) and replacing Y'_ϵ, Y_ϵ with Z'_ϵ, Z_ϵ. In summary,

$$L^1(Z_\epsilon,\ell)\xleftarrow{\mathbf{D}_{Z'_\epsilon,\epsilon}}L^1(Z'_\epsilon,\ell)\xleftarrow{\mathbf{P}_{T_-}}L^1(X_\epsilon,\ell)\xleftarrow{\mathbf{D}_{X,\epsilon}}L^1(X,\ell)\xrightarrow{\mathbf{D}_{X,\epsilon}}L^1(X_\epsilon,\ell)\xrightarrow{\mathbf{P}_{T_+}}L^1(Y'_\epsilon,\ell)\xrightarrow{\mathbf{D}_{Y'_\epsilon,\epsilon}}L^1(Y_\epsilon,\ell) \tag{9.19}$$

We wish to measurably partition $X = A \cup A^c$, $Y_\epsilon = B \cup B^c$, and $Z_\epsilon = C \cup C^c$ such that

(i) $\mathbf{L}_{\epsilon,+}\mathbf{1}_A \approx \mathbf{1}_B$ and $\mathbf{L}_{\epsilon,+}\mathbf{1}_{A^c} \approx \mathbf{1}_{B^c}$,
(ii) $\mathbf{L}_{\epsilon,-}\mathbf{1}_A \approx \mathbf{1}_C$ and $\mathbf{L}_{\epsilon,-}\mathbf{1}_{A^c} \approx \mathbf{1}_{C^c}$,
(iii) $\mu(A) = \nu_\epsilon(B) = \eta_\epsilon(C)$ and $\mu(A^c) = \nu_\epsilon(B^c) = \eta_\epsilon(C^c)$.

We would also like $\mu(A) \approx \mu(A^c)$, $\nu_\epsilon(B) \approx \nu_\epsilon(B^c)$, and $\eta_\epsilon(C) \approx \eta_\epsilon(C^c)$, to avoid highly unbalanced partitions where one of the sets contains almost all of the measure.

We consider

$$\rho(A,B,C) = \frac{1}{2}\left(\left(\frac{\langle \mathbf{L}_{\epsilon,+}\mathbf{1}_A, \mathbf{1}_B\rangle_{\nu_\epsilon}}{\mu(A)} + \frac{\langle \mathbf{L}_{\epsilon,+}\mathbf{1}_{A^c}, \mathbf{1}_{B^c}\rangle_{\nu_\epsilon}}{\mu(A^c)}\right) + \left(\frac{\langle \mathbf{L}_{\epsilon,-}\mathbf{1}_A, \mathbf{1}_C\rangle_{\eta_\epsilon}}{\mu(A)} + \frac{\langle \mathbf{L}_{\epsilon,-}\mathbf{1}_{A^c}, \mathbf{1}_{C^c}\rangle_{\eta_\epsilon}}{\mu(A^c)}\right)\right). \tag{9.20}$$

The four terms in (9.20) are concerned with the four properties in points (i) and (ii) above. Using arguments identical to the autonomous case above, one can show that in the 'advection only' setting, one has

$$\rho_0(A,B,C) = \frac{1}{2}\left(\left(\frac{\mu(A\cap T_+^{-1}B)}{\mu(A)} + \frac{\mu(A^c\cap T_+^{-1}B^c)}{\mu(A^c)}\right) + \left(\frac{\mu(A\cap T_-^{-1}C)}{\mu(A)} + \frac{\mu(A^c\cap T_-^{-1}C^c)}{\mu(A^c)}\right)\right).$$

Consider the optimisation problem

$$\sup_{A\subset X, B\subset Y_\epsilon, C\subset Z_\epsilon} \{\rho(A,B,C) - 1 : \mu(A) = \nu_\epsilon(B) = \eta_\epsilon(C)\}, \tag{9.21}$$

where we have subtracted a constant 1 for convenience. Using the expression (9.20) for $\rho(A,B,C)$ we have (using (9.17) twice)

$$\rho(A,B,C) - 1$$
$$= \frac{1}{2}\left(\left(\frac{\langle \mathbf{L}_{\epsilon,+}\mathbf{1}_A, \mathbf{1}_B\rangle_{\nu_\epsilon}}{\mu(A)} + \frac{\langle \mathbf{L}_{\epsilon,+}\mathbf{1}_{A^c}, \mathbf{1}_{B^c}\rangle_{\nu_\epsilon}}{\mu(A^c)}\right) + \left(\frac{\langle \mathbf{L}_{\epsilon,-}\mathbf{1}_A, \mathbf{1}_C\rangle_{\eta_\epsilon}}{\mu(A)} + \frac{\langle \mathbf{L}_{\epsilon,-}\mathbf{1}_{A^c}, \mathbf{1}_{C^c}\rangle_{\eta_\epsilon}}{\mu(A^c)}\right)\right) - 1 \tag{9.22}$$
$$= \frac{1}{2}\left(\left\langle \mathbf{L}_{\epsilon,+}\left(\sqrt{\frac{\mu(A^c)}{\mu(A)}}\mathbf{1}_A - \sqrt{\frac{\mu(A)}{\mu(A^c)}}\mathbf{1}_{A^c}\right), \sqrt{\frac{\nu_\epsilon(B^c)}{\nu_\epsilon(B)}}\mathbf{1}_B - \sqrt{\frac{\nu_\epsilon(B)}{\nu_\epsilon(B^c)}}\mathbf{1}_{B^c}\right\rangle_{\nu_\epsilon}\right. \tag{9.23}$$
$$\left. + \left\langle \mathbf{L}_{\epsilon,-}\left(\sqrt{\frac{\mu(A^c)}{\mu(A)}}\mathbf{1}_A - \sqrt{\frac{\mu(A)}{\mu(A^c)}}\mathbf{1}_{A^c}\right), \sqrt{\frac{\eta_\epsilon(C^c)}{\eta_\epsilon(C)}}\mathbf{1}_C - \sqrt{\frac{\eta_\epsilon(C)}{\eta_\epsilon(C^c)}}\mathbf{1}_{C^c}\right\rangle_{\eta_\epsilon}\right). \tag{9.24}$$

Using the shorthand $\Psi_A = \sqrt{\frac{\mu(A^c)}{\mu(A)}}\mathbf{1}_A - \sqrt{\frac{\mu(A)}{\mu(A^c)}}\mathbf{1}_{A^c}$, $\Psi_B = \sqrt{\frac{\nu_\epsilon(B^c)}{\nu_\epsilon(B)}}\mathbf{1}_B - \sqrt{\frac{\nu_\epsilon(B)}{\nu_\epsilon(B^c)}}\mathbf{1}_{B^c}$, and $\Psi_C = \sqrt{\frac{\eta_\epsilon(C^c)}{\eta_\epsilon(C)}}\mathbf{1}_C - \sqrt{\frac{\eta_\epsilon(C)}{\eta_\epsilon(C^c)}}\mathbf{1}_{C^c}$, it is straightforward to verify that $||\Psi_A||_\mu = ||\Psi_B||_{\nu_\epsilon} = ||\Psi_C||_{\eta_\epsilon} = 1$ and $\langle \Psi_A, \mathbf{1}_X\rangle_\mu = \langle \Psi_B, \mathbf{1}_{Y_\epsilon}\rangle_{\nu_\epsilon} = \langle \Psi_C, \mathbf{1}_{Z_\epsilon}\rangle_{\eta_\epsilon} = 0$. Thus, a relaxed form of (9.21), where we remove the restriction that the arguments of $\mathbf{L}_{\epsilon,+}$ and $\mathbf{L}_{\epsilon,-}$ are differences of characteristic functions, but retain the orthogonality properties of this ansatz, is

$$\max_{f\in L^2(X,\mu), g\in L^2(Y_\epsilon,\nu_\epsilon), h\in L^2(Z_\epsilon,\eta_\epsilon)} \left\{\frac{1}{2}\left(\frac{\langle \mathbf{L}_{\epsilon,+}f, g\rangle_{\nu_\epsilon}}{||f||_\mu ||g||_{\nu_\epsilon}} + \frac{\langle \mathbf{L}_{\epsilon,-}f, h\rangle_{\eta_\epsilon}}{||f||_\mu ||h||_{\eta_\epsilon}}\right)\right.$$
$$\left. : \langle f, \mathbf{1}_X\rangle_\mu = \langle g, \mathbf{1}_{Y_\epsilon}\rangle_{\nu_\epsilon} = \langle h, \mathbf{1}_{Z_\epsilon}\rangle_{\eta_\epsilon} = 0\right\} \tag{9.25}$$

We now rewrite this relaxed optimisation problem in terms of the self-adjoint operator $\mathbf{Q}_\epsilon = (\mathbf{L}_{\epsilon,+}^*\mathbf{L}_{\epsilon,+} + \mathbf{L}_{\epsilon,-}^*\mathbf{L}_{\epsilon,-})/2$.

$$
\begin{aligned}
(9.25) &= \max_{f\in L^2(X,\mu)} \left\{ \frac{1}{2}\left(\frac{\langle \mathbf{L}_{\epsilon,+}f, \mathbf{L}_{\epsilon,+}f\rangle_{\nu_\epsilon}}{\|f\|_\mu \|\mathbf{L}_{\epsilon,+}f\|_{\nu_\epsilon}} + \frac{\langle \mathbf{L}_{\epsilon,-}f, \mathbf{L}_{\epsilon,-}f\rangle_{\eta_\epsilon}}{\|f\|_\mu \|\mathbf{L}_{\epsilon,-}f\|_{\eta_\epsilon}} \right) : \langle f, \mathbf{1}_X\rangle_\mu = 0 \right\} \\
&= \max_{f\in L^2(X,\mu)} \left\{ \frac{1}{2}\left(\frac{\langle \mathbf{L}_{\epsilon,+}^*\mathbf{L}_{\epsilon,+}f, f\rangle_\mu}{\|f\|_\mu \langle \mathbf{L}_{\epsilon,+}^*\mathbf{L}_{\epsilon,+}f, f\rangle_\mu^{1/2}} + \frac{\langle \mathbf{L}_{\epsilon,-}^*\mathbf{L}_{\epsilon,-}f, f\rangle_\mu}{\|f\|_\mu \langle \mathbf{L}_{\epsilon,-}^*\mathbf{L}_{\epsilon,-}f, f\rangle_\mu^{1/2}} \right) : \langle f, \mathbf{1}_X\rangle_\mu = 0 \right\} \\
&= \max_{f\in L^2(X,\mu)} \left\{ \frac{1}{2}\left(\left(\frac{\langle \mathbf{L}_{\epsilon,+}^*\mathbf{L}_{\epsilon,+}f, f\rangle_\mu}{\|f\|_\mu^2}\right)^{1/2} + \left(\frac{\langle \mathbf{L}_{\epsilon,-}^*\mathbf{L}_{\epsilon,-}f, f\rangle_\mu}{\|f\|_\mu^2}\right)^{1/2} \right) : \langle f, \mathbf{1}_X\rangle_\mu = 0 \right\} \\
&\le \max_{f\in L^2(X,\mu)} \left\{ \left(\frac{1}{2}\left(\frac{\langle \mathbf{L}_{\epsilon,+}^*\mathbf{L}_{\epsilon,+}f, f\rangle_\mu}{\|f\|_\mu^2} + \frac{\langle \mathbf{L}_{\epsilon,-}^*\mathbf{L}_{\epsilon,-}f, f\rangle_\mu}{\|f\|_\mu^2} \right)\right)^{1/2} : \langle f, \mathbf{1}_X\rangle_\mu = 0 \right\} \\
&= \max_{f\in L^2(X,\mu)} \left\{ \left(\frac{\langle \mathbf{Q}_\epsilon f, f\rangle_\mu}{\|f\|_\mu^2} \right)^{1/2} : \langle f, \mathbf{1}_X\rangle_\mu = 0 \right\}.
\end{aligned}
$$

The operator $\mathbf{Q}_\epsilon$ is self-adjoint, compact (by Theorem 9.1, noting that duals, compositions, and sums of compact operators are compact), and positive (since $\mathbf{L}_{\epsilon,+}^*\mathbf{L}_{\epsilon,+}$ and $\mathbf{L}_{\epsilon,-}^*\mathbf{L}_{\epsilon,-}$ are positive, and sums of positive operators are positive). Assuming λ_1 is simple, by Corollary 9.1 the value of (9.25) is $\lambda_2^{1/2}$, the square root of second largest eigenvalue of $\mathbf{Q}_\epsilon := (\mathbf{L}_{\epsilon,+}^*\mathbf{L}_{\epsilon,+} + \mathbf{L}_{\epsilon,-}^*\mathbf{L}_{\epsilon,-})/2$, and the maximising f is u_2, the second eigenvector of $\mathbf{Q}_\epsilon$. The corresponding maximising g and h are $\mathbf{L}_{\epsilon,+}u_2/\|\mathbf{L}_{\epsilon,+}u_2\|_{\nu_\epsilon}$ and $\mathbf{L}_{\epsilon,-}u_2/\|\mathbf{L}_{\epsilon,-}u_2\|_{\eta_\epsilon}$, respectively. As (9.25) is a relaxation of (9.23), the equalities (9.22)–(9.23) prove:

Theorem 9.5.

$$\max_{A\subset X, B\subset Y_\epsilon, C\subset Z_\epsilon} \{\rho(A,B,C) : \mu(A) = \nu_\epsilon(B) = \eta_\epsilon(C)\} \le 1 + \lambda_2^{1/2}.$$

To construct a partition A, A^c from the solution to (9.25), we set $A = A_\beta := \{x \in X : u_2(x) > \beta\}$, $A^c = A_\beta^c := \{x \in X : u_2(x) \le \beta\}$, $B = B_{\beta'} := \{y \in Y_\epsilon : \mathbf{L}_{\epsilon,+}u_2(y) > \beta'\}$, $B^c = B_{\beta'}^c := \{y \in Y_\epsilon : \mathbf{L}_{\epsilon,+}u_2(y) \le \beta'\}$, $C = C_{\beta''} := \{z \in Z_\epsilon : \mathbf{L}_{\epsilon,-}u_2(z) > \beta''\}$, $C^c = C_{\beta''}^c := \{z \in Z_\epsilon : \mathbf{L}_{\epsilon,-}u_2(z) \le \beta''\}$, where β is chosen to maximise $\rho(A_\beta, B_{\beta'}, C_{\beta''})$ and $\beta' = \beta'(\beta)$, $\beta'' = \beta''(\beta)$ are chosen so that $\mu(A_\beta) = \nu_\epsilon(B_{\beta'}) = \eta_\epsilon(C_{\beta''})$. In practice, the optimal value of β is found via a line search; see Sect. 9.6.5 for details.

9.5 Further Discussion

Having completed the description of the three dynamical set-ups, we now discuss some further properties, focussing mainly on the similarities and differences of Sects. 9.4.2 and 9.4.3.

9.5.1 Single- vs. Bidirectional Coherence

One of the main features that we demonstrate in the two-dimensional numerical case studies is that coherent sets in the sense of Sect. 9.4.2 typically have boundaries that are approximately aligned along (time-dependent) stable and unstable manifolds of organising trajectories of the dynamics. In particular, the boundaries of the coherent sets at the initial time are aligned with stable directions and those at the final time are aligned with unstable directions. Why should this be the case? If the advective dynamics is invertible, the only way that finite-time mixing can occur is via diffusion. Therefore, an efficient way for a set to be rapidly mixed over a finite-time interval would be for the set to be stretched into long filaments, thus greatly increasing the length of its boundary, allowing diffusion to have a much greater effect. Alternatively, choosing a set at the initial time that already has a very long boundary would also enhance mixing as the initial diffusion would have a large effect.

Optimally coherent sets resist this diffusive mixing by having short boundaries at both the initial and final times. Intuitively this is accomplished by the boundary of the initial set being mostly roughly aligned with stable directions; thus, under forward-time evolution, these parts of the boundary are not stretched much, and when the final time is reached, they have evolved so as to roughly align with unstable directions. Symmetrically, one can take the backward-time viewpoint: coherent sets at the final time should have small boundary and evolve backwards in time to sets with small boundary; this is achieved by the sets at the final time being roughly aligned with unstable directions. The length of the finite-time interval under consideration governs how well the boundaries are aligned with stable/unstable manifolds (the latter are time-asymptotic objects); the longer the interval, the stronger the alignment. In fact, we will show in Sect. 9.7 that the optimally coherent sets depend on the time interval considered, as they should.

We have specifically chosen case studies that have regions of strong hyperbolicity to illustrate this point. When the coherence is due to elliptic-type dynamics, as, e.g. in the polar vortex example in [19] or the Agulhas rings in [20], this phenomenon is not observed.

9.5.2 Creating a Sequence of Finite-Time Coherent Sets

In applications, one may be interested in sets that are coherent over a finite-time duration of length τ and in *tracking* such sets over a time horizon of *several multiples of* τ. If one were to compute coherent sets using the techniques of Sect. 9.4.2 from time $t-\tau$ to t, and then from t to $t+\tau$, the sets obtained at t would not match. This is because the discussion in the previous subsection indicates that the first experiment would yield a set at t with boundary roughly aligned with unstable directions, while the second experiment would yield a set at t with boundary roughly aligned with stable directions. The construction of Sect. 9.4.3, on the other hand, computing over the window $t - \tau$ to $t + \tau$ finds a 'central' set at time t with the property that its boundary does not grow large in either backward or forward time. By applying the method of Sect. 9.4.3 to a series of windows of length 2τ, one obtains a sequence of such sets which should vary continuously with time. This is one of the main motivations for the constructions in Sect. 9.4.3.

A similar effect result could be achieved using the method of Sect. 9.4.2 in the following way:

(i) Compute the transfer operator for the period $t - \tau$ to $t + \tau$, and establish optimally coherent sets at $t - \tau$ and $t + \tau$ via thresholding.
(ii) Push the left singular vector (at $t - \tau$) forward with the transfer operator from $t - \tau$ to t and threshold according to the conservation of mass principle.

By pushing forward to the intermediate time t, the sets obtained should also have the property that they have short boundaries when pushed forward to $t + \tau$ or pulled back to $t - \tau$. We will explore this possibility in future work.

9.5.3 *Further Mathematical Properties of the Coherent Set Framework*

In [14] several properties of the operator $\mathbf{L}_\epsilon$ from Sect. 9.4.2 are proved. We briefly mention some of these properties here in the situation where one uses ϵ-ball diffusion for $\alpha_{X,\epsilon}$ and $\alpha_{Y,\epsilon}$. These properties are dealt with in greater detail and generality in [14].

(i) The analytical framework for identifying finite-time coherent sets based on the second singular vectors of $\mathbf{L}_\epsilon$ is *frame-invariant* or *objective*. This means if the framework is applied in a general time-dependent rotating and translating frame, the coherent sets obtained will be the same (except rotated and translated) as those obtained in a static frame. The issue of frame invariance does not apply to Sect. 9.4.1 as the dynamics should not be time-varying. The arguments in [14] could be adapted to the setting of Sect. 9.4.3.
(ii) If T is a diffeomorphism and $X = Y_\epsilon = M$, the leading singular value $\sigma_1 = 1$ of $\mathbf{L}_\epsilon$ is simple. Arguments similar to those in [14] could be applied to the material in Sect. 9.4.1 and 9.4.3 to demonstrate simplicity of λ_1.
(iii) If T is a diffeomorphism and $X = Y_\epsilon = M$, a lower bound on the second singular value $\sigma_{2,\epsilon} < 1$ is given in [14], depending on ϵ. The techniques in [14] could be adapted to the constructions in Sect. 9.4.1 and 9.4.3.
(iv) If T is a diffeomorphism and $X = Y_\epsilon = M$, subdominant singular vectors of $\mathbf{L}_\epsilon$ are 1/2-Hölder regular, with the Hölder constant having an explicit dependence on ϵ (larger ϵ, smaller constant). In particular, this places some limitations on the geometric shapes of the optimally coherent sets. One could apply the techniques in [14] to the constructions in both Sects. 9.4.1 and 9.4.3.

Regarding point (iii) above, in the case of autonomous two-dimensional area-preserving maps T, Junge *et al.* [29] state that the probability to map out of a T*-invariant* set by an ϵ-perturbed systems is bounded to first order from above by ϵ (i.e. the amplitude of perturbation) multiplied by the ratio of set boundary length to set volume. They also state a lower bound on the second largest eigenvalue λ_2 of an operator $\mathbf{R}_\epsilon$; thus for a fixed ϵ, if the invariant set of the unperturbed system has small boundary and its (normalised) Lebesgue measure is close to 1/2, then λ_2 is closer to 1. In [29], the self-adjoint operator $\mathbf{R}_\epsilon$ is constructed similarly to our $\mathbf{Q}_\epsilon$, but using a one-sided diffusion only (i.e. akin to $(\mathbf{D}_{X,\epsilon}\mathbf{P} + (\mathbf{D}_{X,\epsilon}\mathbf{P})^*)/2$).

9.6 Numerical Representations of Transfer Operators

In order to apply the theory developed in the previous sections to specific mathematical models, we require a computer representation of the operators $\mathbf{P}, \mathbf{L}, \mathbf{P}_\epsilon$, and $\mathbf{L}_\epsilon$. We recall here the standard approach of Ulam [49], adapted to our specific operator constructions. We represent these operators as a projected action on a finite-cardinality basis of characteristic functions. Let $\{B_1, \ldots, B_m\}$ denote a partition of X and define $\pi_{X,m} : L^1(X, \ell) \to \mathrm{sp}\{\mathbf{1}_{B_1}, \ldots, \mathbf{1}_{B_m}\}$ by

$$\pi_{X,m} f = \sum_{i=1}^{m} \left(\frac{1}{\ell(B_i)} \int_{B_i} f \, d\ell \right) \mathbf{1}_{B_i}.$$

Similarly, let $\{C_1, \ldots, C_n\}$ denote a partition of Y_ϵ and define $\pi_{Y_\epsilon,n} : L^1(Y_\epsilon, \ell) \to \mathrm{sp}\{\mathbf{1}_{C_1}, \ldots, \mathbf{1}_{C_n}\}$ by

$$\pi_{Y_\epsilon,n} f = \sum_{j=1}^{n} \left(\frac{1}{\ell(C_j)} \int_{C_j} f \, d\ell \right) \mathbf{1}_{C_j},$$

where ℓ is Lebesgue measure. If μ, ν_ϵ are absolutely continuous with respect to ℓ, and the maximal diameter of the partition elements decreases to zero as $m, n \to \infty$, then $\pi_{X,m}, \pi_{Y_\epsilon,n}$ converge strongly to the identity operator in $L^p(X, \mu), L^p(Y_\epsilon, \nu_\epsilon)$, $p = 1, 2$ (see, e.g. Prop. 9, Chap. 6 [43]).

9.6.1 *Numerically Representing* P *and* L

We consider the operator $\pi_{Y_\epsilon,n}\mathbf{P}\pi_{X,m} : \mathrm{sp}\{\mathbf{1}_{B_1}, \ldots, \mathbf{1}_{B_m}\} \to \mathrm{sp}\{\mathbf{1}_{C_1}, \ldots, \mathbf{1}_{C_n}\}$, which has matrix representation[12]

[12] Li [32] contains the first statement of this result in the context of interval maps, but it is straightforward to derive using the property that for each measurable $A \subset Y_\epsilon$, $\int_A \mathbf{P} f \, d\ell = \int_{T^{-1}A} f \, d\ell$ for all $f \in L^1(X, \ell)$.

$$P_{ij} = \frac{\ell(B_i \cap T^{-1}(C_j))}{\ell(C_j)}. \tag{9.26}$$

In what follows, it will be useful to consider the related matrix

$$\bar{P}_{ij} = \frac{\ell(B_i \cap T^{-1}(C_j))}{\ell(B_i)}, \tag{9.27}$$

which may be considered as a discrete action of T on measures. $\bar{P}$ is row-stochastic, has leading eigenvalue 1, and has an interpretation as a transition matrix where the entry $\bar{P}_{ij}$ represents the conditional probability that a randomly chosen point in B_i lands in C_j after one application of T.

In what follows, we use the shorthand $\ell^B = [\ell(B_1), \ldots, \ell(B_m)]$ and $\ell^C = [\ell(C_1), \ldots, \ell(C_n)]$. Given μ, let $p_i = \mu(B_i)$ and we approximate h_μ as $\sum_{i=1}^m (p_i/\ell_i^B)\mathbf{1}_{B_i}$; from now on for brevity, we drop this functional representation for densities and measures and write them as vectors. The image density $h_{\nu_\epsilon} = \mathbf{P}_\epsilon h_\mu$ is estimated as $\sum_{i=1}^m (p_i/\ell_i^B)P_{ij}$ and the image measure ν_ϵ is estimated as $\sum_{i=1}^m (p_i/\ell_i^B)P_{ij}\ell_j^C = \sum_{i=1}^m p_i \bar{P}_{ij} =: q_j$. Thus, to construct an approximate matrix representation for $\mathbf{L}$, we use the definition (9.2) to obtain

$$\begin{aligned} L_{ij} &= \frac{(p_i/\ell_i^B)P_{ij}}{\sum_{i=1}^m (p_i/\ell_i^B)P_{ij}} \\ &= \frac{(p_i/\ell_i^B)\bar{P}_{ij}(\ell_i^B/\ell_j^C)}{\sum_{i=1}^m (p_i/\ell_i^B)\bar{P}_{ij}(\ell_i^B/\ell_j^C)} \\ &= \frac{p_i \bar{P}_{ij}}{\sum_{i=1}^m p_i \bar{P}_{ij}}. \end{aligned} \tag{9.28}$$

This latter expression appeared in [19]. It is clear that $\mathbf{1}L = \mathbf{1}$. Denoting the inner products $\langle x, y\rangle_p$ and $\langle x, y\rangle_q$ by $\sum_{i=1}^m x_i y_i p_i$ and $\sum_{i=1}^n x_i y_i q_i$, respectively, it is straightforward to check that L^*, the matrix dual satisfying $\langle xL, y\rangle_q = \langle x, yL^*\rangle_p$, is $\bar{P}^\top$ and that $\mathbf{1}L^* = \mathbf{1}$.

In the autonomous setting, recall one has $X = Y_\epsilon$ and $\mu = \nu_\epsilon$ and μ should be T-invariant. To estimate the T-invariant μ, we use the leading eigenvector of $\bar{P}$, i.e. choose p to be the (assumed unique) vector satisfying $p = p\bar{P}$. The expression for L now becomes

$$L_{ij} = \frac{p_i \bar{P}_{ij}}{p_j}, \tag{9.29}$$

and is a discrete approximation of $\mathbf{L} : L^2(X, \mu) \to L^2(X, \mu)$; in fact, in this setting L is nothing but a discrete approximation of the Perron-Frobenius operator $\mathbf{P}_\mu : L^1(X, \mu) \to L^1(X, \mu)$ defined with respect to μ, rather than ℓ. In the autonomous setting, there have been a number of papers that discuss choosing partitions in a way that Ulam's method produces the most accurate estimate of the physical invariant measure μ for T [23, 28, 38]. For the purposes of finding almost-invariant and coherent sets, as numerical diffusion plays an important role, we advise choosing partition sets that are approximately spherically symmetric (e.g. squares or cubes) so that the implicit numerical diffusion that results is approximately isotropic.

We remark that the matrix representation (9.28) is not the same as the matrix representation of $\pi_{X,m}\mathbf{L}_0\pi_{Y,n}$; that is, L is not exactly a Galerkin projection of $\mathbf{L}_0$. We have chosen this alternative formulation for numerical convenience and do not believe that the numerical impact is great. To estimate the entry $\bar{P}_{ij}$ numerically, one may sample test points $x_{i,k}$, $k = 1, \ldots, K$ uniformly distributed over B_i (e.g. on a uniform grid) and then compute $T(x_{i,k})$ and count how many fall in C_j; that is,

$$\bar{P}_{ij} \approx \frac{\#\{k : T(x_{i,k}) \in C_j\}}{K}. \tag{9.30}$$

The software package GAIO [9] is used to estimate the transition matrix entries. GAIO uses generalised rectangles (*boxes*) as partition elements and, using a multilevel tree-like data structure, can efficiently find which boxes contain image points.

9.6.2 *Numerical Representation of* $\mathbf{P}_\epsilon$ *and* $\mathbf{L}_\epsilon$

Although in the previous section we have constructed a matrix representation of $\mathbf{L}_0$, not $\mathbf{L}_\epsilon$ for some $\epsilon > 0$, in fact, as a consequence of the discretisation we have already implicitly incorporated a low level of numerical diffusion of the order of the diameter of the partition elements. We now discuss two ways to construct a matrix representation of $\mathbf{L}_\epsilon$, explicitly including diffusion of the type governed by $\mathbf{D}_{X,\epsilon}$ and $\mathbf{D}_{Y'_\epsilon,\epsilon}$:

(i) For each test point $x_{i,k} \in B_i$, $k = 1, \ldots, K$ represent the diffusion over an ϵ-ball centred at $x_{i,k}$ by a second grid of points $y_{i,k,k'}$, $k' = 1, \ldots, K'$ uniformly spread over $B_\epsilon(x_{i,k})$. For example, one could select $y_{i,k,k'}$ on a grid centred at $x_{i,k}$; in terms of computer code, the most efficient approach is to create a 'mask' of such points for an ϵ-ball centred at zero and merely add the vector $x_{i,k}$ to translate this fixed 'mask'. Now compute $T(y_{i,k,k'})$, $k = 1, \ldots, K$, $k' = 1, \ldots, K'$; in total this represents $K \cdot K'$ points for each box B_i. Finally, for each image point $T(y_{i,k,k'})$ we again create K'' points uniformly distributed in a ball of radius ϵ, centred at $T(y_{i,k,k'})$; call this final set of $K \cdot K' \cdot K''$ points $z_{i,k,k',k''}$. This final set of points can again be created via the 'mask' procedure described above; now one adds the vector $T(y_{i,k,k'})$. Finally, we estimate

$$\bar{P}_{\epsilon,i,j} = \frac{\#\{z_{i,k,k',k''} \in C_j\}}{K \cdot K' \cdot K''}.$$

This was the approach taken in [14]. It has a high accuracy because it directly simulates the concatenation $\mathbf{D}_{Y'_\epsilon,\epsilon}\mathbf{P}\mathbf{D}_{X,\epsilon}$ via test points and effectively applies a discretisation only once.

(ii) While the above approach is cheap from a memory point of view, there is an overhead to computing T-images of $K \cdot K'$ points. A faster (but somewhat less accurate) approach would be to discretise each of the three operators $\mathbf{D}_{Y'_\epsilon,\epsilon}, \mathbf{P}, \mathbf{D}_{X,\epsilon}$ separately and then estimate their product by matrix multiplication. Such an approach is less accurate because the effects of the three discretisations are multiplied together; however, if ϵ is much larger than the box diameters, the error should be comparatively small. An advantage to separately discretising is that one can try different diffusion amplitudes without having to recompute the discretisation $\mathbf{P}$.

Let $\{B_1, \ldots B_m\}$ denote a partition of X. We first set up the matrix $\bar{P}$ without diffusion as described in Sect. 9.6.1. We construct a matrix $D_{X,\epsilon}$ representing a discretised version of $\mathbf{D}_{X,\epsilon}$ as follows. In each box B_i, we choose K test points $x_{i,k} \in B_i$, $k = 1, \ldots, K$. For each $x_{i,k} \in B_i$ we represent the diffusion over an ϵ-ball centred at $x_{i,k}$ by a second grid of points $y_{i,k,k'}$, $k' = 1, \ldots, K'$, uniformly spread over $B_\epsilon(x_{i,k})$, and estimate

$$D_{X,\epsilon,i,j} = \frac{\#\{y_{i,k,k'} \in B_j\}}{K \cdot K'}.$$

We similarly construct a matrix $D_{Y'_\epsilon,\epsilon}$ based on sets $\{C_1, \ldots, C_n\}$. The matrix P_ϵ is then obtained as $P_\epsilon = D_{Y'_\epsilon,\epsilon} P D_{X,\epsilon}$.

In this paper we take an even faster and coarser approach whereby a ball of radius ϵ is approximated as a square or cube of side-length 2ϵ. This approach is faster as it makes use of the internal data structure of GAIO [9].

In the following sections we briefly describe how to appropriately put together the matrices L. From now on, we drop the ϵ subscript for the matrices P, Q, and L.

9.6.3 Autonomous Setting

In the autonomous setting, we construct $L_{ij} = p_i \bar{P}_{ij} / p_j$, where the map T used to construct $\bar{P}$ represents the dynamics over the fixed time duration we are interested in. As we are in the autonomous setting, only the duration matters, not the initial time. We consider

$$Q := (L + L^*)/2; \tag{9.31}$$

note that Q is self-adjoint in $\langle \cdot, \cdot \rangle_p$ (but is not a symmetric matrix in general) and that $\mathbf{1}Q = \mathbf{1}Q^* = \mathbf{1}$. By Corollary 9.1 the solution to (9.12) is given by the second left eigenvector u_2 of $\mathbf{Q}$; we will numerically estimate u_2 as the second left eigenvector of Q, normalised so that $\langle u_2, u_2 \rangle_p = 1$. If there is strong almost-invariance present, we expect the entries of the vector u_2 to be around ± 1. Note that Froyland [12] proposed the use of the second *right* eigenvector of $(\bar{P} + \hat{P})/2$ to obtain almost-invariant sets (where $\hat{P}$ denotes the transition matrix for the time-reversed Markov chain governed by $\bar{P}$). As $(\bar{P} + \hat{P})/2 = Q^\top$, our use of the second *left* eigenvector of Q yields an identical result. Here we have incorporated the autonomous and nonautonomous constructions under a single unified set of constructions and also have demonstrated how to naturally incorporate diffusive aspects of dynamics.

The aim is to find an optimal partition of X. We restate the algorithm as used in [16] for finding almost-invariant sets using R.

Algorithm 1 (Almost-invariant sets).

(i) *Partition the state space X into a collection of connected sets $\{B_1, \ldots, B_m\}$ of small diameter.*
(ii) *Construct the Ulam matrix $\bar{P}$ using (9.30) and compute the (assumed unique) fixed left eigenvector p of $\bar{P}$. If explicit diffusion is to be added, use one of the methods in Sect. 9.6.2.*
(iii) *Construct the matrix L using (9.29) and Q using (9.31).*
(iv) *Compute the second largest eigenvalue $\lambda_2 < 1$ of Q and corresponding left eigenvector u_2, normalised so that $\langle u_2, u_2 \rangle_p = 1$.*
(v) *Denote $I(b) = \{i : u_{2,i} > b\}$, $I^c(b) = \{i : u_{2,i} \leq b\}$. Perform a line search on b to maximise*[13]

$$\frac{\sum_{i \in I(b), j \in I(b)} p_i \bar{P}_{ij}}{\sum_{i \in I(b)} p_i} + \frac{\sum_{i \in I^c(b), i \in I^c(b)} p_i \bar{P}_{ij}}{\sum_{i \in I^c(b)} p_i}. \tag{9.32}$$

(vi) *Denote by $\hat{b}$ the optimal b and set $\hat{A} := \cup_{i \in I(\hat{b})} B_i$, $\hat{A}^c := \cup_{i \in I^c(\hat{b})} B_i$.*

9.6.4 Nonautonomous Setting: Single Direction

In this setting we wish to study transport over the time interval $[t_0, t_1]$ and build the matrix $\bar{P}$ using a map T that describes the dynamics over this interval. We then construct L using (9.28). Proposition 1 [14] shows that the value of (9.18) is $\lambda_2^{1/2}$, the square root of the second largest eigenvalue of $\mathbf{Q}_\epsilon := \mathbf{L}_\epsilon^* \mathbf{L}_\epsilon$, and the maximising f (resp. g) is u_2 (resp. v_2), the corresponding left eigenvector of $\mathbf{Q}_\epsilon$ (resp. $\mathbf{Q}_\epsilon^*$). One could define a matrix approximation of $\mathbf{Q}_\epsilon$ as $Q := LL^*$ and a matrix approximation of $\mathbf{Q}_\epsilon^*$ as $Q^* := L^*L$; however, it is more efficient numerically to find left and right singular vectors of L directly. The reason for this is that L is reasonably sparse, especially for small ϵ, and LL^* may be significantly more dense than the matrices involved in a calculation of singular value of L. The following algorithm was put forward in [19].

[13] The expression (9.32) is a discrete form of (9.8).

Algorithm 2 (Finite-time coherent sets, single time direction).

(i) *Partition the domains* X *and* Y_ϵ *into a collection of connected sets* $\{B_1, \ldots, B_m\}$ *and* $\{C_1, \ldots, C_n\}$*, respectively, of small diameter. If* Y_ϵ *is not known precisely, then set* Y_ϵ *to be a neighbourhood of* $T(X)$ *that contains all possible perturbed image points.*
(ii) *Select the reference measure* μ *as the mass distribution to be tracked, and set* $p_i = \mu(B_i)$*.*
(iii) *Construct the Ulam matrix* $\bar{P}$ *as in (9.30), and compute* $q = p\bar{P}$*. If additional explicit diffusion is used, then use one of the approaches in Sect. 9.6.2.*
(iv) *Define diagonal matrices* $(\Pi_p)_{ii} = p_i$, $i = 1, \ldots, m$, *and* $(\Pi_q)_{jj} = q_j$, $j = 1, \ldots, n$, *compute the second largest singular value* $\sigma_2 < 1$ *of* $\Pi_p^{1/2} \bar{P} \Pi_q^{-1/2}$ *and corresponding left and right singular vectors* $\tilde{u}_2, \tilde{v}_2$*, and set* $u_2 := \tilde{u}_2 \Pi_p^{-1/2}$, $v_2 := \tilde{v}_2 \Pi_q^{-1/2}$*.*
(v) *Denote* $I(b) = \{i : u_{2,i} > b\}, I^c(b) = \{i : u_{2,i} \le b\}, J(b') = \{j : v_{2,j} > b'\}, J^c(b') = \{j : v_{2,j} \le b'\}$*. Perform a line search on* b *to maximise*[14]

$$\frac{\sum_{i \in I(b), j \in J(b')} p_i \bar{P}_{ij}}{\sum_{i \in I(b)} p_i} + \frac{\sum_{i \in I^c(b), j \in J^c(b')} p_i \bar{P}_{ij}}{\sum_{i \in I^c(b)} p_i}, \tag{9.33}$$

selecting $b' = b'(b)$ *so that* $|\sum_{i \in I(b)} p_i - \sum_{j \in J(b')} q_j|$ *is minimised*[15] *for each choice of* b*.*
(vi) *Denote by* $\hat{b}, \hat{b}'$ *the optimal* b, b' *and set* $\hat{A} := \cup_{i \in I(\hat{b})} B_i$, $\hat{A}^c := \cup_{i \in I^c(\hat{b})} B_i$, $\hat{B} := \cup_{j \in J(\hat{b}')} C_j$, $\hat{B}^c := \cup_{j \in J^c(\hat{b}')} C_j$*.*

In the numerical case studies section we will frequently plot the output of item (iv) above, namely, the vectors u_2 and v_2, and also the output of item (vi) above, namely the sets $\hat{A}, \hat{A}^c, \hat{B}$, and $\hat{B}^c$. If the vectors $\tilde{u}_2, \tilde{v}_2$ have ℓ^2-norm[16] 1, then the vectors u_2, v_2 will be normalised so that $\langle u_2, u_2 \rangle_p = 1$ and $\langle v_2, v_2 \rangle_q = 1$. If there is strong coherence present, we expect the entries of the vectors u_2 and v_2 to be around ± 1.

9.6.5 Nonautonomous Setting: Both Directions

In this setting we are interested in sets at time point t that remain coherent both in forward and backward time. We build matrices $\bar{P}_+$ and $\bar{P}_-$ using maps T_+ and T_- that describe the dynamics from t to $t + \tau_2$ and from t to $t - \tau_1$, respectively. We then construct L_+, L_- using (9.28) and consider the matrix approximation to $\mathbf{Q}_\epsilon$:

---

[14] The expression (9.33) is a discrete form of (9.13).

[15] This is the discrete version of insisting that $\mu(A(b)) = \nu_\epsilon(B(b'))$.

[16] This is the default output normalisation for MATLAB, for example.

$Q := (L_+L_+^* + L_-L_-^*)/2$ (recall we always use left multiplication). We propose the following algorithm:

Algorithm 3 (Finite-time coherent sets, both time directions).

(i) *Partition the domains* X, Y_ϵ, *and* Z_ϵ *into a collection of connected sets* $\{B_1,\ldots,B_m\}$, $\{C_1,\ldots,C_n\}$, *and* $\{E_1,\ldots,E_o\}$, *respectively, of small diameter. If* Y_ϵ *and* Z_ϵ *are not known precisely, then set* Y_ϵ *to be a neighbourhood of* $T_+(X)$ *that contains all possible perturbed image points, likewise* Z_ϵ *a neighbourhood of* $T_-(X)$.

(ii) *Select the reference measure* μ *at time* t *as the mass distribution to be tracked, and set* $p_i = \mu(B_i)$.

(iii) *Construct the Ulam matrix* $\bar{P}_+$ *as in (9.30) using* $T = T_+$ *and compute* $q^+ = p\bar{P}_+$. *To construct* $\bar{P}_-$, *use* $T = T_-$ *and replace* C_j *with* E_j *in (9.30). Set* $q^- = p\bar{P}_-$. *If additional explicit diffusion is used, then use one of the approaches in Sect. 9.6.2.*

(iv) *Construct* L_+ *as in (9.28) using* $\bar{P}_+$ *and* p *and* L_- *using* $\bar{P}_-$ *and* p. *Form* $Q = (L_+L_+^* + L_-L_-^*)/2$ *by matrix multiplication where* $L_+^* = \bar{P}_+^\top$ *and* $L_-^* = \bar{P}_-^\top$.

(v) *Compute the second largest eigenvalue* $\lambda_2 < 1$ *of* Q *and corresponding left eigenvector* u_2. *Set* $v_2^+ = u_2L_+$ *and* $v_2^- = u_2L_-$. *Normalise* u_2 *so that* $\langle u_2,u_2\rangle_p = 1$ *and normalise* $v_2^\pm$ *so that* $\langle v_2^\pm, v_2^\pm\rangle_{q^\pm} = 1$.

(vi) *Denote* $I(b) = \{i : u_{2,i} > b\}$, $I^c(b) = \{i : u_{2,i} \le b\}$, $J(b') = \{j : v_{2,j}^+ > b'\}$, $J^c(b') = \{j : v_{2,j}^+ \le b'\}$ *and* $K(b'') = \{l : v_{2,l}^- > b''\}$, $K^c(b'') = \{l : v_{2,l}^- \le b''\}$. *Perform a line search on* b *to maximise*[17]

$$\frac{\sum_{i\in I(b),j\in J(b')} p_i\bar{P}_{+,ij}}{\sum_{i\in I(b)} p_i} + \frac{\sum_{i\in I^c(b),j\in J^c(b')} p_i\bar{P}_{+,ij}}{\sum_{i\in I^c(b)} p_i} + \frac{\sum_{i\in I(b),k\in K(b'')} p_i\bar{P}_{-,ik}}{\sum_{i\in I(b)} p_i} + \frac{\sum_{i\in I^c(b)k\in K^c(b'')} p_i\bar{P}_{-,ik}}{\sum_{i\in I^c(b)} p_i}, \tag{9.34}$$

selecting $b' = b'(b)$ *and* $b'' = b''(b)$ *so that* $|\sum_{i\in I(b)} p_i - \sum_{j\in J(b')} q_j^+|$ *and* $|\sum_{i\in I(b)} p_i - \sum_{k\in K(b'')} q_k^-|$ *are minimised*[18] *for each choice of* b.

(vii) *Denote by* $\hat{b},\hat{b}',\hat{b}''$ *the optimal* b,b',b'' *and set* $\hat{A} := \bigcup_{i\in I(\hat{b})} B_i$, $\hat{A}^c := \bigcup_{i\in I^c(\hat{b})} B_i$, $\hat{B} := \bigcup_{j\in J(\hat{b}')} C_j$, $\hat{B}^c := \bigcup_{j\in J^c(\hat{b}')} C_j$ *and* $\hat{C} := \bigcup_{k\in K(\hat{b}'')} E_k$, $\hat{C}^c := \bigcup_{k\in K^c(\hat{b}'')} E_k$.

In the numerical case studies section we will frequently plot the output of item (v) above, namely the vectors u_2, v_2^+, and v_2^-, and also the output of item (vii) above, namely the sets $\hat{A}$, $\hat{A}^c$, $\hat{B}$, $\hat{B}^c$, $\hat{C}$, and $\hat{C}^c$. If there is strong coherence present, we again expect the entries of the vectors u_2, v_2^+, and v_2^- to be around ± 1.

---

[17] The expression (9.34) is a discrete form of (9.23).

[18] This is the discrete version of insisting that $\mu(A(b)) = \nu_\epsilon(B(b')) = \eta_\epsilon(C(b''))$.

9.7 Numerical Examples

In this section we will apply the different constructions to two well-known example systems. First we consider a periodically driven double gyre flow [45], which has frequently been used as a test bed for different tools for the numerical analysis of transport. Due to the system's periodicity we will analyse the system both with respect to almost-invariant sets and finite-time coherent sets and point out the differences of the constructions. We discuss the effects of diffusion for almost-invariant sets, and, for coherent sets, additionally time direction and flow duration.

As a second example system we consider the transitory double gyre flow as introduced in [37]. Here the dynamics is only nonautonomous on a finite-time interval, but autonomous outside. Therefore the system is well suited to analysing finite-time coherent structures. Special emphasis will be placed on how the structure of finite-time coherent sets depends on the time direction and flow duration.

While in these examples the domain at the initial time and final time remains the same, the coherent set framework also easily handles situations where the initial domain and final domain do not intersect at all, as in, e.g. [20].

9.7.1 *Case Study 1: Periodically Driven Double Gyre Flow*

We consider the time-dependent system of differential equations [45]

$$\begin{aligned} \dot{x} &= -\pi A \sin(\pi f(x,t)) \cos(\pi y) \\ \dot{y} &= \pi A \cos(\pi f(x,t)) \sin(\pi y) \frac{df}{dx}(x,t), \end{aligned} \tag{9.35}$$

where $f(x,t) = \delta \sin(\omega t) x^2 + (1 - 2\delta \sin(\omega t))x$.

For detailed discussions of the system, we refer to [16, 45]. As in [16], we fix parameter values $A = 0.25$, $\delta = 0.25$, and $\omega = 2\pi$ and obtain a flow of period $p = 1$. The system preserves Lebesgue measure on $[0,2] \times [0,1]$. We partition the domain[19] $M = [0,2] \times [0,1]$ in $n = 32768 = 2^{15}$ square boxes. Here we will identify $M = X = Y_\epsilon = Z_\epsilon$. We form matrices $\bar{P}$ by integrating with a fourth-order Runge-Kutta scheme with constant stepsize $h = 0.01$ from $t = 0$ over different time spans, i.e. over one period ($\tau = 1$) and $\tau = \pm 2.5$, using $K = 400$ uniformly distributed test points per box (inner grid[20] points). In this set-up, with box diameter 0.0078, we have a numerically induced diffusion of about $\epsilon \approx 0.0039$. In addition, we consider explicit diffusion via left and right multiplication of $\bar{P}$ with diffusion matrices $D_{X,\epsilon}$, $D_{Y'_\epsilon,\epsilon}$. Here we choose 100 inner grid points per box, for

[19] Note that the boundaries of M are invariant.

[20] See [9] for a description of inner grid points.

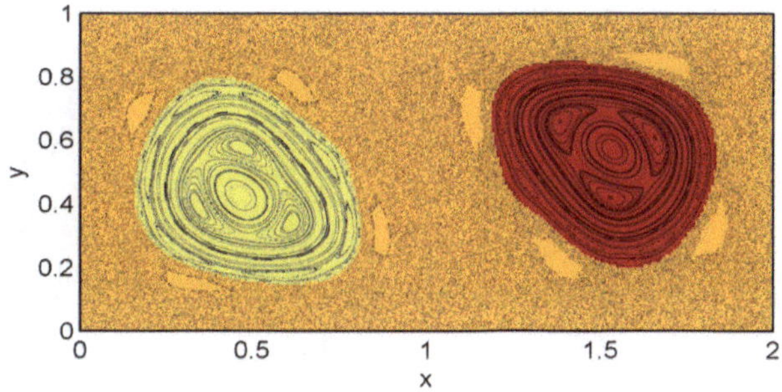

Fig. 9.1 Almost-invariant sets in the double gyre flow on the time interval $[0, 1]$ ($\tau = 1$) – no explicit diffusion ($\epsilon = 0$). Due to symmetry in the system, both $\hat{b} = -0.7259$ and $\hat{b} = 0.7259$ are optimal thresholds in Algorithm 1. We obtain $\rho(\hat{A}) = 1.9910$, where $\hat{A}$ is the light ($\hat{b} = -0.7259$) or dark set ($\hat{b} = 0.7259$) and $\hat{A}^c = M \setminus \hat{A}$. The *black dots* correspond to typical orbits of the time-1 map T, visualising regular and chaotic phase space structures. In this setting, the almost-invariant sets correspond to regular (invariant) structures in phase space

each test point we approximate a ball of radius $\epsilon = 0.02$ by 25 inner grid points of a box of diameter 2ϵ centred in the respective point.

In the following, we will demonstrate the usage of the different constructions and discuss effects of diffusion, flow duration, and time direction.

9.7.1.1 Almost-Invariant Sets

As the system is 1-periodic by construction, the time-1 flow map T describes an autonomous dynamical system. Thus, we want to determine fixed regions in phase space that are almost-invariant under the dynamics of T. To this end, we compute[21] $p = p\bar{P}$ and form matrices L and Q. Following Algorithm 1 we consider the second left eigenvector u_2 to eigenvalue $\lambda_2 = 0.9998$, normalised such that $\langle u_2, u_2 \rangle_p = 1$ (Fig. 9.2a). Optimal almost-invariant sets are obtained for $\hat{b} = \pm 0.7259$, leading to two equally optimal almost-invariant sets shown in pale yellow and red in Fig. 9.1; for both of these sets, one has $\rho(\hat{A}) = 1.9910$. In Fig. 9.1 we have also plotted typical orbits of the map T, visualising the typical phase space structure of area-preserving map consisting of regular islands and chaotic regions. As indicated, regular phase space structures, i.e. truly invariant sets, are picked up as optimal almost-invariant sets in our approach. We note that Fig. 9.1 does not change visibly when more iterates of the time-1 map T (i.e. longer flow times) are considered; for example, for $\tau = 2$ we obtained $\lambda_2 = 0.9998$ and $\rho(\hat{A}) = 1.9909$

[21] As Lebesgue measure is preserved, p should give equal weight $1/m$ to each of the m partition sets. However, in order to account for possible sampling-induced numerical inaccuracies when setting up $\bar{P}$, we will use the numerically obtained p, which is very close to a constant vector in practice.

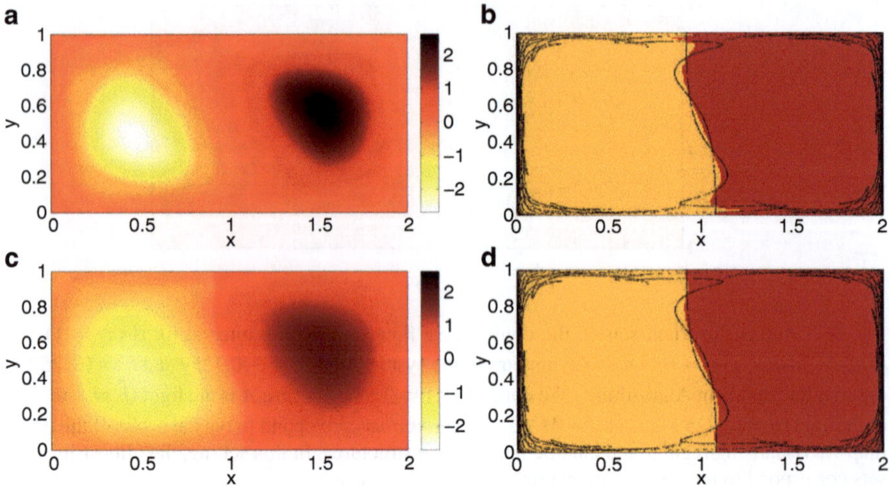

Fig. 9.2 Second eigenvectors and corresponding almost-invariant sets in the double gyre flow on the finite-time interval [0, 1] ($\tau = 1$), with and without explicit diffusion. (**a**) Second eigenvector u_2 of Q ($\lambda_2 = 0.9998$), $\epsilon = 0$. (**b**) Partition by sign results in $\rho(A_0) = 1.9803$. Such a partition would also be obtained when restricting the domain to the chaotic region as visualised in Fig. 9.1; see [16] for a related case study. (**c**), (**d**) Effects of explicit diffusion on second eigenvectors and almost-invariant sets in the double gyre flow. (**c**) Second eigenvector u_2 to eigenvalue $\lambda_2 = 0.9974$ of Q in the diffusive case, $\epsilon = 0.02$. (**d**) Optimal partition into almost-invariant sets based on u_2 in (**c**). Here $\hat{b} = 0$ is the optimal threshold, resulting in $\rho(\hat{A}) = 1.9758$. In (**b**) and (**d**), we have plotted approximations of the stable (resp. unstable) manifolds of hyperbolic periodic orbits (i.e. fixed points of T) located on the $y = 0$ (resp. $y = 1$) axes. The partitions into almost-invariant sets are influenced by these manifolds

and for $\tau = 5$ we obtained $\lambda_2 = 0.9997$ and $\rho(\hat{A}) = 1.9903$. This is understandable because we are approximating *invariant* sets and $\epsilon = 0$: the only mass loss occurs from those points in boxes on the boundary of the pale yellow and red sets that sit outside the true invariant sets, and taking additional iterates does not increase the mass loss.

In order to find a partition into almost-invariant sets that is not determined by these invariant structures, one may restrict the box covering and transition matrix to the chaotic region. Here one finds $\hat{b} = 0$, i.e. a partition by sign, is now the global optimum, but only a local optimum when the entire domain is considered. A similar experiment was carried out in [16]. We show the partition by sign in Fig. 9.2b, obtaining $\rho(A_0) = 1.9803$. Note that for the optimal partition, $\rho(\hat{A})$ is within the upper $(1 + \lambda_2 = 1.9998)$ and lower bounds $(2 - 2\sqrt{2(1 - \lambda_2)} = 1.96)$; this applies also to the suboptimal partition by sign.

In Fig. 9.2c, d we show the results of explicit diffusion, incorporated by left and right multiplication of diffusion matrices $D_{X,\epsilon}, D_{Y'_\epsilon,\epsilon}$ ($\epsilon = 0.02$) with $\bar{P}$ as described above. Applying Algorithm 1 to the diffused matrices results in a visibly smoother eigenvector u_2 (Fig. 9.2c) compared to the case without explicit diffusion (Fig. 9.2a). Moreover, the optimal partitions now are given by $\hat{b} = 0$ and the boundaries are smoother and shorter. Figure 9.2d compared to Fig. 9.2b. As expected, diffusion regularises eigenvectors of Q; this in turn usually shortens and smoothens the boundaries of almost-invariant sets.

Another important phenomenon that has been illustrated here is an apparent bifurcation of the almost-invariant sets with variation in noise amplitude. When only numerical diffusion was present, the invariant sets shown in Fig. 9.1 were selected by Algorithm 1. The addition of explicit diffusion resulted in Algorithm 1 selecting the sets shown in Fig. 9.2d. The reason for this switch is that with very low diffusion, advective flux dominates diffusive flux and the invariant sets in Fig. 9.1 minimise transport across their boundaries. As the diffusion amplitude is raised, the optimal almost-invariant structures change to those shown in Fig. 9.2d, which have a shorter, non-invariant boundary, leading to a reduction in diffusive flux, but an increase in advective flux; the net effect is, however, lower flux than the invariant sets in Fig. 9.1 under the $\epsilon = 0.02$ diffusion regime. We will further investigate bifurcation aspects in future work. We note that bifurcations of almost-invariant and almost-cyclic sets when changing a system parameter have been studied in [29] and [22], respectively; qualitative bifurcations of the invariant density under varying noise amplitude have been observed numerically in [5].

In Fig. 9.2b, d we have also plotted approximations[22] to stable and unstable manifolds of hyperbolic periodic orbits (oscillating around $x = 1$ on the $y = 0$ and $y = 1$ axes). Apparently, as already discussed in [16], the transfer operator approach finds a decomposition into almost-invariant sets, which is influenced by the underlying manifold structure, but tries to find a more optimal decomposition than a geometrical approach such as lobe dynamics [41, 42] would suggest. While the stable and unstable manifolds concern time-asymptotic dynamics, the optimal almost-invariant sets identified are tuned to a finite flow time of the dynamics.

9.7.1.2 Finite-Time Coherent Sets: Single Time Direction

In this section we are no longer interested in spatially fixed sets, but in possibly dynamical regions in phase space that remain coherent during some finite time span. We will first study finite-time coherent sets obtained from considering the dynamics in a single time direction. We restrict ourselves to the case without explicit diffusion.

As a first case study, we will consider the dynamics on the time interval [0, 1], i.e. using flow time $\tau = 1$, such as in the previous paragraph, but apply Algorithm 2 to the respective matrix $\bar{P}$. The resulting (normalised) singular vectors u_2 and v_2 with respect to the singular value $\sigma_2 = 1 - 1.6 \times 10^{-5}$ are shown in Fig. 9.3a, c. The thresholds $\hat{b} = 0$ and $\hat{b}' = 0$ turn out to define optimal coherent pairs shown in Fig. 9.3b, d. We obtain $\rho(\hat{A}, \hat{B}) = 1.9976$, which is bounded from above by

[22]We refer the reader to [16] for more details.

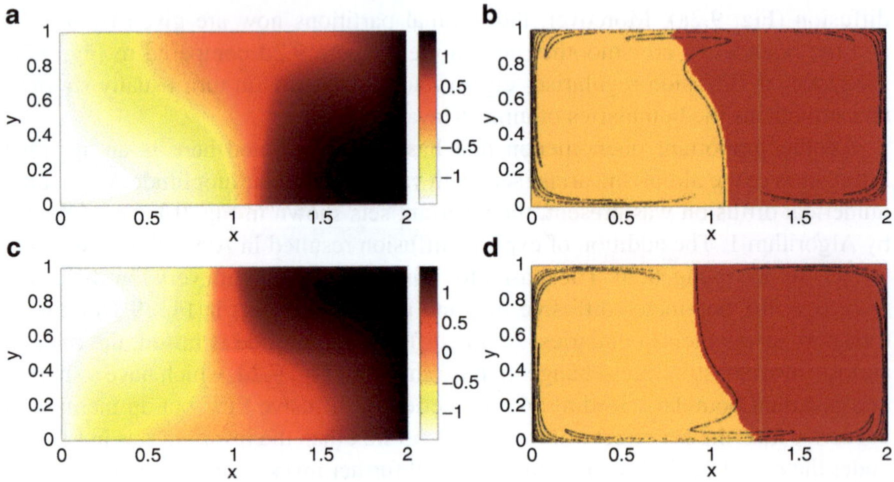

Fig. 9.3 Singular vectors and finite-time coherent pairs ($\hat{b} = \hat{b}' = 0$ in Algorithm 2) in the double gyre flow on the time interval [0, 1] ($\tau = 1$) for the case without explicit diffusion ($\epsilon = 0$). (**a**) *Left* (normalised) singular vector u_2 to singular value $\sigma_2 = 1 - 1.6 \times 10^{-5}$ as obtained from Algorithm 2. (**b**) Finite-time coherent sets $(\hat{A}, \hat{A}^c)$ at $t = 0$ from u_2. Superimposed is an approximation of the stable manifold of a hyperbolic periodic orbit on the $y = 0$ axis. (**c**) *Right* (normalised) singular vector v_2. (**d**) Finite-time coherent sets $(\hat{B}, \hat{B}^c)$ at $t = 1$ from v_2 and approximation of the unstable manifold of a hyperbolic periodic orbit on the $y = 1$ axis. We obtain $\rho(\hat{A}, \hat{B}) = 1.9976$, which is bounded from above by $1 + \sigma_2$

$1 + \sigma_2$. In addition, we have overlaid the optimal partition in Fig. 9.3b with an approximation to the stable manifold of the periodic orbit on the $y = 0$ axis and the optimal partition in Fig. 9.3d with an approximation to the unstable manifold of the periodic orbit on the $y = 1$ axis. The (asymptotic) geometric structures influence the shape of the coherent sets, but due to the short finite-time horizon on which their computation is based on, we do not get a nearly exact correspondence.

We repeat this study by considering the dynamics on the time interval [0, 2.5] as well as on the time interval [−2.5, 0], i.e. longer time intervals ($\tau = 2.5$) which are not integer multiples of the period of the flow (though the latter property is not so important here). Figure 9.4a, c show the outcome of an application of Algorithm 2 on the time interval [0, 2.5] and Fig. 9.4b, d the results for the interval [−2.5, 0].

Again $\hat{b} = \hat{b}' = 0$ turn out to be the optimal thresholds for a decomposition into finite-time coherent sets[23]. The decompositions $(\hat{A}, \hat{A}^c)$ at $t = 0$ and $(\hat{B}, \hat{B}^c)$ at $t = 2.5$ are shown in Fig. 9.4a, c, whereas the respective decompositions $(\hat{A}, \hat{A}^c)$ at $t = -2.5$ and $(\hat{B}, \hat{B}^c)$ at $t = 0$ can be seen in Fig. 9.4 b, d. For both settings, we obtain $\sigma_2 = 0.9999$ and $\rho(\hat{A}, \hat{B}) = 1.9951$, bounded from above by $1 + \sigma_2$.

[23]Varying b and b' around $b = b' = 0$, one obtains partitions that are very close to optimal. Because of the symmetry in the system, we concentrate on the partition by sign.

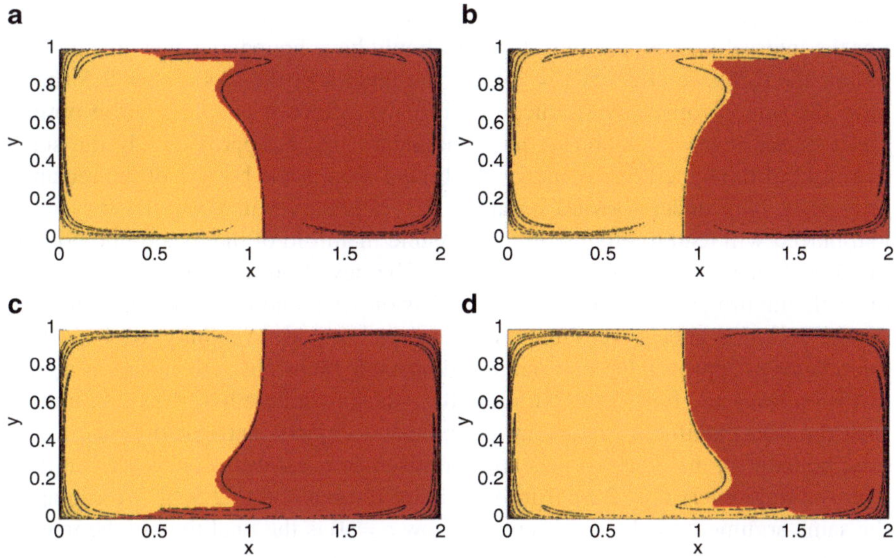

Fig. 9.4 Finite-time coherent pairs in the double gyre flow on the time intervals $[0, 2.5]$ and $[-2.5, 0]$, i.e. $\tau = 2.5$, for the case without explicit diffusion ($\epsilon = 0$). (**a**) Finite-time coherent sets $(\hat{A}, \hat{A}^c)$ at $t = 0$ from left (normalised) singular vector u_2 as obtained from Algorithm 2. Here the dynamics on the time interval $[0, 2.5]$ is considered. (**b**) Finite-time coherent sets $(\hat{A}, \hat{A}^c)$ at $t = -2.5$ from left (normalised) singular vector u_2 as obtained from Algorithm 2. Here the dynamics on the time interval $[-2.5, 0]$ is considered. (**c**) Finite-time coherent sets $(\hat{B}, \hat{B}^c)$ at $t = 2.5$ from right (normalised) singular vector v_2 on the time interval $[0, 2.5]$. (**d**) Finite-time coherent sets $(\hat{B}, \hat{B}^c)$ at $t = 0$ from right (normalised) singular vector v_2 from dynamics on $[-2.5, 0]$. For both settings we obtain $\rho(\hat{A}, \hat{B}) = 1.9951$, which is bounded from above by $1 + \sigma_2$, with singular value $\sigma_2 = 0.9999$. The optimal thresholds in Algorithm 2 are $\hat{b} = \hat{b}' = 0$. The partitions are overlaid with approximations of the stable (**a**),(**b**) and unstable manifolds (**c**), (**d**) of hyperbolic periodic orbits

We compare Fig. 9.4a with Fig. 9.3b to explain the effect of flow duration. Both of these images describe coherent sets at $t = 0$; the only difference is the flow duration. One can see that the boundary in Fig. 9.4a is longer than the boundary in Fig. 9.3b and is also closer to the stable manifold of the hyperbolic fixed point for the time-1 map on the $x = 0$ axis. The reason for this is that with increasing flow time, advective flux effects increase, relative to diffusive flux. Thus, a short boundary is less important for the longer flow duration; instead a boundary that grows in length at a slower rate over the longer time interval is more important. This ensures that the boundary in Fig. 9.4c (at $t = 2.5$) is shorter than the push-forward of the boundary in Fig. 9.3b to time $t = 2.5$ would be. In each case (flows times 1 and 2.5), the coherent sets are tuned to their particular finite-time duration to resist mixing over that period.

Secondly, we notice that although the finite-time coherent sets in Fig. 9.4a, d are both defined at $t = 0$, the different time spans under consideration ($[0, 2.5]$

and $[-2.5, 0]$) produce markedly different results. This can be explained as follows. For the interval $[0, 2.5]$, the sets at $t = 0$ should have boundaries that are small (to reduce the diffusive effect at $t = 0$) and the images of the sets at $t = 2.5$ should have the same property (to reduce the diffusive effect at $t = 2.5$). The optimal coherent set at $t = 0$ therefore has a boundary that is approximately the stable manifold of the hyperbolic point for the time-1 map at the bottom of the rectangle at $t = 0$; then under forward iteration, this boundary will not grow very much and indeed will tend to align with the unstable manifold of the hyperbolic point of the time-1 map at the top of the rectangle. The deviation from the true stable and unstable manifolds is a function of the diffusion level and the finite-time duration. As the diffusion goes to zero and the time duration goes to infinity, we expect the boundaries to approach the true stable and unstable manifolds. However, for a given diffusion level and finite-time duration, the sets shown here have less flux transfer than the true manifolds (which are complicated objects that would create a very long boundary); the coherent sets are optimised to resist diffusion-assisted mixing over the time interval $[0, 2.5]$. Turning now to the interval $[-2.5, 0]$, we can apply the same argument as above, except that now $t = 0$ is the final time in the interval and by the above argument one expects the coherent set boundaries to be roughly aligned along unstable directions, rather than stable directions. This is indeed what we see in Fig. 9.4.

9.7.1.3 Finite-Time Coherent Sets: Both Time Directions

Finally, we consider the coherent set framework that takes into account both time directions. For this we study again both the dynamics on the time interval $[0, 2.5]$ and the backward-time dynamics on the time interval $[-2.5, 0]$ as in the previous paragraph and apply Algorithm 3, i.e. flow times $\tau = \pm 2.5$. We obtain the second left eigenvector u_2 (after normalisation) at $t = 0$, corresponding to the eigenvalue $\lambda_2 = 0.9998$ as well as corresponding optimal vectors v_2^+ at $t = 2.5$ and v_2^- at $t = -2.5$. The optimal thresholds in Algorithm 3 are given by $\hat{b} = \hat{b}' = \hat{b}'' = 0$, defining finite-time coherent sets $(\hat{A}, \hat{A}^c)$ at $t = 0$, as shown in Fig. 9.5a $(\hat{B}, \hat{B}^c)$ at $t = 2.5$ (Fig. 9.5c) and $(\hat{C}, \hat{C}^c)$ at $t = -2.5$ (Fig. 9.5e). One obtains $\rho(\hat{A}, \hat{B}, \hat{C}) = 1.9878$, which is bounded from above by $1 + (\lambda_2)^{1/2} = 1.9999$.

While the boundaries of the coherent sets at $t = 0$ obtained when considering only one time direction are roughly aligned along stable or unstable manifolds (see Fig. 9.4a, d), the result of the triple construction using both time directions simultaneously is a dynamical compromise, influenced by both stable and unstable directions at $t = 0$ (see Fig. 9.5a).

We also remark that the longer flow time (5 time units in Fig. 9.5 compared to 2.5 time units in Fig. 9.4) result in an even stronger correspondence of Fig. 9.5c–f with stable/unstable manifolds (a zoom is shown in Fig. 9.5g).

We also consider explicit diffusion with $\epsilon = 0.02$ and obtain coherent triples $\rho(\hat{A}, \hat{B}, \hat{C}) = 1.9628$ and $(\lambda_2)^{1/2} = 0.9972$; see Fig. 9.5 b, d, and f. The boundaries

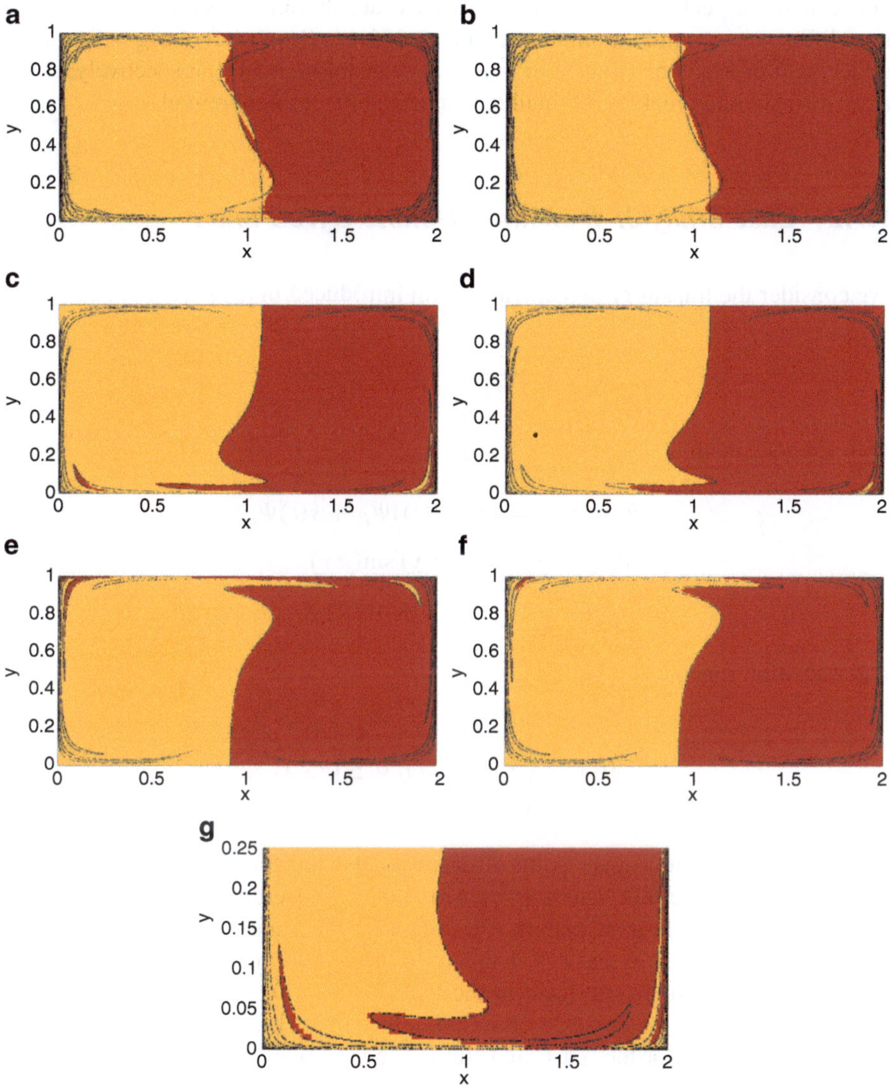

Fig. 9.5 Finite-time coherent sets using forward- and backward-time dynamics on $[0, 2.5]$ ($\tau = 2.5$) and $[-2.5, 0]$ ($\tau = -2.5$) based on non-diffusive setting ($\epsilon = 0$, *left*) and for the case with explicit diffusion ($\epsilon = 0.02$, *right*). (**a**) Finite-time coherent sets $(\hat{A}, \hat{A}^c)$ at $t = 0$ obtained from Algorithm 3, no explicit diffusion. (**b**) Same as (**a**) but for $\epsilon = 0.02$. (**c**) Corresponding finite-time coherent sets $(\hat{B}, \hat{B}^c)$ at $t = 2.5$ ($\epsilon = 0$); (**g**) shows a close-up. (**d**) Same as (**c**) but for $\epsilon = 0.02$. (**e**) Corresponding finite-time coherent sets $(\hat{C}, \hat{C}^c)$ at $t = -2.5$ ($\epsilon = 0$). (**f**) Same as (**e**) but for $\epsilon = 0.02$. Again we have overlaid the respective stable and unstable manifolds of hyperbolic periodic orbits. The coherence of the triples for the case $\epsilon = 0$ can be estimated as $\rho(\hat{A}, \hat{B}, \hat{C}) = 1.9878$, which is bounded from above by $1 + (\lambda_2)^{1/2} = 1.9999$. For $\epsilon = 0.02$ one obtains $\rho(\hat{A}, \hat{B}, \hat{C}) = 1.9628$ and $(\lambda_2)^{1/2} = 0.9972$

between the respective finite-time coherent sets are shorter and smoother than in the non-diffusive case, and the small disconnected pieces (lower left of Fig. 9.5c (dark), upper right of Fig. 9.5e (light)) no longer appear in Fig. 9.5d, f, respectively, as the extra diffusive flux makes the small disconnected pieces nonoptimal.

9.7.2 Case Study 2: Transitory Double Gyre Flow

We consider the transitory dynamical system introduced in [37]

$$\dot{x} = -\frac{\partial}{\partial y}\Psi, \quad \dot{y} = \frac{\partial}{\partial x}\Psi,$$

with stream function

$$\begin{aligned} \Psi(x,y,t) &= (1-s(t))\Psi_P + s(t)\Psi_F \\ \Psi_P(x,y) &= \sin(2\pi x)\sin(\pi y) \\ \Psi_F(x,y) &= \sin(\pi x)\sin(2\pi y) \end{aligned}$$

and transition function

$$s(t) = \begin{cases} 0, & t < 0, \\ t^2(3-2t), & 0 \le t \le 1, \\ 1, & t > 1. \end{cases}$$

A horizontal double gyre pattern (described by the 'past' system with stream function Ψ_P) on the unit square is rotated anticlockwise during times $0 < t < 1$ into a vertical double gyre pattern (described by the 'future' system with stream function Ψ_F). For $t \le 0$ and $t \ge 1$ the system is autonomous; thus the interesting finite-time behaviour is restricted to the time interval $[0, 1]$. Figure 9.6 illustrates the complex mixing processes on the time interval $[0, 1]$. Of particular importance are the separatrices for the 'past' and 'future' autonomous systems, i.e. the vertical line at $x = 0.5$ for $t \le 0$ (Fig. 9.6a) and the horizontal line at $y = 0.5$ for $t \ge 1$ (Fig. 9.6f). Their images and preimages in the transitory time interval $[0, 1]$ completely describe the transport mechanism as can be seen in Fig. 9.6. We refer to [37] for a detailed discussion and analysis of this transitory dynamical system.

For our numerical study we partition the domain $M = [0,1] \times [0,1]$ in $n = 16384 = 2^{14}$ square boxes and form matrices $\bar{P}$ by integrating with a fourth-order Runge-Kutta scheme with constant stepsize $h = 0.01$ for different flow times $\tau = 1$ and $\tau = \pm 0.5$ on $[0, 1]$, using $K = 400$ uniformly distributed test points per box (inner grid points). In this set-up, again with box diameter 0.0078, we have a numerically induced diffusion of about $\epsilon \approx 0.0039$, but we do not take into account any explicit additional diffusion from the operators $\mathbf{D}_{X,\epsilon}$ or $\mathbf{D}_{Y_\epsilon,\epsilon}$.

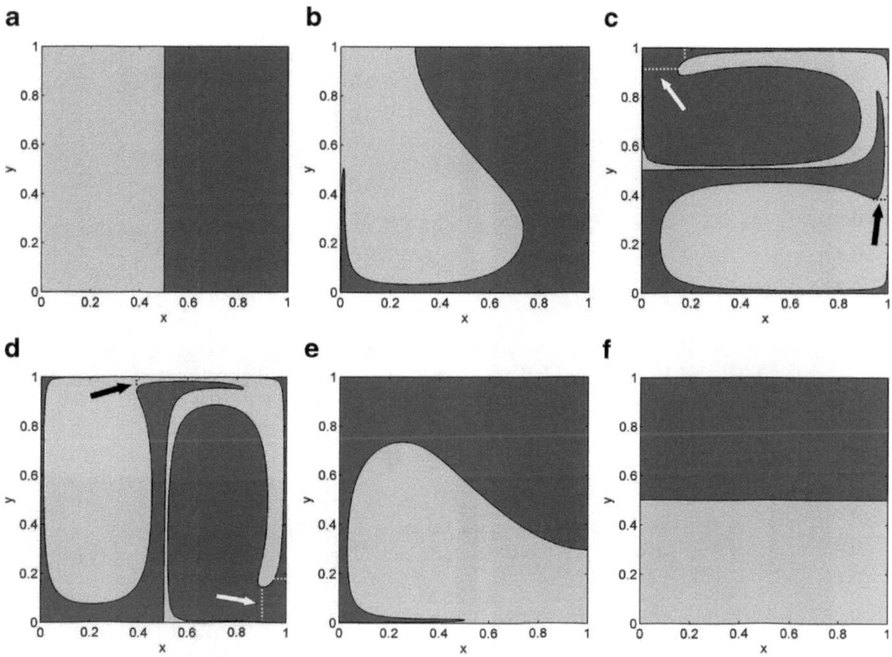

Fig. 9.6 Complex mixing dynamics of the transitory double gyre system. (**a**) Two vertical sets, corresponding to invariant sets for the autonomous dynamics for $t \leq 0$, are initialized at time $t = 0$ and evolved forward under the dynamics. The sets are bounded by a heteroclinic orbit connecting two saddle point equilibria at $(0.5, 0)$ and $(0.5, 1)$ in the 'past' autonomous system. (**b**) Image sets of (**a**) at $t = 0.5$ (i.e. after flowing time $\tau = 0.5$) and (**c**) at $t = 1$ (flow time $\tau = 1$). (**d**) Preimage at $t = 0$ (i.e. after flowing backwards $\tau = -1$ from $t = 1$) of the two horizontal sets in (f) that are bounded by the separatrix of the 'future' autonomous system. (**e**) same as (**d**) but at $t = 0.5$ (i.e. $\tau = -0.5$). (**e**) Two sets corresponding to invariant sets for the autonomous dynamics for $t \geq 1$ are initialized at time $t = 1$ and evolved backwards under the dynamics. The sets are bounded by a heteroclinic orbit connecting two saddle point equilibria at $(0, 0.5)$ and $(1, 0.5)$ in the 'future' autonomous system. For an explanation of the *dashed lines* in (**c**) and (**d**) (indicated by *arrows*), see main text

First we consider the flow on the entire transition interval $[0, 1]$ ($\tau = 1$). The results of an application of Algorithm 2 are shown in Fig. 9.7. The (normalised) singular vectors u_2 and v_2 with respect to the singular value $\sigma_2 = 0.9997$ are shown in Fig. 9.7a, c. The thresholds $\hat{b} = -0.0525$, $\hat{b}' = -0.0530$ turn out to define optimal coherent sets $(\hat{A}, \hat{A}^c)$ at $t = 0$ and $(\hat{B}, \hat{B}^c)$ at $t = 1$, shown in Fig. 9.7b, d. The coherence can be estimated as $\rho(\hat{A}, \hat{B}) = 1.9885$, which is bounded from above by $1 + \sigma_2 = 1.9997$. The finite-time coherent sets appear to pick up the dominant light structures in Fig. 9.6c, f, whose boundaries match those of the sets considerably.

However, one may wonder why the left hand 'blob' is picked up as the optimal coherent set in Fig. 9.7b and not the right hand 'blob'. An inspection of Fig. 9.6c reveals that a small horizontal 'cut' across the thin light filament at approximately

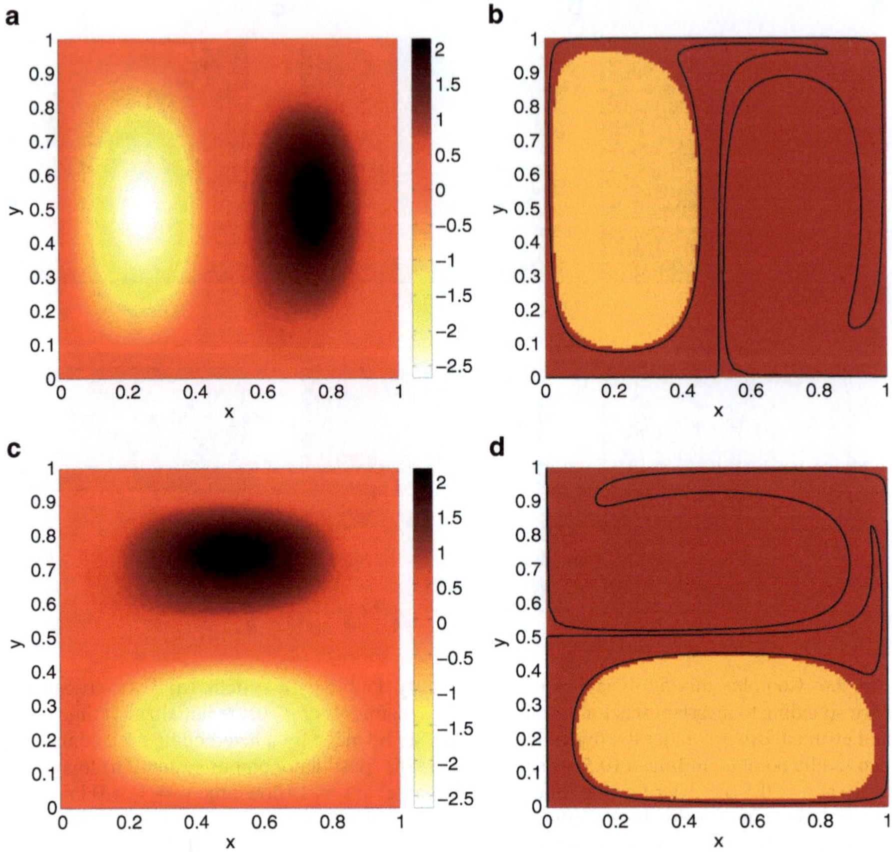

Fig. 9.7 Singular vectors and finite-time coherent pairs in the transitory double gyre flow on [0, 1], $\tau = 1$; no explicit diffusion ($\epsilon = 0$). (**a**) *Left* (normalised) singular vector u_2 to singular value $\sigma_2 \approx 0.9997$ obtained from Algorithm 2. (**b**) Finite-time coherent sets $(\hat{A}, \hat{A}^c)$ at $t = 0$ from u_2. (**c**) *Right* (normalised) singular vector v_2. (**d**) Finite-time coherent sets $(\hat{B}, \hat{B}^c)$ at $t = 1$ from v_2. We obtain $\rho(\hat{A}, \hat{B}) = 1.9885$, which is well bounded from above by $1 + \sigma_2$. The transport barriers from Fig. 9.6c and (**f**) are overlaid, delineating considerable parts of the boundaries of the finite-time coherent sets

$(x, y) = (1, 0.4)$ (indicated by black dotted line) will separate the light image blob covering most of the lower half of Fig. 9.6c. This also holds for the light preimage blob covering most of the left half of Fig. 9.6d (here cutting at approximately $(x, y) = (0.4, 1)$). Such a cut will lead to only a small advective flux across the cut and both the light blob and its image have relatively small boundary, to reduce diffusive flux. On the other hand, to similarly separate the dark image blob covering most of the top half of Fig. 9.6c would require either a longer horizontal cut around $(x, y) = (0.1, 0.9)$, leading to a larger advective flux, or a small vertical cut near, e.g. $(x, y) = (0.2, 1)$ (see white dotted lines), leading to small advective flux, but slightly more diffusive flux as the boundary of the dark image blob has been lengthened, similarly in Fig. 9.6d. Thus, there is some slight asymmetry in the system, which is clearly picked up by the coherent set calculations.

In a second set of numerical experiments we consider transport and mixing on the time subintervals $[0, 0.5]$ and $[0.5, 1]$, i.e. for flow times $\tau = 0.5$. The results of applying Algorithm 2 are shown in Fig. 9.8. The decompositions $(\hat{A}, \hat{A}^c)$ at $t = 0$ and $(\hat{B}, \hat{B}^c)$ at $t = 0.5$ w.r.t. the dynamics on the time interval $[0, 0.5]$ can be seen in Fig. 9.8a, c, whereas $(\hat{A}, \hat{A}^c)$ at $t = 0.5$ and $(\hat{B}, \hat{B}^c)$ at $t = 1$ w.r.t. the dynamics on the time interval $[0.5, 1]$ are shown in Fig. 9.8b, d.

On both time intervals we get the same second singular value $\sigma_2 = 0.9998$ and $\hat{b} = 0.1488$, $\hat{b}' = 0.1511$ turn out to be the optimal thresholds for a decomposition into finite-time coherent sets. The coherence values $\rho(\hat{A}, \hat{B}) = 1.9919$ for both settings are bounded by the theoretical upper bound of $1 + \sigma_2$. While the shapes of the sets are visibly influenced by the respective geometric structures, the boundaries of the sets do not fully match the images of $\{x = 1/2\}$ and $\{y = 1/2\}$. The reason is again the small level of diffusion. The black curves in Fig. 9.8a–d are all much longer than the boundaries of the coherent sets. Thus, again, the coherent sets are tuned to a particular flow duration and small diffusion level.

Finally, we consider again the triple construction considering the dynamics on the time intervals $[0, 0.5]$ and $[0.5, 1]$ from the previous paragraph ($\tau = \pm 0.5$). The results of an application of Algorithm 3 are shown in Fig. 9.9. In particular, Fig. 9.9a shows the optimal decomposition into finite-time coherent sets $(\hat{A}, \hat{A}^c)$ at $t = 0.5$, (b) the sets $(\hat{B}, \hat{B}^c)$ at $t = 1$, and (c) the corresponding partition $(\hat{C}, \hat{C}^c)$ at $t = 0$. With thresholds $\hat{b} = -0.2532$, $\hat{b}' = \hat{b}'' = -0.2464$, we obtain $\rho(\hat{A}, \hat{B}, \hat{C}) = 1.9883$ bounded by $1 + (\lambda_2)^{1/2}$, where $\lambda_2 = 0.9994$.

While the optimal sets do not exactly match the boundaries of the geometric structures in Fig. 9.6, the partitions at $t = 0$ and $t = 1$ are very similar to the ones in Fig. 9.7. Moreover, the central sets at $t = 0.5$ (Fig. 9.9a) are influenced both by stable and unstable directions and thus form, as expected, a dynamical compromise which accounts for both the forward- and backward-time dynamics. As we have observed before, geometrical structures such as invariant manifolds may not necessarily bound sets of minimal leakage. This is again due to the fact that deterministic time-asymptotic objects do not account for the finite-time and diffusive effects that are central to our transfer operator construction.

9.8 Summary

In this chapter we proposed a unified setting for finite-time almost-invariant and coherent set constructions. The constructions were based around a building block operator that combined advective dynamics via a Perron-Frobenius operator with small amplitude diffusive dynamics, developed in [14]. This building block operator was then manipulated to set up suitable optimisation problems for the autonomous and nonautonomous settings. The unified setting clarified the similarities and

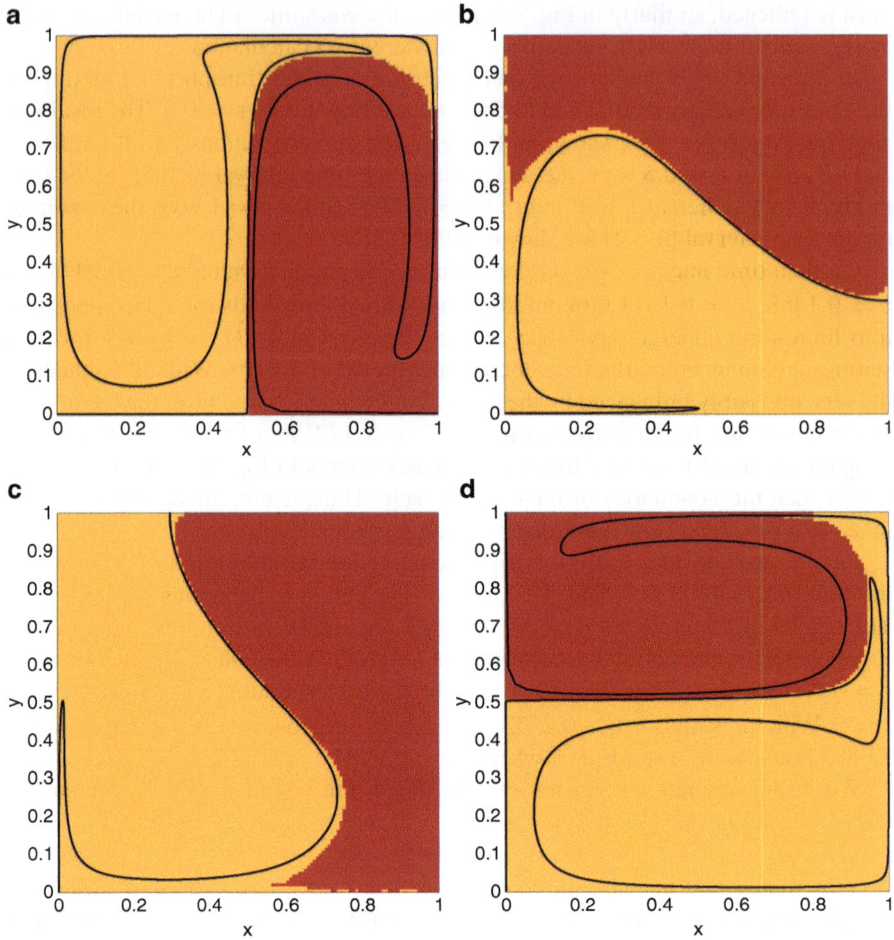

Fig. 9.8 Finite-time coherent sets in the transitory double gyre flow on the time interval $[0, 0.5]$ (*left*) as well as on the time interval $[0.5, 1]$ (*right*) using the one-sided construction in Algorithm 2. Here $\tau = 0.5$ and $\epsilon = 0$. (**a**) Finite-time coherent sets $(\hat{A}, \hat{A}^c)$ at $t = 0$ from dynamics on the time interval $[0, 0.5]$. (**b**) Finite-time coherent sets $(\hat{A}, \hat{A}^c)$ at $t = 0.5$ from dynamics on the time interval $[0.5, 1]$. (**c**) Finite-time coherent sets $(\hat{B}, \hat{B}^c)$ at $t = 0.5$ from dynamics on the time interval $[0, 0.5]$. (**d**) Finite-time coherent sets $(\hat{B}, \hat{B}^c)$ at $t = 1$ from dynamics on the time interval $[0.5, 1]$. For both time intervals $\rho(\hat{A}, \hat{B}) = 1.9919 < 1 + \sigma_2 = 1.9998$. The optimal thresholds in Algorithm 2 are $\hat{b} = 0.1488$, $\hat{b}' = 0.1511$. The corresponding transport barriers from Fig. 9.6 are overlaid

differences of the dynamical problems being solved in the autonomous and nonautonomous cases. These optimisation problems made use of the fact that the underlying transfer operator was compact and self-adjoint, leading to a simple analytic solution given by the eigenfunction corresponding to the second largest eigenvalue.

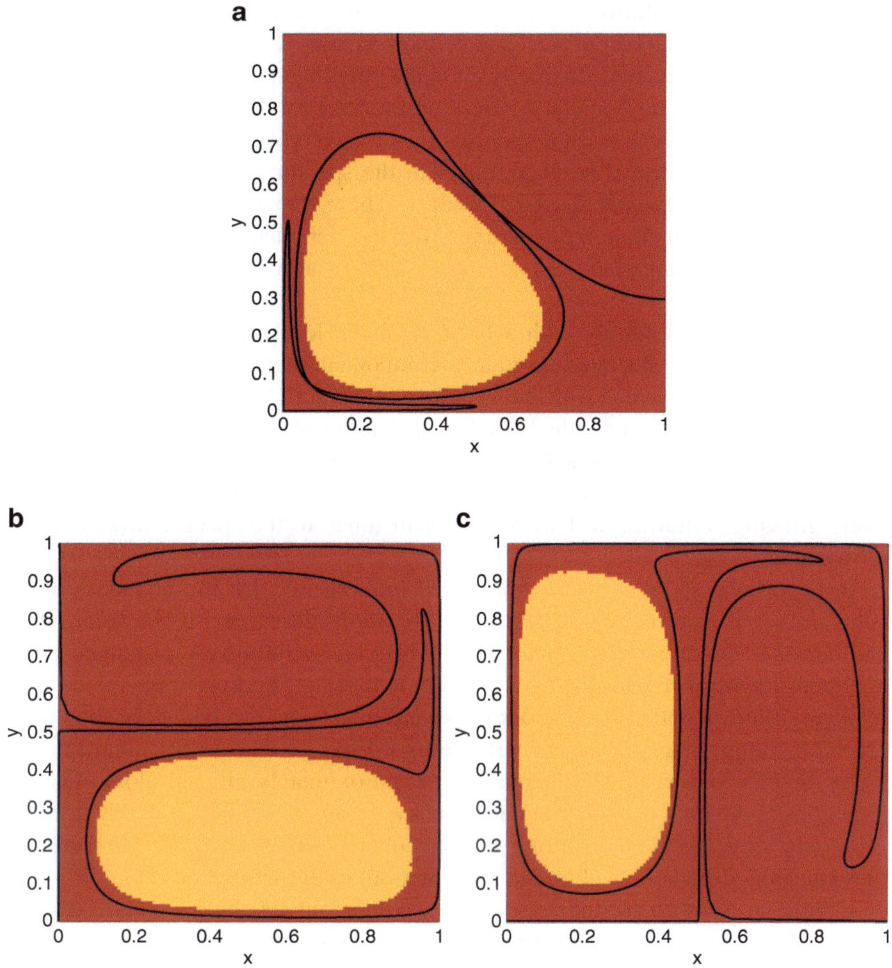

Fig. 9.9 Finite-time coherent sets for transitory double gyre flow on the time intervals $[0, 0.5]$ and $[0.5, 1]$ ($\tau = \pm 0.5$) using the two-sided construction (Algorithm 3, no explicit diffusion ($\epsilon = 0$). (**a**) Optimal finite-time coherent sets $(\hat{A}, \hat{A}^c)$ at $t = 0.5$. (**b**) $(\hat{B}, \hat{B}^c)$ at $t = 1$. (**c**) $(\hat{C}, \hat{C}^c)$ at $t = 0$. Here $\hat{b} = -0.2532$, $\hat{b}' = \hat{b}'' = -0.2464$, and $\rho(\hat{A}, \hat{B}, \hat{C}) = 1.9883$. The corresponding transport barriers from Fig. 9.6 are overlaid

Via two detailed case studies we investigated the dependence of almost-invariant and coherent sets on three aspects of the dynamics: the level of diffusion, the flow duration, and the time direction of the dynamics. We also compared the boundaries of the coherent sets with the time-dependent stable and unstable manifolds of organising hyperbolic points in flows.

As proved formally in [14], for fixed flow times, we showed that increased levels of diffusion produce more regular eigenfunctions and coherent sets with shorter boundaries. Intuitively, this is because diffusive flux is proportional to the boundary lengths of coherent sets, and so the boundary would like to be shortened to minimise the effects of diffusive flux. On the other hand, shortening the boundary usually means that the boundary is less equivariant (or in the autonomous case, invariant) in an advective sense, increasing the advective flux. The result is that an optimal compromise is reached, tuned to the specific diffusion level and flow duration, with a somewhat shorter boundary. These remarks apply equally to the autonomous and nonautonomous settings. We also demonstrated an instance where the optimal almost-invariant sets appear to undergo a bifurcation as the diffusion level is increased.

Increasing flow duration, with a fixed diffusion level, has the opposite effect. Now, advective flux becomes a greater component of the overall flux, and the boundary tries to move in a way that is more aligned with 'stable' directions. This is so that at the final flow time, the boundary has not grown very much (which would lead to high diffusive flux at the final time). Again, an optimal balance is reached; the boundary of the coherent set at the initial time moves close to the stable direction, likely growing somewhat in length, the result tuned to the specific flow time and diffusion level.

In hyperbolic settings, we demonstrated numerically that the boundary of the coherent set is approximately aligned with *stable* directions at the *initial* time and *unstable* directions at the *final* time. This can be intuitively explained by an argument identical to those above; the coherent set at the final time should have relatively short boundary, and under backward advection, the boundary should not grow very long; otherwise, there will be high diffusive flux at the initial time. In order for this to occur, the coherent set is approximately aligned along unstable directions at the final time so that under backward advection the boundary does not grow very long; the exact positioning of the coherent sets are tuned to the particular flow time and diffusion level to minimise flux out of the sets.

Finally, in order to produce a sequence of coherent sets over a sliding window of fixed finite-time duration, we proposed a new construction where the focus is on coherent sets in the middle of the finite-time window; these sets remain coherent in both forward and backward time and can be used to create a natural time-dependent sequence of coherent sets over several translated finite-time windows.

Future work will include further investigation of bifurcation phenomena and computational improvements to the bidirectional coherent set calculations.

Acknowledgements GF and KPG thank the Banff International Research Station for hospitality during the workshop on 'Open Dynamical Systems: Ergodic Theory, Probabilistic Methods and Applications', which provided the catalyst for this work. GF's research was partially supported by an ARC Discovery Project (DP110100068) and an ARC Future Fellowship.

9.9 Appendix

9.9.1 *Proof of Theorem 9.3*

Our building block operator $\mathbf{L}$ may be written as $\mathbf{L}f(y) = \int_X k(x,y)f(x)\,d\mu(x)$, where $k \in L^2(X, Y_\epsilon)$, satisfies $k \geq 0$. Thus $\mathbf{Q} = (\mathbf{L} + \mathbf{L}^*)/2$ is a self-adjoint operator $\mathbf{Q} : L^2(X,\mu) \to L^2(X,\mu)$ defined by $\mathbf{Q}f(y) = \int \kappa(x,y)f(x)\,d\mu(x)$, with $\mathbf{Q1} = \mathbf{1}$ (see [14] for the specific forms of k and κ). From self-adjointness it follows that $\kappa(x,y) = \kappa(y,x)$ and thus

$$\int \kappa(x,y)\,d\mu(y) = 1 = \int \kappa(x,y)\,d\mu(x). \tag{9.36}$$

9.9.1.1 Lower Bound

The proof of Lemma 9.2 draws heavily on the proof of Theorem 2.1 [31] (which is effectively a continuous time version of Lemma 9.2) and includes techniques from Theorem 4.3, Chap. 6 [6].

Lemma 9.2. *Let λ_2 denote the second largest eigenvalue of $\mathbf{Q}$. Then $1-\lambda_2 \geq c^2/8$ where $c = \inf_A(\int \mathbf{Q1}_A \cdot \mathbf{1}_{A^c}\,d\mu)/(\mu(A)\mu(A^c))$.*

Proof. We know that $\sup\{\langle \mathbf{Q}f, f\rangle_\mu/\langle f,f\rangle_\mu : \langle f,\mathbf{1}\rangle_\mu = 0\} = \lambda_2$, where λ_2 is the second largest eigenvalue of $\mathbf{Q}$. Thus $\inf\{\langle (I-\mathbf{Q})f, f\rangle_\mu/\langle f,f\rangle_\mu : \langle f,\mathbf{1}\rangle_\mu = 0\} = 1-\lambda_2$.

Now,

$$\begin{aligned}
&\langle (I-\mathbf{Q})f, f\rangle_\mu \\
&= \int f^2\,d\mu - \int \kappa(x,y)f(y)\,d\mu(y)\cdot f(x)\,d\mu(x) \\
&= \int f^2\,d\mu - \int \kappa(x,y)f(y)\,d\mu(y)\cdot f(x)\,d\mu(x) \\
&= \int f(x)\left(\int \kappa(x,y)(f(x)-f(y))\,d\mu(y)\right)d\mu(x) \quad \text{using (9.36)} \\
&= \int f(y)\left(\int \kappa(x,y)(f(y)-f(x))\,d\mu(y)\right)d\mu(x) \quad \text{interchanging } x \text{ and } y \\
&= (1/2)\int \left(\int \kappa(x,y)(f(x)-f(y))^2\,d\mu(y)\right)d\mu(x) \quad \text{combining previous 2 lines.}
\end{aligned}$$

Now

$$\int \kappa(x,y)(f(x)+f(y))^2\,d\mu(x)d\mu(y)$$
$$\leq 2\int \kappa(x,y)(f(x)^2+f(y)^2)\,d\mu(x)d\mu(y) \quad \text{using } (a+b)^2 \leq 2(a^2+b^2)$$
$$= \int 4\kappa(x,y)f(x)^2\,d\mu(x)d\mu(y) \quad \text{by symmetry of } \kappa$$
$$= \int 4f(x)^2\,d\mu(x) \quad \text{by (9.36).}$$

So

$$(1/2)\int\left(\int \kappa(x,y)(f(x)-f(y))^2\,d\mu(y)\right)d\mu(x)$$
$$\geq (1/2)\int\left(\int \kappa(x,y)(f(x)-f(y))^2\,d\mu(y)\right)d\mu(x)\cdot\frac{\int \kappa(x,y)(f(x)+f(y))^2\,d\mu(x)d\mu(y)}{\int 4f(x)^2\,d\mu(x)}$$
$$\geq (1/8)\frac{\left(\int \kappa(x,y)|(f(x)^2-f(y)^2)|\,d\mu(x)d\mu(y)\right)^2}{\int f^2\,d\mu} \quad \text{by Hölder.} \tag{9.37}$$

Now

$$\int \kappa(x,y)|(f(x)^2-f(y)^2)|\,d\mu(x)d\mu(y)$$
$$= 2\int \kappa(x,y)\mathbf{1}_{\{f(x)^2-f(y)^2>0\}}(f(x)^2-f(y)^2)\,d\mu(x)d\mu(y) \quad \text{by symmetry of } \kappa$$
$$= 2\int_0^\infty d\alpha\int \kappa(x,y)\mathbf{1}_{\{f(x)^2>\alpha\geq f(y)^2\}}\,d\mu(x)d\mu(y)$$
$$= 2\int_0^\infty d\alpha\int \kappa(x,y)\mathbf{1}_{A_\alpha}(x)\cdot\mathbf{1}_{A_\alpha^c}(y)\,d\mu(x)d\mu(y) \quad \text{where } A_\alpha = \{f^2>\alpha\}$$
$$= 2\int_0^\infty d\alpha\int \mathbf{Q}\mathbf{1}_{A_\alpha}(y)\cdot\mathbf{1}_{A_\alpha^c}(y)\,d\mu(y)$$
$$\geq 2c\int_0^\infty d\alpha\,\mu(A_\alpha)\mu(A_\alpha^c), \quad \text{where } c = \inf_A\left(\int \mathbf{Q}\mathbf{1}_A\cdot\mathbf{1}_{A^c}\,d\mu\right)/(\mu(A)\mu(A^c))$$
$$= 2c\int_0^\infty d\alpha\int \mathbf{1}_{\{f(x)^2>\alpha\geq f(y)^2\}}d\mu(x)d\mu(y)$$
$$= 2c\int_0^\infty d\alpha\int \mathbf{1}_{\{f(x)^2-f(y)^2>0\}}(f(x)^2-f(y)^2)d\mu(x)d\mu(y)$$
$$= c\int |f(x)^2-f(y)^2|\,d\mu(x)d\mu(y).$$

By (9.37) we have

$$\begin{aligned}
\langle (I-\mathbf{Q})f, f\rangle_\mu &\geq (1/8)\frac{c^2\left(\int |f(x)^2 - f(y)^2|\, d\mu(x)d\mu(y)\right)^2}{\left(\int f^2\, d\mu\right)} \\
&\geq (1/8)c^2\kappa' \int f^2\, d\mu \\
&\geq (1/8)c^2 \int f^2\, d\mu \quad \text{by [31], Prop. 2.2 (see for a definition of } \kappa').
\end{aligned}$$

Thus $1-\lambda_2 \geq c^2/8$.

Define

$$\rho(A) = \frac{\int \mathbf{Q1}_A \cdot \mathbf{1}_A\, d\mu}{\mu(A)} + \frac{\int \mathbf{Q1}_{A^c}\cdot \mathbf{1}_{A^c}\, d\mu}{\mu(A^c)} \quad \text{and} \quad c(A) = \frac{\int \mathbf{Q1}_A \cdot \mathbf{1}_{A^c}\, d\mu}{\mu(A)\mu(A^c)}.$$

Lemma 9.3. $\rho(A) = 2 - c(A)$.

Proof.

$$\begin{aligned}
&\frac{\int \mathbf{Q1}_A \cdot \mathbf{1}_A\, d\mu}{\mu(A)} + \frac{\int \mathbf{Q1}_{A^c}\cdot \mathbf{1}_{A^c}\, d\mu}{\mu(A^c)} \\
&= \frac{\mu(A^c)(\int \mathbf{Q1}_A \cdot \mathbf{1}\, d\mu - \int \mathbf{Q1}_A\cdot \mathbf{1}_{A^c}\, d\mu)}{\mu(A)\mu(A^c)} + \frac{\mu(A)(\int \mathbf{Q1}_{A^c}\cdot \mathbf{1}\, d\mu - \int \mathbf{Q1}_{A^c}\cdot \mathbf{1}_A\, d\mu)}{\mu(A)\mu(A^c)} \\
&= \frac{\mu(A^c)\mu(A) - \mu(A^c)\int \mathbf{Q1}_A\cdot \mathbf{1}_{A^c}\, d\mu)}{\mu(A)\mu(A^c)} + \frac{\mu(A)\mu(A^c) - \mu(A)\int \mathbf{Q1}_{A^c}\cdot \mathbf{1}_A\, d\mu)}{\mu(A)\mu(A^c)} \\
&= 2 - \mu(A^c)c(A) - \mu(A)c(A^c) = 2 - c(A) \quad \text{as } c(A) = c(A^c).
\end{aligned}$$

Corollary 9.2. $\rho := \sup_A \rho(A) \geq 2 - \sqrt{8(1-\lambda_2)}$.

Proof. $\sup_A \rho(A) = 2 - \inf_A c(A) = 2 - c = 2 - \sqrt{8(1-\lambda_2)}$. □

9.9.1.2 Upper Bound

Lemma 9.4. $\rho \leq \lambda_2 + 1$.

Proof. We know that $\langle \mathbf{Q}f, f\rangle_\mu \leq \lambda_2$ for all $\langle f, \mathbf{1}\rangle_\mu = 0$. Consider the test function $f = \sqrt{\mu(A^c)/\mu(A)}\mathbf{1}_A - \sqrt{\mu(A)/\mu(A^c)}\mathbf{1}_{A^c}$. We have

$$\begin{aligned}
&\langle \mathbf{Q}f, f\rangle_\mu \\
&= \langle \sqrt{\mu(A^c)/\mu(A)}\mathbf{Q1}_A, \sqrt{\mu(A^c)/\mu(A)}\mathbf{1}_A\rangle_\mu - \langle \sqrt{\mu(A^c)/\mu(A)}\mathbf{Q1}_A, \sqrt{\mu(A^c)/\mu(A)}\mathbf{1}_{A^c}\rangle_\mu \\
&\quad - \langle \sqrt{\mu(A^c)/\mu(A)}\mathbf{Q1}_{A^c}, \sqrt{\mu(A^c)/\mu(A)}\mathbf{1}_A\rangle_\mu + \langle \sqrt{\mu(A^c)/\mu(A)}\mathbf{Q1}_A^c, \sqrt{\mu(A^c)/\mu(A)}\mathbf{1}_{A^c}\rangle_\mu
\end{aligned}$$

$$
\begin{aligned}
&= \frac{\mu(A^c)}{\mu(A)}\langle \mathbf{Q1}_A, \mathbf{1}_A\rangle_\mu - 2\langle \mathbf{Q1}_A, \mathbf{1}_{A^c}\rangle_\mu + \frac{\mu(A)}{\mu(A^c)}\langle \mathbf{Q1}_{A^c}, \mathbf{1}_{A^c}\rangle_\mu \\
&= \frac{\mu(A^c)}{\mu(A)}\langle \mathbf{Q1}_A, \mathbf{1}_A\rangle_\mu + \frac{\mu(A)}{\mu(A^c)}\langle \mathbf{Q1}_{A^c}, \mathbf{1}_{A^c}\rangle_\mu \\
&\quad -(\langle \mathbf{Q1}_A, \mathbf{1}\rangle_\mu - \langle \mathbf{Q1}_A, \mathbf{1}_A\rangle_\mu + \langle \mathbf{Q1}_{A^c}, \mathbf{1}\rangle_\mu - \langle \mathbf{Q1}_{A^c}, \mathbf{1}_{A^c}\rangle_\mu) \\
&= \frac{\langle \mathbf{Q1}_A, \mathbf{1}_A\rangle_\mu}{\mu(A)} + \frac{\langle \mathbf{Q1}_{A^c}, \mathbf{1}_{A^c}\rangle_\mu}{\mu(A^c)} - \mu(A) - \mu(A^c) \\
&\leq \lambda_2.
\end{aligned}
$$

As A is arbitrary, the result follows.

References

1. Allshouse, M., Thiffeault, J.L.: Detecting coherent structures using braids. Phys. D **241**(2), 95–105 (2012)
2. Aref, H.: The development of chaotic advection. Phys. Fluid. **14**(4), 1315–1325 (2002)
3. Billings, L., Schwartz, I.B.: Identifying almost invariant sets in stochastic dynamical systems. Chaos **18**, 023,122 (2008)
4. Birman, M.S., Solomjak, M.Z.: Spectral Theory of Self-Adjoint Operators in Hilbert Space. D. Reidel Publishing Company, Inc, Boston (1986)
5. Bollt, E.M., Billings, L., Schwartz, I.B.: A manifold independent approach to understanding transport in stochastic dynamical systems. Phys. D **173**, 153–177 (2002)
6. Brémaud, P.: Markov chains. Gibbs fields, Monte Carlo simulation, and queues. Texts in Applied Mathematics, vol. 31. Springer, New York (1999)
7. Conway, J.: A course in functional analysis. Graduate Texts in Mathematics, vol. 96, 2nd edn. Springer, New York (1990)
8. Dellnitz, M., Junge, O.: Almost invariant sets in Chua's circuit. Int. J. Bif. Chaos **7**(11), 2475–2485 (1997)
9. Dellnitz, M., Froyland, G., Junge, O.: The Algorithms behind GAIO–Set Oriented Numerical Methods for Dynamical Systems, pp. 145–174. Springer, New York (2001)
10. Dellnitz, M., Junge, O.: On the approximation of complicated dynamical behaviour. SIAM J. Numer. Anal. **36**(2), 491–515 (1999)
11. Deuflhard, P., Huisinga, W., Fischer, A., Schütte, C.: Identification of almost invariant aggregates in nearly uncoupled Markov chains. Lin. Algebra Appl. **315**, 39–59 (2000)
12. Froyland, G.: Statistically optimal almost-invariant sets. Phys. D **200**, 205–219 (2005)
13. Froyland, G.: Unwrapping eigenfunctions to discover the geometry of almost-invariant sets in hyperbolic maps. Phys. D **237**, 840–853 (2008)
14. Froyland, G.: An analytic framework for identifying finite-time coherent sets in time-dependent dynamical systems. Phys. D **250**, 1–19 (2013)
15. Froyland, G., Dellnitz, M.: Detecting and locating near-optimal almost-invariant sets and cycles. SIAM J. Sci. Comput. **24**(6), 1839–1863 (2003)
16. Froyland, G., Padberg, K.: Almost-invariant sets and invariant manifolds—connecting probabilistic and geometric descriptions of coherent structures in flows. Phys. D **238**, 1507–1523 (2009)
17. Froyland, G., Lloyd, S., Quas, A.: Coherent structures and isolated spectrum for Perron-Frobenius cocycles. Ergod. Theor. Dyn. Syst.**30**, 729–756 (2010)
18. Froyland, G., Lloyd, S., Santitissadeekorn, N.: Coherent sets for nonautonomous dynamical systems. Phys. D **239**, 1527–1541 (2010)

19. Froyland, G., Santitissadeekorn, N., Monahan, A.: Transport in time-dependent dynamical systems: Finite-time coherent sets. Chaos **20**, 043,116 (2010)
20. Froyland, G., Horenkamp, C., Rossi, V., Santitissadeekorn, N., Gupta, A.S.: Three-dimensional characterization and tracking of an Agulhas ring. Ocean Model. **52–53**, 69–75 (2012)
21. Gouillart, E., Thiffeault, J.L., Finn, M.: Topological mixing with ghost rods. Phys. Rev. E **73**(3), 036,311 (2006)
22. Grover, P., Ross, S., Stremler, M., Kumar, P.: Topological chaos, braiding and bifurcation of almost-cyclic sets. Chaos: An Interdisciplinary Journal of Nonlinear Science, AIP Publishing **22**(4), 043135 (2012)
23. Guder, R., Dellnitz, M., Kreuzer, E.: An adaptive method for the approximation of the generalized cell mapping. Chaos, Solitons & Fractals **8**(4), 525–534 (1997)
24. Haller, G.: Finding finite-time invariant manifolds in two-dimensional velocity fields. Chaos **10**, 99–108 (2000)
25. Haller, G.: A variational theory of hyperbolic Lagrangian Coherent Structures. Phys. D **240**, 574–598 (2011)
26. Haller, G., Beron-Vera, F.: Geodesic theory of transport barriers in two-dimensional flows. Phys. D **241**(20), 1680–1702 (2012)
27. Huisinga, W., Schmidt, B.: Metastability and dominant eigenvalues of transfer operators. In: Leimkuhler, B., Chipot, C., Elber, R., Laaksonen, A., Mark, A., Schlick, T., Schütte, C., Skeel, R., Barth, T.J., Griebel, M., Keyes, D.E., Nieminen, R.M., Roose, D., Schlick, T. (eds.) New Algorithms for Macromolecular Simulation, Lecture Notes in Computational Science and Engineering, vol. 49, pp. 167–182. Springer, Berlin (2006)
28. Junge, O.: An adaptive subdivision technique for the approximation of attractors and invariant measures: proof of convergence. Dyn. Syst. **16**(3), 213–222 (2001)
29. Junge, O., Marsden, J.E., Mezic, I.: Uncertainty in the dynamics of conservative maps. Proceedings 43rd IEEE Conference December Control, pp. 2225–2230 (2004)
30. Lasota, A., Mackey, M.C.: Chaos, Fractals, and Noise: Stochastic Aspects of Dynamics.Applied Mathematical Sciences, vol. 97, 2nd edn. Springer, New York (1994)
31. Lawler, G.F., Sokal, A.D.: Bounds on the L^2 spectrum for Markov chains and Markov processes: a generalization of Cheeger's inequality. Trans. Amer. Math. Soc. **309**, 557–580 (1988)
32. Li, T.Y.: Finite approximation for the Frobenius-Perron operator. a solution to Ulam's conjecture. J. Approx. Theor. **17**, 177–186 (1976)
33. Liu, W., Haller, G.: Strange eigenmodes and decay of variance in the mixing of diffusive tracers. Phys. D **188**, 1–39 (2004)
34. Meiss, J.D.: Symplectic maps, variational principles, and transport. Rev. Mod. Phys. **64**(3), 795–848 (1992)
35. Mezic, I.: On the geometrical and statistical properties of dynamical systems: theory and applications. Ph.D. thesis, California Institute of Technology (1994)
36. Mezić, I., Wiggins, S.: A method for visualization of invariant sets of dynamical systems based on the ergodic partition. Chaos **9**(1), 213–218 (1999)
37. Mosovsky, B., Meiss, J.: Transport in transitory dynamical systems. SIAM J. Dyn. Syst. **10**, 35–65 (2011)
38. Murray, R.: Optimal partition choice for invariant measure approximation for one-dimensional maps. Nonlinearity **17**(5), 1623–1644 (2004)
39. Pikovsky, A., Popovych, O.: Persistent patterns in deterministic mixing flows. Europhys. Lett. **61**(5), 625–631 (2003)
40. Popovych, O., Pikovsky, A., Eckhardt, B.: Abnormal mixing of passive scalars in chaotic flows. Phys. Rev. E **75**, 036,308 (2007)
41. Rom-Kedar, V., Leonard, A., Wiggins, S.: An analytical study of transport, mixing and chaos in an unsteady vortical flow. J. Fluid Mech. **214**, 347–394 (1990)
42. Rom-Kedar, V., Wiggins, S.: Transport in two-dimensional maps. Arch. Rational Mech. Anal. **109**, 239–298 (1990)
43. Royden, H.: Real Analysis, 3 edn. Pearson, London (1988)

44. Schütte, C., Huisinga, W., Deuflhard, P.: Transfer operator approach to conformational dynamics in biomolecular systems. In: Fiedler, B. (ed.) Ergodic Theory, Analysis, and Efficient Simulation of Dynamical Systems, pp. 191–223. Springer, Berlin (2001)
45. Shadden, S.C., Lekien, F., Marsden, J.E.: Definition and properties of Lagrangian coherent structures from finite-time Lyapunov exponents in two-dimensional aperiodic flows. Phys. D **212**, 271–304 (2005)
46. Sinclair, A., Jerrum, M.: Approximate counting, uniform generation and rapidly mixing Markov chains. Inform. Comput. **82**(1), 93–133 (1989)
47. Singh, M., Speetjens, M., Anderson, P.: Eigenmode analysis of scalar transport in distributive mixing. Phys. Fluid. **21**, 093,601–093,601 (2009)
48. Stremler, M., Ross, S., Grover, P., Kumar, P.: Topological chaos and periodic braiding of almost-cyclic sets. Phys. Rev. Lett. **106**(11), 114, 101 (2011)
49. Ulam, S.: Problems in Modern Mathematics. Interscience, New York (1964)
50. Wiggins, S.: Chaotic Transport in Dynamical Systems. Springer, New York (1992)
51. Wiggins, S.: The dynamical systems approach to Lagrangian transport in oceanic flows. Annu. Rev. Fluid Mech. **37**, 295–328 (2005)

Chapter 10
Return-Time Statistics, Hitting-Time Statistics and Inducing

Nicolai T.A. Haydn, Nicole Winterberg, and Roland Zweimüller

Abstract In the framework of abstract ergodic probability-preserving transformations, we prove that the limiting return-time statistics and hitting-time statistics persist if we pass from the original system to a first-return map and vice versa.

10.1 Introduction

The asymptotic behaviour of return-time and hitting-time distributions of small sets H in an ergodic probability-preserving dynamical system $(X, \mathscr{A}, \mu, T)$ is a well-studied circle of questions. For the open dynamical system obtained by regarding H as a *hole* in X, this is of obvious interest since the hitting time of H represents the survival time of orbits in the open system.

In the present note, we shall always assume that $(X, \mathscr{A}, \mu)$ is a probability space and that $T : X \to X$ is an ergodic measure-preserving map thereon. Also, H, H_l and Y will always denote measurable sets of positive measure. By ergodicity and the Poincaré recurrence theorem, the measurable *(first) hitting-time* function of H, $\varphi_H : X \to \overline{\mathbb{N}} := \{1, 2, \ldots, \infty\}$ with $\varphi_H(x) := \inf\{n \geq 1 : T^n x \in H\}$, is finite a.e. on X. When restricted to H it is called the *(first) return-time* function of our set, and it satisfies Kac' formula $\int_H \varphi_H \, d\mu_H = 1/\mu(H)$, where $\mu_H(A) := \mu(H \cap A)/\mu(H)$, $A \in \mathscr{A}$. That is, when regarded as a random variable on the probability space

N.T.A. Haydn (✉)
Department of Mathematics, University of Southern California, Los Angeles, CA 90089, USA
e-mail: nhaydn@usc.edu

N. Winterberg
FB Mathematik, Universität Salzburg, Hellbrunner Strasse 34, 5020 Salzburg, Austria
e-mail: nicole.winterberg@sbg.ac.at

R. Zweimüller
Fakultät für Mathematik, Universität Wien, Oskar-Morgenstern-Platz 1, 1090 Vienna, Austria
e-mail: rzweimue@member.ams.org

W. Bahsoun et al. (eds.), *Ergodic Theory, Open Dynamics, and Coherent Structures*, Springer Proceedings in Mathematics & Statistics 70, DOI 10.1007/978-1-4939-0419-8_10,

© Springer Science+Business Media New York 2014

$(H, \mathscr{A} \cap H, \mu_H)$, the return time has a distribution with expectation $1/\mu(H)$, and we will often normalize our variable accordingly, thus passing to $\mu(H)\,\varphi_H$.

In the following, our goal is to obtain further information about this *return-time distribution* of H, in particular when H is really small, that is, we are going to study sequences $(H_l)_{l\geq 1}$ of *asymptotically rare events*, meaning that[1] $\mu(H_l) \to 0$. The sequence is said to have *(asymptotic) return-time statistics* given by a random variable $\tilde{\mathsf{R}}$ which takes values in $[0, \infty]$, if its normalized return-time distributions converge, in the usual sense, to that of $\tilde{\mathsf{R}}$, so that $\tilde{F}_l(t) := \mu_{H_l}(\mu(H_l)\,\varphi_{H_l} \leq t) \to \tilde{F}(t) := \Pr[\tilde{\mathsf{R}} \leq t]$ as $l \to \infty$ whenever $t > 0$ is a continuity point of $\tilde{F}$. In standard probabilistic notation, this is expressed, on the level of distribution functions, as $\tilde{F}_l \Longrightarrow \tilde{F}$.

Similarly, one may study the *hitting-time distribution* of H, that is, the law of φ_H on the original probability space $(X, \mathscr{A}, \mu)$, and ask, for sequences $(H_l)_{l\geq 1}$ as above, for *(asymptotic) hitting-time statistics* given by some $[0, \infty]$-valued random variable R. This means that $F_l(t) := \mu(\mu(H_l)\,\varphi_{H_l} \leq t) \to F(t) := \Pr[\mathsf{R} \leq t]$ for all continuity points $t > 0$ of F, that is $F_l \Longrightarrow F$. This situation is somewhat simpler, as the underlying measure μ remains the same.

A large amount of material on return- and hitting-time statistics for specific types of dynamical systems is available. We refer the reader to [8,9,12] and the references cited there. (For systems with some hyperbolicity, and reasonably well behaved sequences $(H_l)_{l\geq 1}$ one typically gets convergence to an exponential law.)

Moreover, certain fundamental questions can be posed and answered in an abstract ergodic-theoretical setup. Most important for us, this is the case for the relation between the two types of limits introduced above, which has been clarified in [6]. We will recall it below; see Theorem 10.1.

The focus of the present paper is on the behaviour of the two limiting relations under the standard operation of *inducing* on a suitable reference set $Y \subseteq X$. The technique of inducing is a basic tool, both for the analysis of specific systems and for abstract ergodic theory. In the former sense, it has been used to identify return-time statistics of certain non-uniformly hyperbolic systems in [4], where it is shown that return-time statistics persist under inducing, at least in situations in which X is a Riemannian manifold, (H_l) is a sequence of ε-neighbourhoods of a typical point, and where the limit law has no mass at zero. This has been exploited in further papers; see, e.g. [5].

Here, we extend this principle to the general measure-theoretical setup and dispose of these extra conditions; see Theorem 10.3 below. We do so by first proving a corresponding abstract statement for the easier case of hitting times (Theorem 10.2) and then transfer it to return times via the aforementioned universal correspondence between the two. The results below were obtained independently in [7,13].

[1] This is obviously a property of the sequence and not of the individual events H_l. Still, we take the liberty of following this imprecise but common terminology.

10.2 Preliminaries

10.2.1 Distributional Convergence

A *sub-probability distribution function* on $[0,\infty)$ is a non-decreasing right-continuous function $F : [0,\infty) \to [0,1]$ (with a canonical extension to $\mathbb{R}$ which vanishes on $(-\infty,0)$). These F are in a one-to-one correspondence with the Borel probability measures Q on $[0,\infty]$, where Q corresponds to the function given by $F(t) := Q([0,t])$, $t \in [0,\infty)$, so that $Q(\{\infty\}) = 1 - \lim_{t\to\infty} F(t)$. These Q, in turn, are the distributions of random variables R taking values in $[0,\infty]$, $Q(B) = \Pr[\mathsf{R} \in B]$, $B \in \mathscr{B}_{[0,\infty]}$. R is (almost surely) real-valued iff $F(t) \to 1$ as $t \to \infty$, i.e. iff F is a proper *probability distribution function* on $[0,\infty)$.

If F, F_l are sub-probability distribution functions on $[0,\infty)$, then $F_l \Longrightarrow F$ means that $F_l(t) \to F(t)$ as $l \to \infty$ for all continuity points $t > 0$ of F. This is equivalent to the usual weak convergence $Q_l \Longrightarrow Q$ (cf. [3]) of the corresponding Borel probabilities on $[0,\infty]$. (This is obvious if, e.g. we map $[0,\infty]$ onto $[0,1]$ by some orientation-preserving homeomorphism and carry the measures and distribution functions along.)

In the present context we are interested in certain measurable functions $R_l : X \to [0,\infty]$, the distributions of which may be taken w.r.t. different probabilities ν_l on $(X,\mathscr{A})$. Convergence in the above sense of these distributions to the law of some random variable R will be denoted by

$$R_l \overset{\nu_l}{\Longrightarrow} \mathsf{R} \qquad \text{as } l \to \infty, \tag{10.1}$$

that is, with $F(t) := \Pr[\mathsf{R} \le t]$ denoting the distribution function of R, (10.1) means

$$\nu_l(R_l \le t) \longrightarrow F(t) \quad \text{for all continuity points } t > 0 \text{ of } F.$$

As a special case, this includes the notation $R_l \overset{\nu}{\Longrightarrow} \mathsf{R}$ for distributional convergence of the R_l when regarded as random variables on the common probability space $(X,\mathscr{A},\nu)$, that is, $\nu(R_l \le t) \longrightarrow F(t)$ for all continuity points $t > 0$ of F.

If, in the setup of the introduction, $R_l := \mu(H_l)\,\varphi_{H_l}$, then return-time statistics refer to convergence $R_l \overset{\nu_l}{\Longrightarrow} \mathsf{R}$ with $\nu_l = \mu_{H_l}$, while hitting-time statistics concern the convergence $R_l \overset{\nu}{\Longrightarrow} \mathsf{R}$ with $\nu := \mu$.

10.2.2 Relation Between Return-Time and Hitting-Time Statistics

Given an arbitrary sequence $(H_l)_{l\geq 1}$ of asymptotically rare events, its return-time statistics and its hitting-time statistics are intimately related to each other. The following fundamental result was first established in [6]. For an alternative proof see [2].

Theorem 10.1 (Hitting-time statistics versus return-time statistics). *Let $(X, \mathscr{A}, \mu, T)$ be an ergodic probability-preserving system and $(H_l)_{l\geq 1}$ a sequence of asymptotically rare events. Then*

$$\mu(H_l)\,\varphi_{H_l} \overset{\mu}{\Longrightarrow} \mathsf{R} \quad \textit{for some random variable } \mathsf{R} \textit{ in } [0,\infty] \tag{10.2}$$

iff

$$\mu(H_l)\,\varphi_{H_l} \overset{\mu_{H_l}}{\Longrightarrow} \tilde{\mathsf{R}} \quad \textit{for some random variable } \tilde{\mathsf{R}} \textit{ in } [0,\infty]. \tag{10.3}$$

In this case, the sub-probability distribution functions F and $\tilde{F}$ of R and $\tilde{\mathsf{R}}$ satisfy

$$\int_0^t (1-\tilde{F}(s))\,ds = F(t) \quad \textit{for } t \geq 0. \tag{10.4}$$

Through this integral equation each of F and $\tilde{F}$ uniquely determines the other. Moreover, F is necessarily continuous and concave with $F(t) \leq t$, while $\tilde{F}$ is a probability distribution function s.t. $\int_0^\infty (1-\tilde{F}(s))\,ds \leq 1$.

It is also known that the properties recorded above [which follow easily from (10.4)] completely determine the class of all F and $\tilde{F}$ which do occur as hitting- and return-time limits in an arbitrary aperiodic ergodic system; see [10, 11].

10.2.3 Strong Distributional Convergence for Hitting Times of Rare Events

It is an interesting but sometimes neglected fact that many distributional limit theorems for ergodic processes automatically hold for large collections of initial probability distributions ν on the underlying space. In Sect. 10.4 below, this principle will serve as an important technical tool. Let $(R_l)_{l\geq 1}$ be a sequence of non-negative measurable functions on $(X, \mathscr{A}, \mu)$ and R some $[0,\infty]$-valued random variable. *Strong distributional convergence* w.r.t. μ of $(R_l)_{l\geq 1}$ to R means that

$$R_l \overset{\nu}{\Longrightarrow} \mathsf{R} \qquad \text{for all probability measures } \nu \ll \mu, \tag{10.5}$$

compare [1]. This type of convergence is denoted by $R_l \overset{\mathscr{L}(\mu)}{\Longrightarrow} \mathsf{R}$. (The probabilistic literature sometimes uses the term *(Rényi-)mixing*.)

A situation in which one always has strong convergence for trivial reasons is that of an a.s. constant limit variable R, i.e. $\Pr[\mathsf{R} = c] = 1$ for some $c \in [0, \infty]$. There are, however, more interesting scenarios in which strong distributional convergence is automatic. A discussion of a natural and widely applicable sufficient condition for this to happen in an ergodic system can be found in [15]. Using it, it is not hard to see that asymptotic hitting-time distributions for rare events always behave in this way: If a limit law shows up under one particular initial distribution $\nu \ll \mu$, then it does so for all $\nu \ll \mu$. The following fact, which will be very useful for the proof of our main results, is contained in Corollary 5 of [15].

Proposition 10.1 (Strong distributional convergence of hitting times). *Let $(X, \mathscr{A}, \mu, T)$ be ergodic and probability preserving and $\nu \ll \mu$ some probability measure. Let $(H_l)_{l\geq 1}$ be a sequence of asymptotically rare events. Then, for any random variable* R *in* $[0, \infty]$,

$$\mu(H_l)\,\varphi_{H_l} \stackrel{\nu}{\Longrightarrow} \mathsf{R} \quad \textit{implies} \quad \mu(H_l)\,\varphi_{H_l} \stackrel{\mathscr{L}(\mu)}{\Longrightarrow} \mathsf{R}. \tag{10.6}$$

As a consequence, we see that proving $\mu(H_l)\,\varphi_{H_l} \stackrel{\nu}{\Longrightarrow} \mathsf{R}$ for some particular ν is as good as proving it for any other probability measure.

We emphasize that the analogous statement for return-time statistics is false. First, for an arbitrarily chosen probability $\nu \ll \mu$, we may have $\nu(H_l) = 0$, so that the statement doesn't even make sense. But even if $\nu(H_l) > 0$, there is no hope for a universality statement like that of the proposition. We illustrate this in the simple setup of disjoint sequences $(H_l)_{l\geq 1}$.

Example 10.1. Let $(X, \mathscr{A}, \mu, T)$ be an ergodic probability-preserving system and $(H_l)_{l\geq 1}$ a sequence of pairwise disjoint asymptotically rare events. Assume that, for some non-constant random variable $\tilde{\mathsf{R}}$, we have $\tilde{R}_l := \mu(H_l)\,\varphi_{H_l} \stackrel{\mu_{H_l}}{\Longrightarrow} \tilde{\mathsf{R}}$. Then there is some probability $\nu \ll \mu$ such that $\nu(H_l) > 0$ for $l \geq l^*$ but $\tilde{R}_l \stackrel{\nu_{H_l}}{\not\Longrightarrow} \tilde{\mathsf{R}}$.

To see this, let $t^* \in (0, \infty)$ be a continuity point of $\tilde{F}(t) := \Pr[\tilde{\mathsf{R}} \leq t]$ such that $0 < \tilde{F}(t^*) < 1$. Define $H_l^* := H_l \cap \{\tilde{R}_l \leq t^*\}$, $l \geq 1$. By assumption, $\mu_{H_l}(H_l^*) \to \tilde{F}(t^*)$, so that $\mu(H_l^*) > 0$ for $l \geq l^*$. Let $Z := \bigcup_{l\geq l^*} H_l^*$ and $\nu := \mu_Z$. Since the H_l are pairwise disjoint, we find that $\nu_{H_l}(\tilde{R}_l \leq t^*) = \nu(H_l^*)/\nu(H_l) = \mu(H_l^*)/\mu(H_l^*) = 1$ for $l \geq l^*$. Our claim follows.

10.3 Hitting-Time Statistics via Inducing

As before, we let $(X, \mathscr{A}, \mu, T)$ be an ergodic probability-preserving system. Now fix some $Y \in \mathscr{A}$, $\mu(Y) > 0$. It is a well-known classical result that the *first-return map* $T_Y : Y \to Y$ defined by

$$T_Y\, x := T^{\varphi_Y(x)} x, \quad x \in Y,$$

is a measure-preserving ergodic map on the probability space $(Y, \mathscr{A} \cap Y, \mu_Y)$. (For measure preservation by more general versions of induced maps, see, e.g., [14].) When studying specific systems, one often tries to find some good reference set Y such that T_Y is more convenient a map than T. In this case, it often pays to prove a relevant property first for T_Y and to transfer it back to T afterwards – if possible. Here, we show that this strategy can be employed to deal with hitting-time statistics. The conclusion is pleasantly simple, as the respective limit laws for T and T_Y coincide.

In the following, we let $\varphi_H^Y : Y \to \overline{\mathbb{N}}$ denote the hitting time of $H \in \mathscr{A} \cap Y$ under the first-return map T_Y, that is,

$$\varphi_H^Y(x) := \inf\{j \geq 1 : T_Y^j x \in H\}, \quad x \in Y. \tag{10.7}$$

The hitting-time functions φ_H and φ_H^Y are naturally related to each other in that

$$\varphi_H = \sum_{j=0}^{\varphi_H^Y - 1} \varphi_Y \circ T_Y^j \quad \text{on } Y. \tag{10.8}$$

This can be exploited in a fairly straightforward manner to obtain

Theorem 10.2 (Hitting-time statistics via inducing). *Let $(X, \mathscr{A}, \mu, T)$ be an ergodic probability-preserving system and $Y \in \mathscr{A}$, $\mu(Y) > 0$. Assume that $(H_l)_{l \geq 1}$ is a sequence of asymptotically rare events in $\mathscr{A} \cap Y$ and that R is any random variable with values in $[0, \infty]$. Then*

$$\mu_Y(H_l)\, \varphi_{H_l}^Y \overset{\mu_Y}{\Longrightarrow} \mathsf{R} \quad \text{as } l \to \infty \tag{10.9}$$

iff

$$\mu(H_l)\, \varphi_{H_l} \overset{\mu}{\Longrightarrow} \mathsf{R} \quad \text{as } l \to \infty. \tag{10.10}$$

Proof. (i) Let $F : [0, \infty) \to [0, 1]$ be the (sub-)distribution function of R. Due to Proposition 10.1 we know that in (10.10) convergence $\overset{\mu}{\Longrightarrow}$ w.r.t. μ is equivalent to convergence $\overset{\mu_Y}{\Longrightarrow}$ w.r.t. μ_Y. It is therefore sufficient to prove that

$$\mu_Y\left(\mu(Y)^{-1}\mu(H_l)\, \varphi_{H_l}^Y \leq t\right) \longrightarrow F(t) \quad \text{for all continuity points } t > 0 \tag{10.11}$$

iff

$$\mu_Y\left(\mu(H_l)\, \varphi_{H_l} \leq t\right) \longrightarrow F(t) \quad \text{for all continuity points } t > 0. \tag{10.12}$$

To this end, we fix any continuity point $t > 0$ of F and any $\varepsilon > 0$. Next, we choose $\delta > 0$ so small that

$$F(t) - \varepsilon/4 < F(e^{-\delta}t) \leq F(e^{\delta}t) < F(t) + \varepsilon/4. \tag{10.13}$$

Since F is continuous on a dense set, we may also assume that both $e^{-\delta}t$ and $e^{\delta}t$ are continuity points of F.

(ii) By the Ergodic theorem and Kac' formula, we have

$$m^{-1}\sum_{j=0}^{m-1}\varphi_Y \circ T_Y^j \longrightarrow \mu(Y)^{-1} \quad \text{a.e. on } Y. \tag{10.14}$$

This implies that the increasing sequence of sets given by

$$E_M := \{\sum_{j=0}^{m-1}\varphi_Y \circ T_Y^j \geq e^{-\delta}\mu(Y)^{-1}m \quad \text{for all } m \geq M\} \in \mathscr{A} \cap Y$$

satisfies $\mu_Y(E_M^c) \to 0$ as $M \to \infty$. Now fix some M such that $\mu_Y(E_M^c) < \varepsilon/4$.

Next, let $F_l := \{\varphi_{H_l}^Y \geq M\} \in \mathscr{A} \cap Y, l \geq 1$. Then $F_l = Y \cap \bigcap_{j=1}^{M-1} T_Y^{-j} H_l^c$. Hence $\mu_Y(F_l^c) \leq \sum_{j=1}^{M-1} \mu_Y(T_Y^{-j} H_l) \leq M\, \mu_Y(H_l) \to 0$ as $l \to \infty$, since T_Y preserves μ_Y, and the sequence $(H_l)_{l\geq 1}$ is asymptotically rare. Therefore there is some $L \geq 1$ such that $\mu_Y(F_l^c) < \varepsilon/4$ for $l \geq L$.

(iii) Now recall (10.8). By definition of E_M and F_l, we have

$$\varphi_{H_l} = \sum_{j=0}^{\varphi_{H_l}^Y - 1} \varphi_Y \circ T_Y^j \geq e^{-\delta}\, \mu(Y)^{-1}\varphi_{H_l}^Y \quad \text{on } F_l \cap E_M.$$

Therefore, for any $s > 0$,

$$F_l \cap E_M \cap \{\mu(H_l)\,\varphi_{H_l} \leq s\} \subseteq F_l \cap E_M \cap \{\mu(Y)^{-1}\mu(H_l)\,\varphi_{H_l}^Y \leq e^{\delta}s\},$$

and hence

$$\begin{aligned}\mu_Y\left(\mu(H_l)\,\varphi_{H_l} \leq s\right) &\leq \mu_Y\left(\mu(Y)^{-1}\mu(H_l)\,\varphi_{H_l}^Y \leq e^{\delta}s\right) + \mu_Y((F_l \cap E_M)^c) \\ &< \mu_Y\left(\mu(Y)^{-1}\mu(H_l)\,\varphi_{H_l}^Y \leq e^{\delta}s\right) + \varepsilon/2 \quad \text{for } l \geq L.\end{aligned} \tag{10.15}$$

(iv) Assume (10.11). Since $e^{\delta}t$ is a continuity point of F, we can pick $L' \geq 1$ s.t. $\mu_Y(\mu(Y)^{-1}\mu(H_l)\,\varphi_{H_l}^Y \leq e^{\delta}t) < F(e^{\delta}t) + \varepsilon/4$ for $l \geq L'$. Combining this with (10.13) and with (10.15) for $s := t$, we obtain

$$\mu_Y\left(\mu(H_l)\,\varphi_{H_l} \leq t\right) < F(t) + \varepsilon \quad \text{for } l \geq \max(L, L'). \tag{10.16}$$

If, on the other hand, we start from (10.12), then $\mu_Y\left(\mu(H_l)\,\varphi_{H_l} \leq e^{-\delta}t\right) > F(e^{-\delta}t) - \varepsilon/4$ for $l \geq L'$ with L' large enough, as $e^{-\delta}t$, too, is a continuity point of F. Combining this with (10.13) and with (10.15) for $s := e^{-\delta}t$, we get

$$\mu_Y\left(\mu(Y)^{-1}\mu(H_l)\,\varphi^Y_{H_l} \leq t\right) > F(t) - \varepsilon \quad \text{for } l \geq \max(L, L'). \tag{10.17}$$

Versions of (10.16) and (10.17) providing the corresponding estimate in the opposite direction are obtained in exactly the same way. As $\varepsilon > 0$ was arbitrary, our claim follows. □

In the proof, Proposition 10.1 was used at the very start to ensure that

$$\mu(H_l)\,\varphi_{H_l} \overset{\mu}{\Longrightarrow} \mathsf{R} \quad \text{iff} \quad \mu(H_l)\,\varphi_{H_l} \overset{\mu_Y}{\Longrightarrow} \mathsf{R}. \tag{10.18}$$

This special case of strong distributional convergence can also be verified directly. We indicate an argument which is somewhat more elementary than the theory behind Proposition 10.1.

Proof (of (10.18), direct version). (i) Let F denote the (necessarily continuous, see Theorem 10.1) sub-probability distribution function of R. To prove that for every $t > 0$ (henceforth fixed),

$$\mu\left(\varphi_{H_l} \leq s_l\right) \to F(t) \quad \text{iff} \quad \mu_Y\left(Y \cap \{\varphi_{H_l} \leq s_l\}\right) \to F(t) \tag{10.19}$$

with $s_l := t/\mu(H_l)$, we show that for $\varepsilon \in (0, t)$ there is some $L_0 \geq 1$ s.t.

$$\mu_Y\left(Y \cap \{\varphi_{H_l} \leq s_l^-\}\right) - \varepsilon < \mu\left(\varphi_{H_l} \leq s_l\right) < \mu_Y\left(Y \cap \{\varphi_{H_l} \leq s_l^+\}\right) + \varepsilon, \tag{10.20}$$

for $l \geq L_0$, where $s_l^{\pm} := (t \pm \varepsilon)/\mu(H_l)$. Below we focus on the second of these estimates. The other one is obtained by an analogous argument.

(ii) To switch from μ on X to its restriction to Y in (10.20), we will show that $\mu\left(\varphi_{H_l} \leq s_l\right) \approx \int_Y \varphi_Y \cdot 1_{\{\varphi_{H_l} \leq s_l\}}\, d\mu$ (which only involves μ on Y). This uses the well-known canonical representation (from the theory of induced transformations)

$$\mu(A) = \textstyle\sum_{n\geq 0}\mu\left(Y \cap \{\varphi_Y > n\} \cap T^{-n}A\right) \quad \text{for } A \in \mathscr{A}. \tag{10.21}$$

As μ is finite, there is some N^* such that

$$\textstyle\sum_{n>N^*}\mu\left(Y \cap \{\varphi_Y > n\}\right) < \varepsilon/4. \tag{10.22}$$

We observe an approximate invariance property of the variables φ_{H_l}. Note that

$$\varphi_{H_l} \circ T^n = \varphi_{H_l} - n \quad \text{on } \{\varphi_{H_l} > n\}. \tag{10.23}$$

As (H_l) is asymptotically rare, there is some L_1 s.t. $\mu(\varphi_{H_l} \leq N^*) < \varepsilon/4(N^*+1)$ for $l \geq L_1$. Moreover, there is some L_2 s.t. $s_l + N^* \leq (t \pm \varepsilon/2)/\mu(H_l) =: s_l^*$ whenever $l \geq L_2$. For $n \in \{0, \dots, N^*\}$ and $l \geq L_1 \vee L_2$ we then find that

$$\begin{aligned}
&\mu\left(Y \cap \{\varphi_Y > n\} \cap T^{-n}\{\varphi_{H_l} \leq s_l\}\right) \\
&\quad \leq \mu\left(Y \cap \{\varphi_{H_l} > N^*\} \cap \{\varphi_Y > n\} \cap \{\varphi_{H_l} \leq s_l + n\}\right) + \mu\left(\{\varphi_{H_l} \leq N^*\}\right) \\
&\quad < \mu\left(Y \cap \{\varphi_Y > n\} \cap \{\varphi_{H_l} \leq s_l^*\}\right) + \varepsilon/4(N^*+1).
\end{aligned}$$

Combining the preceding considerations, we then obtain

$$\begin{aligned}
\mu\left(\varphi_{H_l} \leq s_l\right) &< \textstyle\sum_{n=0}^{N^*} \mu\left(Y \cap \{\varphi_Y > n\} \cap T^{-n}\{\varphi_{H_l} \leq s_l\}\right) + \varepsilon/4 \\
&< \textstyle\sum_{n=0}^{N^*} \mu\left(Y \cap \{\varphi_Y > n\} \cap \{\varphi_{H_l} \leq s_l^*\}\right) + 2\varepsilon/4 \\
&\leq \textstyle\sum_{n\geq 0} \mu\left(Y \cap \{\varphi_Y > n\} \cap \{\varphi_{H_l} \leq s_l^*\}\right) + \varepsilon/2 \\
&= \textstyle\int_Y \varphi_Y \cdot 1_{\{\varphi_{H_l} \leq s_l^*\}}\, d\mu + \varepsilon/2. \qquad (10.24)
\end{aligned}$$

(iii) We now verify that $\int_Y \varphi_Y \cdot 1_{\{\varphi_{H_l} \leq s_l\}}\, d\mu \approx \mu(Y)^{-1} \mu\left(Y \cap \{\varphi_{H_l} \leq s_l\}\right)$. To this end, we use that T_Y is measure preserving and write

$$\textstyle\int_Y \varphi_Y \cdot 1_{\{\varphi_{H_l} \leq s_l^*\}}\, d\mu = N^{-1} \sum_{n=0}^{N-1} \int_Y \varphi_Y \circ T_Y^n \cdot 1_{\{\varphi_{H_l} \leq s_l^*\}} \circ T_Y^n\, d\mu. \qquad (10.25)$$

Due to ergodicity of T_Y and the L^1-ergodic theorem (plus Kac' formula), we can choose this N in such a way that

$$\int_Y \left| N^{-1} \textstyle\sum_{n=0}^{N-1} \varphi_Y \circ T_Y^n - \mu(Y)^{-1} \right| d\mu < \varepsilon/4, \qquad (10.26)$$

and hence, a fortiori,

$$N^{-1} \textstyle\sum_{n=0}^{N-1} \int_Y (\varphi_Y \circ T_Y^n)\, 1_{\{\varphi_{H_l} \leq s_l^+\}}\, d\mu < \mu_Y\left(Y \cap \{\varphi_{H_l} \leq s_l^+\}\right) + \varepsilon/4. \qquad (10.27)$$

The left-hand expression here differs from the right-hand side of (10.25). However, to switch from $1_{\{\varphi_{H_l} \leq s_l^*\}} \circ T_Y^n$ to $1_{\{\varphi_{H_l} \leq s_l^+\}}$, we can again exploit approximate invariance of the φ_{H_l}. We first record that there is some $\delta > 0$ such that

$$\textstyle\int_A \varphi_Y \circ T_Y^n\, d\mu < \varepsilon/4 \quad \text{for all } n \geq 0 \text{ and } A \in \mathscr{A} \cap Y \text{ with } \mu(A) < \delta. \qquad (10.28)$$

Indeed, since T_Y is measure preserving on Y, it is immediate that the sequence $(\varphi_Y \circ T_Y^n)_{n\geq 0}$ is uniformly integrable on Y, whence (10.28).

Reviewing the idea which gave (10.23), we see that

$$\varphi_{H_l} \circ T_Y^n = \varphi_{H_l} - \sum_{j=0}^{n-1} \varphi_Y \circ T_Y^j \quad \text{on } Y \cap \{\varphi_{H_l}^Y > n\}. \tag{10.29}$$

Let $\Phi := \sum_{j=0}^{N-1} \varphi_Y \circ T_Y^j$ which is finite a.e. on Y. Take M so large that $\mu(Y \cap \{\Phi > M\}) < \delta/2$. Next, choose L_3 such that $\mu(Y \cap \{\varphi_{H_l}^Y \leq N\}) < \delta/2$ and $s_l^* + M \leq s_l^+$ whenever $l \geq L_3$. Then $A_l := Y \cap \{\Phi \leq M\} \cap \{\varphi_{H_l}^Y > N\}$ satisfies $\mu(Y \setminus A_l) < \delta$ for $l \geq L_3$. For such l and $n \in \{0, \ldots, N\}$, we thus get

$$\begin{aligned} &\int_Y \varphi_Y \circ T_Y^n \cdot 1_{\{\varphi_{H_l} \leq s_l^*\}} \circ T_Y^n \, d\mu \\ &\leq \int_{A_l} \varphi_Y \circ T_Y^n \cdot 1_{\{\varphi_{H_l} \leq s_l^* + M\}} \, d\mu + \int_{Y \setminus A_l} \varphi_Y \circ T_Y^n \, d\mu \\ &< \int_Y \varphi_Y \circ T_Y^n \cdot 1_{\{\varphi_{H_l} \leq s_l^+\}} \, d\mu + \varepsilon/4. \end{aligned} \tag{10.30}$$

(iv) The proof of the right-hand half of (10.20) is completed by letting $L_0 := L_1 \vee L_2 \vee L_3$ and combining (10.24) with (10.30) and (10.27). □

10.4 Return-Time Statistics via Inducing

We finally extend the applicability of the inducing method to return-time statistics from the setup of [4] to the general abstract framework of measure-theoretic ergodic theory. With Theorem 10.1 and Proposition 10.1 at our disposal, we can easily pass from the fully general Theorem 10.2 to a corresponding result for return-time statistics which does not require any further assumptions. This is a significant improvement of Theorem 2.1 of [4], since the latter (a) only deals with a slightly restricted class of limit laws where the limiting distribution $\tilde{F}$ of the return times satisfies $\lim_{t \to 0^+} \tilde{F}(t) = 0$, i.e. there is no point mass at $t = 0$, (b) assumes that X is a Riemannian manifold and (c) only applies to sequences (H_l) of ε-neighbourhoods of typical (but not arbitrary) points of X.

Theorem 10.3 (Return-time statistics via inducing). *Let $(X, \mathscr{A}, \mu, T)$ be an ergodic probability-preserving system, and $Y \in \mathscr{A}$, $\mu(Y) > 0$. Assume that $(H_l)_{l \geq 1}$ is a sequence of asymptotically rare events in $\mathscr{A} \cap Y$ and that $\tilde{\mathsf{R}}$ is any random variable with values in $[0, \infty]$. Then*

$$\mu_Y(H_l)\, \varphi_{H_l}^Y \overset{\mu_{H_l}}{\Longrightarrow} \tilde{\mathsf{R}} \quad \text{as } l \to \infty \tag{10.31}$$

iff

$$\mu(H_l)\, \varphi_{H_l} \overset{\mu_{H_l}}{\Longrightarrow} \tilde{\mathsf{R}} \quad \text{as } l \to \infty. \tag{10.32}$$

Proof. Applying Theorem 10.1 to T_Y, we see that (10.31) is equivalent to

$$\mu_Y(H_l)\,\varphi_{H_l}^Y \overset{\mu_Y}{\Longrightarrow} \mathsf{R} \quad \text{as } l \to \infty, \tag{10.33}$$

with R and $\tilde{\mathsf{R}}$ related by the integral equation (10.4) for their respective distribution functions F and $\tilde{F}$. Due to Theorem 10.2, (10.33) is equivalent to

$$\mu(H_l)\,\varphi_{H_l} \overset{\mu}{\Longrightarrow} \mathsf{R} \quad \text{as } l \to \infty. \tag{10.34}$$

But then we can again apply Theorem 10.1, this time to T, to validate that (10.34) is indeed equivalent to (10.32), as claimed. □

References

1. Aaronson, J.: An Introduction to Infinite Ergodic Theory. AMS, New York (1997)
2. Abadi, M., Saussol, B.: Hitting and returning to rare events for all alpha-mixing processes. Stoch. Process. Appl. **121**, 314–323 (2011)
3. Billingsley, P.: Convergence of Probability Measures. 2nd edn. Wiley, New York (1999)
4. Bruin, H., Saussol, B., Troubetzkoy, S., Vaienti, S.: Return time statistics via inducing. Ergod. Theor. Dyn. Syst. **23**, 991–1013 (2003)
5. Bruin, H., Vaienti, S.: Return time statistics for unimodal maps. Fundam. Math. **176**, 77–94 (2003)
6. Haydn, N., Lacroix, Y., Vaienti, S.: Hitting and return times in ergodic dynamical systems. Ann. Probab. **33**, 2043–2050 (2005)
7. Haydn, N.: A note on the limiting entry and return times distributions for induced maps. preprint, arXiv: 1208.6059v1
8. Hirata, M., Saussol, B., Vaienti, S.: Statistics of return times: A general framework and new applications. Commun. Math. Phys. **206**, 33–55 (1999)
9. Keller, G.: Rare events, exponential hitting times and extremal indices via spectral perturbation. Dyn. Syst. **27**, 11–27 (2012)
10. Kupsa, M., Lacroix, Y.: Asymptotics for hitting times. Ann. Probab. **33**, 610–619 (2005)
11. Lacroix, Y.: Possible limit laws for entrance times of an ergodic aperiodic dynamical system. Israel J. Math. **132**, 253–263 (2002)
12. Saussol, B.: An introduction to quantitative Poincaré recurrence in dynamical systems. Rev. Math. Phys. **21**, 949–979 (2009)
13. Winterberg, N.: Return time statistics of ergodic dynamical systems. M.Sc Thesis, Universität Salzburg, June 2012.
14. Zweimüller, R.: Invariant measures for general(ized) induced transformations. Proc. Am. Math. Soc. **133**, 2283–2295 (2005)
15. Zweimüller, R.: Mixing limit theorems for ergodic transformations. J. Theor. Probab. **20**, 1059–1071 (2007)

MIX
Papier aus verantwortungsvollen Quellen
Paper from responsible sources
FSC® C105338

If you have any concerns about our products,
you can contact us on
ProductSafety@springernature.com

In case Publisher is established outside the EU,
the EU authorized representative is:
Springer Nature Customer Service Center GmbH
Europaplatz 3, 69115 Heidelberg, Germany

Printed by Libri Plureos GmbH
in Hamburg, Germany